Lecture Notes in Mathematics

Edited by A. Dold, Heidelberg and B. Eckmann, Zürich

365

Robert P. Gilbert
Indiana University, Bloomington, IN/USA

Constructive Methods for Elliptic Equations

Springer-Verlag
Berlin · Heidelberg · New York 1974

AMS Subject Classifications (1970): 35J15, 35J25, 35J30, 35J40, 35J60, 30A92, 30A93, 30A97, 35D99, 35F99, 35C15

ISBN 3-540-06690-X Springer-Verlag Berlin · Heidelberg · New York
ISBN 0-387-06690-X Springer-Verlag New York · Heidelberg · Berlin

Offsetdruck: Julius Beltz, Hemsbach/Bergstr.

PREFACE

This set of lecture notes was developed, while I was Visiting Unidel Chair Professor at the University of Delaware, during the year 1972 - 1973. They correspond, with minor variations, to the lectures I gave there in the Applied Mathematics Seminar. Primarily these lectures are a report on recent work by the Indiana University group working on Function theoretic methods as applied to the theory of partial differential equations. Since this research was done after the publication of the author's first monograph [Gi 1], in general, only recent works are quoted. We do, however, list as references the important books by Bergman [Be 1], Garabedian [Ga 1], and Vekua [Ve 1,2]. An extensive bibliography may be found by referring to the above four books.

For purposes of exposition it was necessary to include some material of an introductory nature. Hence, some coincidence with the previously mentioned works is unavoidable. Usually these portions are brief and the reader is referred to the originals for further details.

The first chapter is a brief introduction to functions of several complex variables and may be skipped by anyone knowing the rudiments of this topic. Chapter two deals with

NOTE

This research was supported in part by the Air Force Office of Scientific Research through AF-AFOSR Grant No. 71-2205A.

normalized second order elliptic equations in the plane and presents the results of Stefan Bergman [Be 1], Ilya Vekua [Ve 1] and the author [Gi 1,2,3,4]. Chapter three shows how these results may be used for numerical/constructive approaches for solving boundary value problems. Chapters four and five develop the theory of higher order elliptic equations in the plane and give some generalizations to $\mathbb{R}^n$. In chapter six the recent work of Marić-Tjong [MT 1], David Colton [Co 3-6], Colton-Gilbert [CG 2], and Gilbert-Lo [GL 1] concerning elliptic differential equations in $\mathbb{R}^3$ and $\mathbb{R}^4$ is described. Chapter seven contains an introduction to singularity theory for solutions of partial differential equations. A definitive answer is found to the search [Gi 0,1] for necessary and sufficient criteria for a harmonic function to be singular. Chapter eight is a report on the author's work with Dean Kukral [GK 2 - 6] on extending the results of the previous chapter to higher order elliptic equations, and higher dimensional equations. Chapter nine treats the analytic theory of semi-linear, elliptic systems of partial differential equations in the plane. Initial value problems are treated by means of the contraction mapping principle. This part of the book is a presentation of the author's work with James Conlan [CG 4-6].

This completes the discussion of the analytic theory of elliptic differential equations.

The final chapters, ten and eleven, develop the theory of weak solutions of elliptic systems in the plane. This treatment depends on some initial work by Lipman Bers [Be 3] and Ilya Vekua [Ve 2] concerning the generalized Cauchy-Riemann equations, and also Avron Douglis' [Do 1] research into the hyperanalytic function theory. The results reported on here are due to the author with Gerald Hile [GH 1-3].

I would like to thank Professors George Hsiao and Richard Weinacht who discussed this work with the author and made many helpful suggestions, and I would also like to thank Mr. William Moss, Dr. Dennis Quinn, and Mr. Satish Bhatnagar for proofing the manuscript. Special mention is due to Diane McCaw, Kathy Kisor, and Alison Ingham for their careful and accurate typing of the manuscript.

Part of the reason this work progressed so smoothly from its conception to completion was the very pleasant but intellectually stimulating atmosphere provided by the University of Delaware, and in particular, by the very competent Applied Mathematics group there. I must comment also on the fact that their Chairman, Professor Willard Baxter, was most considerate in helping to maintain an environment which was most conducive to my work on this project and my joint research with several members of his Department.

Finally, I would like to dedicate this manuscript to my parents, Ralph and Ruth Gilbert, who have always encouraged me in my scientific career.

August, 1973

TABLE OF CONTENTS

1. Introduction

We consider functions defined on $\mathbb{C}^n \equiv \mathbb{C}^1 \times \ldots \times \mathbb{C}^1$, i.e., the space of complex n-tuples. Let D be a domain in $\mathbb{C}^n$, which unless otherwise stated will be assumed to be simply-connected. A function $f(z) \equiv f(z_1,\ldots,z_n)$ will be called (Weierstrass) W-holomorphic in D iff for each point $\tilde{z} \in D$ $f(z)$ permits a power series representation

$$f(z) = \sum c_m (z-\tilde{z})^m \tag{1.1}$$

which converges in some neighborhood of $\tilde{z}$. Here m is a multi-index, $z^m \equiv z_1^{m_1} \ldots z_n^{m_n}$, and the summation is meant to be taken over all non-negative integer values of the components of m. It is easy to show that if $f(z)$ converges at some point b it converges absolutely and uniformly (independent of the order of summation) for all z contained in the polydisk (the Cartesian product of the disks $|z_k - a_k| < \varepsilon_k$),

$$\Delta(a,\varepsilon) \equiv \{z \,\big|\, |z-a| < \varepsilon\} , \quad \varepsilon_k < |b_k - a_k| .$$

A function $f(z)$ is said to be (Cauchy-Riemann) CR-holomorphic in the above domain if all the first partial derivatives $\frac{\partial f}{\partial \bar{z}_k}$, $(k = 1,\ldots,n)$ exist and are continuous. CR-holomorphic implies that $\frac{\partial f}{\partial \bar{z}_k} = 0$, $(k = 1,\ldots,n)$ and that if $\mathfrak{G} \equiv \mathfrak{G}_1 \times \ldots \times \mathfrak{G}_n \subset \mathfrak{D}$ where the $\partial \mathfrak{G}_k$ are sufficiently smooth, say C^1, then one has a Cauchy representation <u>in</u> $\mathfrak{G}$, namely

$$f(z) = \left(\frac{1}{2\pi i}\right)^n \int_{\mathfrak{b}} \frac{f(\mathfrak{z})\, d\mathfrak{z}}{\mathfrak{z} - z} , \tag{1.2}$$

where $\mathfrak{M} \equiv \partial\mathfrak{G}_1 \times \ldots \partial\mathfrak{G}_n$ is the "distinguished-boundary," and $d\zeta = d\zeta_1 \ldots d\zeta_n$. The proof of this representation formula follows immediately by writing (1.1) over as an iterated integral and applying the one-variable Cauchy representation theorem.

The representation (1.2) permits us to conclude, as in the case of one complex variable, that a sequence of CR-holomorphic functions which converges uniformly in D, converges to a CR-holomorphic function. The proof is the same as in the one-variable case.

It is obvious that all W-holomorphic functions are also CR-holomorphic. On the other hand, if a function is continuous in all its variables (simultaneously), and CR-holomorphic then it is W-holomorphic. The proof is similar to the one-variable case; the representation (1.2) is used and the Cauchy-kernel $(\zeta - z)^{-1}$ is expanded about an interior point of $\mathfrak{G}$. If we do not make the assumption that $f(z)$ is continuous then it is more difficult to show that is is W-holomorphic. The fact that this is true is the content of an important and basic theorem due to Hartogs, which we postpone until later.

As in the one variable case, we are able to show that the Taylor coefficients of a W-holomorphic function (1.1) are given by

$$c_m = \frac{1}{m!}\,(D^m f)(\overset{\circ}{z}) \,,$$

$$m! = m_1! \ldots m_n! \quad , \quad D^m \equiv \frac{\partial^{|m|}}{\partial z_1^{m_1} \ldots \partial z_n^{m_n}} \quad , \quad |m| = m_1 + \ldots + m_n \quad .$$

This is done by differentiating (1.1) successively (Majorization tells us this is possible) and evaluating the derivatives at the center of the ps (power series).

Another result which follows directly from one-variable theory is that if the ps

(1.3) $$f(z) = \sum c_m (z-a)^m$$

converges in a polydisk $\Delta(a,\rho)$ then $f(z)$ is W-holomorphic in $\Delta(a,\rho)$. See, for example, [Gi 1] for the details.

The ps (1.3) and the polydisk $\Delta(a,\rho)$ taken together are termed a function element, $\{f,\Delta(a,\rho)\}$. If we choose a point $z' \in \Delta(a,\rho)$ then one may expand $f(z)$ as a ps about z' and this new ps converges at least in the polydisk $\Delta(z',\tilde{\rho})$, where $\tilde{\rho}_k = \rho_k - |z'_k - a_k|$. If the new ps converges in a larger polydisk then one has obtained a continuation of the original function element.

<u>Hartogs' Lemma</u>: Let $f(z)$ be CR-holomorphic in $\overline{\Delta(0,r)}$, $r = (r_1,r_2)$, and bounded in $\overline{\Delta(0,\hat{r})}$, $\hat{r} = (r_1,\rho)$, $0 < \rho < r_2$. Then $f(z)$ is continuous in $\Delta(0,r)$.

<u>Pf.</u>: Since $f(z)$ is CR-holomorphic, for each fixed z_1, $|z_1| \leq r_1$ we have

(1.4) $$f(z) = \sum_{k=0}^{\infty} f_k(z_1) z_2^k$$

converging uniformly in z_2 for $|z_2| \leq r_2$.

Claim: The $f_k(z_1)$ are holomorphic for $|z_1| \leq r_1$.

Pf.: $f_0(z_1) \equiv f(z_1,0)$, and hence is holomorphic. Define

$$F^{(n)}(z_1,z_2) \equiv \frac{f(z_1,z_2) - \sum_{k=0}^{n-1} f_k(z_1)z_2^k}{z_2^n}$$

$$= \sum_{\ell=0}^{\infty} f_{\ell+n}(z_1)z_2^{\ell} ,$$

$$z_2F^{(n)}(z_1,z_2) \equiv F^{(n-1)}(z_1,z_2)-f_{(n-1)}(z_1) .$$

These functions are clearly CR-holomorphic in $\{|z_1| < r_1\}\times\{0 < |z_2| < r_2\}$. Let $\{z_2^{(i)}\}$ be a sequence such that $\lim z_2^{(i)} \to 0$, and $0<|z_2^{(i)}| \leq r_2$. Then

$$h_i^{(n)}(z_1) \equiv F^{(n)}(z_1,z_2^{(i)})$$

are holomorphic for $|z_1| \leq r_1$, and $\lim h_i^{(n)}(z_1) = f_n(z_1)$. If the convergence is uniform in $|z_1| \leq r_1$ then the $f_n(z_1)$ are holomorphic there. Since f is bounded in $\overline{\Delta(0,\hat{r})}$ the Cauchy estimates tell us that $|f_k(z_1)| \leq M\rho^{-k}$. Then for $\forall$ $|z_1| \leq r_1$

$$\lim \left|F^{(n)}(z_1,z_2^{(i)})-f_n(z_1)\right| \leq \lim \sum_{\ell=1}^{\infty} \left|f_{\ell+n}(z_1)\right|\cdot\left|z_2^{(i)}\right|^{\ell}$$

$$\leq M\rho^{-n} \lim \left\{\frac{|z_2^{(i)}|}{\rho-|z_2^{(i)}|}\right\} = 0 ,$$

which establishes our claim.

Claim: The series (1.4) converges uniformly in any relatively compact subset of $\Delta(0,r)$.

Pf.: Each $f_\ell(z_1)$ has at most a finite number of zeroes in $|z_1| \le r_1$ and, consequently, the set of all points where the $\{f_\ell(z_1)\}$ vanish is denumerable. It follows that we can pass a circle $|z_1| = \tilde{r} < r_1$ arbitrarily close to $|z_1| = r_1$, which does not pass through any zero of the $\{f_\ell(z_1)\}$ $(f \not\equiv 0)$. One has $\left|\frac{f_\ell(z_1)}{M} r_2^\ell\right| \le \left(\frac{r_2}{\rho}\right)^\ell$, for $|z_1| \le \tilde{r}$. If $E_\ell \equiv \{z_1 \,\big|\, |z_1| = \tilde{r},\ f_\ell(z_1) r_2^\ell > M\}$ then it can be shown that $\operatorname{mes}(E_\ell) \to 0$. Since

$$f(z) = \sum_{k=0}^{\infty} f_k(z_1) z_2^k$$

converges for $z_1 = z_1^0$, $|z_1^0| = \tilde{r} < r_1$ and $|z_2| \le r_2$ it follows that $z_1^0 \notin E_\ell$ as $\ell \to \infty$. If $Q_\ell = \bigcup_{k \ge \ell} E_k$ then $\operatorname{mes}(E_\ell) \le \operatorname{mes}(Q_\ell) \to 0$, since $\bigcap Q_\ell = \phi$. Define the subharmonic functions*

$$h_\ell(z_1) = \frac{1}{\ell} \ln \left|\frac{f_\ell(z_1)}{M}\right| + \ln r_2$$

Using the Poisson formula one has

$$h_\ell(z_1) \le H_\ell(z_1) = \frac{1}{2\pi} \int_{-\pi}^{\pi} \frac{g_\ell(\tilde{r}e^{i\phi})(\tilde{r}^2 - |z_1|^2)\,d\phi}{\tilde{r}^2 + |z_1|^2 - 2\tilde{r}|z_1|\cos(\phi - \arg[z_1])}$$

$$\le \frac{1}{2\pi} \frac{\tilde{r} + |z_1|}{\tilde{r} - |z_1|} \left(\ln \frac{r_2}{\rho}\right) \operatorname{mes} E_\ell \to 0 .$$

where $H_\ell(z_1)$ is a harmonic function in $|z_1| \le \tilde{r}$ which takes on the

---

*A function $h(z)$ is said to be subharmonic if it satisfies the following:
1. $h(z)$ is upper semicontinuous
2. if $h(z) = H(z)$ on ∂D where $H(z)$ is harmonic then $h(z) - H(z) \le 0$ in D
3. $-\infty \le h(z) < \infty$.

values $g_\ell(z_1) = \ln \frac{r_2}{\rho}$ for $z_1 \in E_\ell$ and 0 for $z_1 \notin E_\ell$. Hence given an $\varepsilon > 0$ $\exists$ an $N(\varepsilon) \ni$ for $\ell > N(\varepsilon)$ $h_\ell(z_1) < \varepsilon$ and, hence, $|f_\ell(z_1)| < \left(M \frac{e^\varepsilon}{r_2}\right)^\ell$. From this, we conclude that the series (1.4) converges uniformly and absolutely in $\{|z_1| \leq r_1 , |z_2| \leq r_2 e^{-\varepsilon}\}$, and since $\varepsilon > 0$ may be taken arbitrarily small, in $\{|z_1| \leq r_1 , |z_2| < r_2\}$*.

<u>Hartogs' Theorem</u>: If $f(z)$ is CR-holomorphic in the bicylinder $\Delta(a,r)$ it is also W-holomorphic in $\Delta(a,r)$.

<u>Pf.</u>: If we can show that f is bounded in the bicylinder $\tilde{\Delta}(a,\rho) = \{z \big| |z_1 - a_1| \leq r_1 , |z_2 - a_2| \leq \rho < r_2\}$ then we are finished. This fact follows from Osgood's Lemma.

<u>Osgood's Lemma</u>: If $f(z)$ is CR-holomorphic in $\overline{\Delta(a,r)}$, then $f(z)$ is bounded in some bidisk $\{z \big| |z_1 - a_1| \leq r_1 , |z_2 - z_2^0| \leq r_2'\} \subset \Delta(a,r)$.

<u>Pf.</u>: Without loss of generalization take $a = 0$. Define $M(z_2) = \sup_{|z_1| \leq r_1} |f(z_1,z_2)|$, where z_2 is arbitrary and $|z_2| \leq r_2$.

Let

$$\Omega_n \equiv \{z_2 \big| |z_2| \leq r_2 , \text{ and } M(z_2) \leq n\} \ ;$$

then $\Omega_{n+1} \supset \Omega_n$. For each z_2 $\exists$ an $N(z_2) \ni$ for $n \geq N(z_2)$, $z_2 \in \Omega_n$.

Since $M(z_2)$ is a continuous function of z_2 on $\Delta \equiv \{|z_2| \leq r_2\}$ it follows that the Ω_n are closed.

---

*Note: The partial sums $S_n(z) = \sum_{k=0}^{n} f_k(z_1) z_2^k$ are actually continuous in all the variables; hence, the $S_n(t)$ converge uniformly in the above bidisk to a continuous function.

Claim: $\exists$ a set Ω_n containing a disk $\{z_2 \,|\, |z_2 - z_2^0| \leq b\}$.

Pf.: If this were not true then the sets Ω_n are nowhere dense in Δ, and $\exists$ a closed disk $D_1 \subset \Delta \ni D_1 \cap \Omega_1 = \phi$. Likewise $\exists$ a closed disk $D_2 \subset D_1 \ni D_2 \cap \Omega_2 = \phi$. In this manner, we construct a nested sequence of closed disks $\{D_n\} \ni D_n \cap \Omega_n = \phi$. It follows that there exists at least one point in $\cap D_n$ and which consequently must be exterior to $\cup \Omega_n$. This is obviously a contradiction, which establishes our claim, and therefore the Lemma.

Proof of Hartogs' Theorem: If $f(z)$ is CR-holomorphic in $\Delta(0,r)$ then it also is in $\Delta'(0) \equiv \{z \,|\, |z_1| \leq r_1 \,,\ |z_2| \leq r_2/3\}$. From the Osgood lemma $f(z)$ is then bounded in some bidisk $\Delta^2 \equiv \{z \,|\, |z_1| \leq r_1 \,,$ $|z_2 - z_2^0| \leq \rho\} \subset \Delta'(0)$. Using the Hartogs' lemma it is clear that $f(z)$ is continuous in $\Delta^3 \equiv \{z \,|\, |z_1| \leq r_1 \,,\ |z_2 - z_2^0| \leq \frac{2r_2}{3}$, and hence $f(z)$ is bounded in $\Delta'(0) \subset \Delta^3$. Again using Hartogs' lemma, applied to the bidisks $\Delta'(0)$ and $\Delta(0,r)$, $f(z)$ is seen to be continuous in $\Delta(0,r)$. This is sufficient to show $f(z)$ is W-holomorphic.

Pluriharmonic Functions: From one-variable theory we know that the real and imaginary parts of holomorphic functions of a complex variable are harmonic, i.e., if $f(z) = \phi(z) + i\psi(z)$, where ϕ and ψ are the real and imaginary parts respectively, then $\Delta\phi = \Delta\psi = 0$. On the other hand, if $f(z)$ is a function of several complex variables, the situation is more complicated. Setting as before $f(z) = \phi(z) + i\psi(z)$ one has

$$\phi(z) = \frac{1}{2}\,[f(z) + \overline{f(z)}] \quad ,$$

$$\psi(z) = \frac{1}{2i}\,[f(z) - \overline{f(z)}] \ ,$$

which implies $\dfrac{\partial^2\phi}{\partial z_k \partial \bar{z}_\ell} = \dfrac{\partial^2\psi}{\partial z_k \partial \bar{z}_\ell} = 0$, $(k, \ell = 1,\ldots,n)$, since

$$\frac{\partial \bar{f}}{\partial z_k} = \overline{\left(\frac{\partial f}{\partial \bar{z}_k}\right)} = 0 \ .$$

<u>Theorem</u>: A necessary and sufficient condition for a real valued $\mathbb{C}^2$-function. $\phi(z)$, to be (locally) the real part of a holomorphic function is that $\dfrac{\partial^2\phi}{\partial z_k \partial \bar{z}_\ell} = 0$, $(k, \ell = 1,\ldots,n)$.

<u>Pf</u>.: We have already shown necessity. To show sufficiency, we introduce the differential form

$$\omega = \sum \frac{\partial \phi}{\partial z_k}\, dz_k \ .$$

Then

$$d\omega = \frac{\partial^2\phi}{\partial z_k \partial z_\ell}\, dz_\ell \wedge dz_k + \sum \frac{\partial^2\phi}{\partial z_k \partial \bar{z}_\ell}\, d\bar{z}_\ell \wedge dz_k \ .$$

Since $dz_\ell \wedge dz_k = -dz_k \wedge dz_\ell$ the first summation vanishes. The second summation vanishes because of the hypothesis. One has that ω is a closed differential form, which implies also that $d\bar{\omega} = 0$, from which it follows that the real and imaginary parts of ω are closed. Since

$$\frac{\partial\phi}{\partial z_k}\, dz_k = \frac{1}{2}\left(\frac{\partial\phi}{\partial x_k}\, dx_k + \frac{\partial\phi}{\partial y_k}\, dy_k\right) + \frac{i}{2}\left(\frac{\partial\phi}{\partial x_k}\, dy_k - \frac{\partial\phi}{\partial y_k}\, dx_k\right)$$

there exists a locally defined function $\psi(z)$ such that $\dfrac{\partial\psi}{\partial y_k} = \dfrac{\partial\phi}{\partial x_k}$ and $\dfrac{\partial\psi}{\partial x_k} = -\dfrac{\partial\phi}{\partial y_k}$, $(k = 1,2,\ldots,n)$. However, the above equations imply

$\psi(z)$ is locally a conjugate function for $\phi(z)$ and $\phi = \mathrm{Re}\{f(z)\}$ where $f(z) = \phi + i\psi$ is a locally defined holomorphic function.

Definition: The real parts of holomorphic functions are called pluriharmonic functions.

Theorem: For $n \geq 2$ the pluriharmonic functions form a proper subclass of the harmonic functions of $2n$ real variables.

Pf.: The harmonic functions satisfy

$$\sum \frac{\partial^2 u}{\partial z_k \partial \bar{z}_k} = 0 \quad .$$

This equation is satisfied by the pluriharmonic functions which satisfy in addition

$$\frac{\partial^2 u}{\partial z_k \partial \bar{z}_\ell} = 0 \,, \qquad (k, \ell = 1,2,\ldots,n) \quad .$$

Theorem: In general, the Dirichlet problem cannot be solved by pluriharmonic functions.

Pf.: The harmonic functions solve the Dirichlet problem uniquely (i.e., with prescribed real, continuous boundary values on a sufficiently smooth boundary of a domain D). There exists exactly one harmonic function that assumes the given boundary values on ∂D.

Definition: A real valued function $u(z)$ is said to be plurisubharmonic in a domain $D \subset \mathbb{C}^n$ iff the following hold

(i) $-\infty \leq u(z) < \infty$

(ii) $u(z)$ is upper semicontinuous

(iii) If $\lambda \in \mathbb{C}^1$ and z^0, $a \in \mathbb{C}^n$ then $u(z^0 + a\lambda)$ is subharmonic as a function of λ.

This definition is related to the definition of a convex function of several real variables in the same way the subharmonic functions are related to the definition of the convex functions of a single real variable.

Theorem: If $u(z)$ is pluriharmonic in D and $\overline{\Delta(z^0,r)} \subset D$, then the following integral relationships hold:

$$\text{(i)}\quad u(z^0) \leq \frac{1}{(2\pi)^n} \int_0^{2\pi} u(z_1^0 + r_1 e^{i\theta_1}, \ldots, z_n^0 + r_n e^{i\theta_n})\, d\theta \ ,$$

$$\text{(ii)}\quad u(z^0) \leq \frac{1}{\pi^n r^2} \int u(z)\,dxdy \ .$$

Here, $r \equiv r_1, \ldots, r_n$, $dxdy = dx_1, \ldots, dx_n dy_1, \ldots, dy_n$, $d\theta = d\theta_1, \ldots, d\theta_n$.

Pf.: Since $u(z)$ is subharmonic on the analytic plane we have

$$u(z^0) \leq \frac{1}{2\pi} \int^{2\pi} u(z_1^0 + r_1 e^{i\theta_1},\ z_2^0, \ldots, z_n^0)\, d\theta_1 \ ;$$

furthermore,

$$u(z_1^0 + r_1 e^{i\theta_1},\ z_2^0, \ldots, z_n^0) \leq \frac{1}{2\pi} \int_0^{2\pi} u(z_1^0 + r_1 e^{i\theta_1},\ z_2^0 + r_2 e^{i\theta_2}, \ldots, z_n^0)\, d\theta_2 \ ,$$

etc.

The second integral inequality follows from the first by multiplying by rdr and performing the corresponding n-fold integral.

Remark: Since a function $u(z)$ is subharmonic in D iff it is locally subharmonic, it follows that the converse of the above theorem is true.

Theorem: If $\{u_k\}$ is a sequence of plurisubharmonic functions in D with the property

$$u(z) \leq u_{k+1}(z) \leq u_k(z) \ , \quad \text{and} \quad \lim u_k(z) = u(z) \ ,$$

then $u(z)$ is plurisubharmonic.

Pf.: On each analytic plane the $u_k(z)$ are subharmonic. On the restriction to an analytic plane Λ, let us consider the arbitrary domain $\mathfrak{G} \subset D \cap \Lambda$. Now, if $h(z)$ is harmonic in $\mathfrak{G}$, $h(z) \in \mathcal{C}(\mathfrak{G}+\partial\mathfrak{G})$, and on $\partial\mathfrak{G}$ $h(z) \geq u(z)$, then for $\varepsilon > 0$ and k sufficiently large

$$h(z)+\varepsilon \geq u_k(z) \quad \text{on} \ \partial\mathfrak{G} \ .$$

Hence, for every ε one has in $\mathfrak{G}$, that $h(z)+\varepsilon \geq u(z)$, which implies $h(z) \geq u(z)$ in $\mathfrak{G}$ or $u(z)$ is subharmonic in $\mathfrak{G}$. Since we have established this result for an arbitrary domain on the intersection of an arbitrary analytic plane Λ with D, this implies that $u(z)$ is plurisubharmonic.

Theorem: A real valued function $u(z)$ of class $\mathcal{C}^2(D)$ is plurisubharmonic iff the hermitian differential form $\sum_{k,\ell} \frac{\partial^2 u}{\partial z_k \partial \bar{z}_\ell} dz_k d\bar{z}_\ell$ is positive semidefinite in D.

Pf.: Suppose $u(z)$ is plurisubharmonic, then on each analytic plane intersecting D, $u(z)$ is subharmonic. This implies since $u \ \mathcal{C}^2(D)$, that

$\frac{\partial^2 u(z^0+a\lambda)}{\partial\lambda\partial\bar{\lambda}} \geq 0$. Hence,

$$\frac{\partial^2 u(z^0+a\ )}{\partial\lambda\partial\bar{\lambda}}\, d\lambda d\bar{\lambda} = \sum_{k,\ell} \frac{\partial^2 u}{\partial z_k \partial \bar{z}_\ell} \frac{\partial z_k}{\partial\lambda} \frac{\partial \bar{z}}{\partial\bar{\lambda}}\, d\lambda d\bar{\lambda} \geq 0 \quad ,$$

which implies that

$$\sum_{\ell,k} \frac{\partial^2 u}{\partial z_k \partial \bar{z}_\ell}\, dz_k\, d\bar{z}_\ell \geq 0 \quad \text{in} \quad D \ .$$

The converse statement follows by reversing our argument.

Remark: If $\sum_{k,\ell} \frac{\partial^2 u}{\partial z_k \partial \bar{z}_\ell}\, dz_k\, d\bar{z}_\ell \equiv 0$, then clearly $\frac{\partial^2 u}{\partial z_k \partial \bar{z}_\ell} = 0$ $(k, \ell = 1,\ldots,n)$ and $u(z)$ is seen to be pluriharmonic.

If we restrict ourselves to functions of class $\mathcal{C}^2$, then it is easy to show that the plurisubharmonic functions are a subclass of the subharmonic functions of $2n$ real variables, $n \geq 2$. The subharmonic functions satisfy $\sum_k \frac{\partial^2 u}{\partial z_k \partial \bar{z}_k} \geq 0$, whereas the plurisubharmonic functions satisfy for instance, $\frac{\partial^2 u}{\partial z_k \partial \bar{z}_k} \geq 0$, $(k = 1,\ldots,n)$, and are therefore also subharmonic. The subclass is proper, since there are functions which satisfy $\sum_k \frac{\partial^2 u}{\partial z_k \partial \bar{z}_k} \geq 0$ but not $\sum_{k,\ell} \frac{\partial^2 u}{\partial z_k \partial \bar{z}_\ell}\, dz_k\, d\bar{z}_\ell \geq 0$.

Problem: Find some such functions.

Theorem: The property of plurisubharmonicity remains invariant under holomorphic transformations [Br 1] .

Pf.: Let $z^* = F(z)$ be a mapping of D onto $D^* \ni$ the Jacobian does not vanish in D. Then we claim $(u \circ F^{-1})(z^*)$ exists in D^* and is plurisubharmonic there. We show this just for functions of class $\mathcal{C}^2(D)$ and to this end we note that since the Jacobian does not vaish $F^{-1}(z^*)$ exists and is holomorphic. If $u(z) \in \mathcal{C}^2(D)$ then

$$\sum_{\ell,k} \frac{\partial^2 u}{\partial z^*_\ell \partial \bar{z}^*_k} dz^*_\ell \, d\bar{z}^*_k = \sum_{\substack{p,m \\ \ell,k}} \frac{\partial^2 u}{\partial z_p \partial \bar{z}_m} \frac{\partial z_p}{\partial z^*_\ell} \frac{\partial \bar{z}_m}{\partial \bar{z}^*_k} dz^*_\ell \, d\bar{z}^*_k$$

$$= \sum_{p,m} \frac{\partial^2 u}{\partial z_p \partial \bar{z}_m} dz_p \, d\bar{z}_m \ ,$$

which implies $(u \circ F^{-1})(z^*)$ is plurisubharmonic in D^* if $u(z)$ is plurisubharmonic in D .

II. Elliptic Partial Differential Equations in Two Variables with Analytic Coefficients

II. 1. Linear Equations of Second Order

We consider equations of the form

$$\underset{\sim}{e}[u] \equiv \Delta u+a(x,y)u_x+b(x,y)u_y+c(x,y)u = 0 \tag{2.1}$$

where the coefficients are analytic functions of the real variables x, y for $(x,y) \in \bar{D} \subset \mathbb{R}^2$. (Without loss of generality we assume that $(0,0) \in D$.) This means that for each $(x_0,y_0) \in D$ $\exists$ a rectangle $R(x_0,y_0) \equiv \{(x,y) \mid x_0-h \leq x \leq x+h\ ,\ y-k \leq y \leq y+k\}$ in which the coefficients have convergent Taylor series expansions. For example

$$a(x,y) = \sum_{m=0}^{\infty} \sum_{n=0}^{\infty} a_{mn} (x-x_0)^m (y-y_0)^n \quad .$$

However, this means that $a(x,y)$ converges for complex values of x, y contained in the bidisk $(z_0;\rho) \equiv \{z \mid |z-z_0| \leq \rho\ ;\ \rho_1 = h\ ,\ \rho_2 = k\}$, i.e., $a(x,y)$ has a holomorphic extension to a complex neighborhood of $\bar{D}$, $\mathcal{N}(\bar{D})$. We now insist on the stronger hypothesis, namely, if one makes the substitution

$$z = x+iy$$

$$\zeta = x-iy$$

(which is a linear transformation from $\mathbb{C}^2$ onto $\mathbb{C}^2$), the coefficients $\tilde{a}(z,\zeta) \equiv a\left(\frac{z+\zeta}{2}, \frac{z-\zeta}{2i}\right)$, $\tilde{b}(z,\zeta) \equiv b\left(\frac{z+\zeta}{2}, \frac{z-\zeta}{2i}\right)$, etc. have holomorphic extensions to $(z,\zeta) \in D \times D^*$, where $D^* \equiv \{z \mid \bar{z} \in D\}$.

Note: $\zeta^* = \bar{z}$ iff x and y are real.

Under the above transformation the elliptic equation (1.1) becomes

(2.2) $$\underset{\sim}{E}[U] \equiv U_{z\zeta}+A(z,\zeta)U_z+B(z,\zeta)U_\zeta+C(z,\zeta)U = 0 \quad ,$$

where

$$A(z,\zeta) = \frac{1}{4}\,(\tilde{a}(z,\zeta)+i\tilde{b}(z,\zeta)) \quad ,$$

$$B(z,\zeta) = \frac{1}{4}\,(\tilde{a}(z,\zeta)-i\tilde{b}(z,\zeta)) \quad ,$$

(2.3) $$C(z,\zeta) = \frac{1}{4}\,\tilde{c}(z,\zeta) \quad ,$$

and

$$U(z,\zeta) = u\left(\frac{z+\zeta}{2}\,,\ \frac{z-\zeta}{2i}\right) \quad .$$

<u>Note</u>: If the coefficients of (2.1) are real then for real x, y $A(z,\bar{z}) = \overline{B(z,\bar{z})} \equiv \bar{B}(\bar{z},z)$, and also, $C(z,\bar{z})$ is real.

By introducing the function

(2.4) $$V(z,\zeta) = U(z,\zeta)\,\exp\left\{\int_0^\zeta A(z,\zeta)\,d\zeta - \alpha(z)\right\} \quad ,$$

with $\alpha(z)$ an arbitrary holomorphic function of z, (2.2) may be put in the reduced form

(2.5) $$\underset{\sim}{L}[V] \equiv V_{z\zeta}+D(z,\zeta)V_\zeta+F(z,\zeta)V = 0 \quad ,$$

where

$$F(z,\zeta) = -(A_z+AB-C) \quad ,$$

and

$$(2.6) \qquad D(z,\zeta) = \alpha'(z) - \int_0^{\zeta} A_z(z,\zeta)d\zeta + B(z,\zeta) \quad .$$

The idea of Bergman is to seek a solution of (2.5) which may be represented in the form

$$(2.7) \qquad V(z,\zeta) = (\underset{\sim}{b}_2' f)(z,\zeta) \quad ,$$

$$(2.8) \qquad (\underset{\sim}{b}_2' f)(z,\zeta) \equiv \int_{-1}^{1} E(z,\zeta,t) f\left(\frac{z}{2}(1-t^2)\right) \frac{dt}{\sqrt{1-t^2}} \quad ,$$

where the path of integration is usually chosen so as to not pass through the origin. The generating function, E, may be seen to satisfy the equation [Be 1]

$$(2.9) \qquad (1-t^2)E_{\zeta t} - t^{-1}E_{\zeta} + 2tz(E_{z\zeta} + DE_{\zeta} + FE) = 0 \quad .$$

Formal substitution of (2.7) into (2.5) and integration by parts yields

$$(2.10) \qquad \underset{\sim}{L}[V] = -\left[E_{\zeta} \frac{\sqrt{1-t^2}}{2zt} f\right]_{-1}^{1} + \int_{-1}^{1} \left\{ \frac{E_{z\zeta} + DE_{\zeta} + FE}{\sqrt{1-t^2}} + \left(E_{\zeta} \frac{\sqrt{1-t^2}}{2zt}\right)_t \right\} f\left(\frac{z}{2}(1-t^2)\right) dt \quad .$$

<u>Problem:</u> Verify (2.10) and (2.9) under the assumption that $t^{-1}E_{\zeta}$ is continuous in $D \times D^* \times T$, $T \equiv \{t: |t| \leq 1\}$. What conditions on E are necessary to make the integrated term vanish?

Bergman [ ] showed that it was possible to construct E-functions of the form

(2.11) $$E(z,\zeta,t) = 1+\sum_{n=1}^{\infty} t^{2n}z^{n}\int_{0}^{\zeta} P^{(2n)}(z,\zeta)d\zeta \quad ,$$

where the series converges in the polydisk $\Delta(0,r)\times T \quad D\times D^{*}\times T$. The coefficients $P^{(2n)}$ are defined recursively by

(2.12) $$P^{(2n)}(z,\zeta) = -2F(z,\zeta) \quad ,$$

$$(2n+1)P^{(2n+2)}(z,\zeta) = -2P_{z}^{(2n)}+DP^{2n}+F\int_{0}^{\zeta} P^{(2n)}d\zeta \quad , \quad n \geq 1 \quad .$$

<u>Problem:</u> Obtain the recursive scheme (2.12).

If $D(z,\zeta)$, $F(z,\zeta)$ are holomorphic in $D\times D^{*}$, then for $\Delta(0,r) \subset D\times D^{*}$ these coefficients are dominated by $M(1-z/r)^{-1}(1-\zeta/r)^{-1}$. Dominants for the $P^{(2n)}$, namely the functions $\tilde{P}^{(2n)} \gg P^{(2n)}$, can be given as follows:

(2.13) $$\tilde{P}^{(0)}(z,\zeta) = 1 \ , \quad \tilde{P}^{(2)}(z,\zeta) = 2M(1-z/r)^{-1}(1-\zeta/r)^{-1} \quad ,$$

$$(2n+1)\tilde{P}^{(2n+2)}(z,\zeta)$$

$$= 2\left\{\tilde{P}_{z}^{(2n)}+M(1-z/r)^{-1}(1-\zeta/r)^{-1}\tilde{P}^{(2n)}+M(1-z/r)^{-1}(1-\zeta/r)^{-1}\int_{0}^{\zeta}\tilde{P}^{(2n)}d\zeta\right\} \quad ,$$

$$n \geq 1$$

This system of majorants may be reduced by introducing the recursively defined functions, $Q^{(2n)}(\zeta)$, by

$$\tilde{P}^{(0)}(z,\zeta) = \tilde{Q}^{(0)}(\zeta) = 1 , \quad \tilde{P}^{(2)}(z,\zeta) = \frac{\tilde{Q}^{(2)}(\zeta)}{1-z/r} ,$$

$$\tilde{P}^{(2n)}(z,\zeta) = \frac{2^{n-1}\tilde{Q}^{(2n)}(\zeta)}{\left(1-\frac{z}{r}\right)^n 1\cdot 3\cdot 5\cdot \ldots \cdot(2n-1)} , \quad n \geq 2 .$$

Then one obtains the reduced system for the $\tilde{Q}^{(2n)}(\zeta)$, $\tilde{Q}^{(2)}(\zeta) = 2M(1-\zeta/r)^{-1}$,

$$\tilde{Q}^{(2n+2)}(\zeta) = \tilde{Q}^{(2n)}(\zeta)\left[\frac{n}{r} + \frac{M}{1-\zeta/r}\right] + \frac{M}{1-\zeta/r}\int_0^\zeta \tilde{Q}^{(2n)}(\zeta)\, d\zeta ,$$

which because $\int_0^\zeta \tilde{Q}^{(2n)}(\zeta)d\zeta << \tilde{Q}^{(2n)}(\zeta)$, yields

$$\tilde{Q}^{(2n+2)}(\zeta) << \tilde{Q}^{(2n)}(\zeta)\,\frac{2(n+Mr)}{r(1-\zeta/r)} ,$$

or

$$\tilde{P}^{(2n)}(z,\zeta) << \frac{2^{(2n-2)}(n-1+Mr)(n-2+Mr)\ldots(1+Mr)Mr}{\left(1-\frac{z}{r}\right)^n r^{n-1}\left(1-\frac{\zeta}{r}\right)^{n-1} 1\cdot 3\cdot\ldots\cdot(2n-1)} .$$

Consequently, one has that

$$|E(z,\zeta,t)| < 1 + \frac{2|t|^2 M}{\left(1-\frac{|z|}{r}\right)\left(1-\frac{|\zeta|}{r}\right)} + \frac{Mr^2}{4}\left(1-\frac{|\zeta|}{r}\right)\sum_{n=2}^{\infty} |t|^{2n}\left|\frac{4z}{r}\right|^n$$

$$\cdot \frac{(n-1+Mr)\ldots(1+Mr)}{\left(1-\frac{|z|}{r}\right)^n\left(1-\frac{|\zeta|}{r}\right)^n 1\cdot 3\cdot\ldots\cdot(2n-1)}$$

Since the series $\sum \frac{(\alpha+1)_n n!}{(2n)!} (2x)^n$, $\alpha > 0$ converges for $|x| < 1$, it can be shown that the majorant series converges uniformly and absolutely in the polycylinder $\Delta(0,r\theta)\times T$, for some θ $0 < \theta < 1$.

Problem: Obtain an estimate for θ .

An alternate representation for the solution (2.7) has also been given by Bergman [Be 1] [Gi 1] pg 114. If we define the function $g(z)$ by

$$g(z) = \sum_{n=0}^{\infty} A_n z^n = \int_{-1}^{1} f\left(\frac{z}{2}[1-t^2]\right) \frac{dt}{\sqrt{1-t^2}} ,$$

with $f(z) = \sum_{n=0}^{\infty} a_n z^n$, then the Taylor coefficients of $g(z)$ are given by $A_n = \frac{\Gamma(n+1/2)\Gamma(1/2)}{2^n\Gamma(n+1)} a_n$. The reciprocal transform for $f(z)$ is given by

(2.14) $$f(z) = \frac{1}{2\pi} \int_{-1}^{1} g(2z[1-t^2]) \frac{dt}{t^2} ,$$

where the integration path is understood not to pass through the origin.

Problem: Use the integral representation for the beta function $\beta(p,q) \equiv \frac{\Gamma(p)\Gamma(q)}{\Gamma(p+q)} = \int_0^1 t^{p-1}(1-t)^{q-1}dt$ to obtain the relationship between the Taylor coefficients a_n and A_n .

Problem: Using contour integration techniques verify the inversion formula (2.14).

The general term in the representation (2.7-8) with E given by (2.11) involves the term,

$$\int_{-1}^{1} t^{2n} f\left(\frac{z}{2}[1-t^2]\right)\frac{dt}{\sqrt{1-t^2}} = \sum_{m=0}^{\infty} a_m\left(\frac{z}{2}\right)^m \int_{-1}^{1}(1-t^2)^{m-1/2}\, t^{2n}dt$$

$$= \sum_{m=0}^{\infty} a_m\left(\frac{z}{2}\right)^m \frac{\Gamma(m+1/2)\Gamma(n+1/2)}{\Gamma(m+n+1)}$$

$$= \frac{1}{2}\cdot\frac{3}{2}\cdots\frac{(2n-1)}{2}\, z^{-n}\underbrace{\int_0^z\cdots\int_0^z}_{n\text{-times}} g(z)(dz)^n \quad .$$

<u>Problem:</u> Fill in the details of the above computation. Consequently, we have that b_2' may be represented in the form

$$(\underset{\sim}{b}_2' f)(z,\zeta) = (\underset{\sim}{c}_2' g)(z,\zeta) \equiv g(z) + \sum_{n=1}^{\infty} Q^{(n)}(z,\zeta)\,\frac{(2n-1)!}{2^{2n-1}(n-1)!}\cdot\int_0^z\cdots\int_0^z g(z)(dz)^n ,$$

with $Q^{(n)}(z,\zeta) = \int_0^{\zeta} P^{(2n)}(z,\zeta)d\zeta$. If $g(0) = 0$, then we may replace the n-fold integration by $\frac{1}{n!}\int_0^z (z-\xi)^{n-1}g(\xi)d\xi$, and obtain

$$\text{(2.15)}\quad V(z,\zeta) = (\underset{\sim}{c}_2' g)(z,\zeta) \equiv g(z) + \sum_{n=1}^{\infty}\frac{Q^{(n)}(z,\zeta)}{2^{2n-1}\beta(n,n+1)}\int_0^z (z,\xi)^{n-1}g(\xi)d\xi \quad .$$

<u>Problem:</u> Using $\int_0^z (z-\xi)^k g(\xi)d\xi = \int_0^z (z-\xi)^k \frac{\partial}{\partial\xi}\int_0^{\xi} g(\eta)d\eta d\xi$ integrate by parts to verify the above substitution.

There are some other representations of (2.15) that are interesting. The most important of the ones we will give is due to I. N. Vekua [Ve 1]. His point of view is to consider the equation as a formally hyperbolic equation and to represent solutions of (2.2) using the theory of the Riemann function.

A complex Riemann function for equation (2.2) is defined as follows [Ga 1]:

$$\text{i)} \quad \hat{\underset{\sim}{E}}[R(z,\zeta;t,\tau)] \equiv 0 \text{ *},$$

(2.16)

$$\text{ii)} \quad \left[\frac{\partial}{\partial z} + B(z,\tau)\right] R(z,\tau;t,\tau) = 0 \ ,$$

$$\text{iii)} \quad \left[\frac{\partial}{\partial \zeta} + A(t,\zeta)\right] R(t,\zeta;t,\tau) = 0 \ ,$$

$$\text{iv)} \quad R(t,\tau;t,\tau) = 1 \ .$$

Here $t = \xi+i$, $\tau = \xi-i$. The characteristic conditions ii), and iii) may be written, using iv), as

(2.17)

$$\text{ii')} \quad R(z,\tau;t,\tau) = \exp\left[-\int_t^z B(\sigma,\tau)d\sigma\right] \ ,$$

and

$$\text{iii')} \quad R(t,\zeta;t,\tau) = \exp\left[-\int_\tau^\zeta A(t,\sigma)d\sigma\right] \ .$$

An equivalent definition for the Riemann function is that it is the normalized coefficient (condition iv) above) of the singular term in a fundamental solution, $S(z,\zeta;t,\tau)$, for (2.2), namely

—————————
*$\hat{\underset{\sim}{E}}$ is the formal adjoint to $\underset{\sim}{E}$.

$$S(z,\zeta;t,\tau) = \frac{1}{2} R(z,\zeta;t,\tau)\{\log(z-t)+\log(\zeta-\tau)\}+H(z,\zeta;t,\tau) \quad . \tag{2.18}$$

Here the function H is regular, and moreover, satisfies the nonhomogeneous equation

$$\underset{\sim}{L}[H(z,\zeta;t,\tau)] = \rho(z,\zeta) \quad ,$$

where

$$\rho(z,\zeta) \equiv \underset{\sim}{L}[R(z,\zeta;t,\tau)\log r] \quad , \quad r^2 = |z-t|\cdot|\zeta-\tau| \ .$$

We list the classical Riemann representation formula for $\underset{\sim}{E}[U] = f(z,\zeta)$ ([Gi 1] pg. 141),

$$\begin{aligned} U(z,\zeta) &= U(t,\tau)R(z,\zeta;t,\tau) \\ &+ \int_t^z R(z,\zeta;s,\tau)[U_s(s,\tau)+B(s,\tau)U(s,\tau)]ds \\ &+ \int_\tau^\zeta R(z,\zeta;t,\sigma)[U_\sigma(t,\sigma)+A(t,\sigma)U(t,\sigma)]d\sigma \\ &+ \int_\tau^\zeta d\sigma \int_t^z R(z,\zeta;s,\sigma)f(s,\sigma)ds \quad . \end{aligned} \tag{2.19}$$

For the special case of equation (2.5) this reduces to

$$\begin{aligned} V(z,\zeta) &= V(t,\tau)R(z,\zeta;t,\tau) \\ &+ \int_t^z R(z,\zeta;s,\tau)[V_s(s,\tau)+D(s,\tau)V(s,\tau)]ds \\ &+ \int_\tau^\zeta R(z,\zeta;t,\sigma)U_\sigma(t,\sigma)d\sigma \quad . \end{aligned} \tag{2.20}$$

The representation (2.20) solves the Goursat problem, i.e., the characteristic boundary value problem with data given on $z = t$ and on $\zeta = \tau$. If we take $t = \tau = 0$, and set

$$V(z,\tau) = \phi(z) ,$$

$$V(t,\zeta) = \phi^*(\zeta) ,$$

and define the functions $\Phi(z) = \phi(z)+\int_0^z D(t,0)\phi(t)dt$, $\Phi^*(\zeta) = \phi^*(\zeta)$ then (2.20) may be put in the form

$$V(z,\zeta) = \Phi(0)R(z,\zeta;0,0)+\int_0^z \Phi'(t)R(z,\zeta;t,0)dt$$

$$+\int_0^\zeta \Phi^{*\prime}(t)R(z,\zeta;0,\tau)d\tau$$

From (2.15) it is clear that $V(0,\zeta) = g(0)$. Recalling that the $Q^{(n)}(z,\zeta) = \int_0^\zeta P^{(2n)}(z,\zeta)d\zeta$, (2.15) also indicates that $V(z,0) = g(z)$. Hence Bergman's solution also has the representation

$$(2.21)\qquad V(z,\zeta) = g(0)R(z,\zeta;0,0)+\int_0^z R(z,\zeta;s,0)[g'(s)+D(s,0)g(s)]ds$$

$$= g(z)R(z,\zeta;z,0)+\int_0^z g(s)[D(s,0)R(z,\zeta;s,0)-R_s(z,\zeta;s,0)]ds.$$

This representation is also given by Vekua [Ve 1] pg. 123.

We next consider the special case of (2.1) where the coefficients are analytic functions of the radius squared, $r = |z|$, i.e.

$$\Delta w + a(r^2) r \frac{\partial w}{\partial r} + c(r^2) w = 0 \quad .$$

This equation may be simplified by the substitution

$$u = w \cdot \exp \left(-\frac{1}{2} \int_0^r a(r^2) r dr \right)$$

to the form

$$\Delta u + F(r^2) u = 0 \quad , \tag{2.22}$$

with $F(r^2) = -\frac{r}{2} a_r - a - \frac{r^2}{4} a^2 + c$. For this equation the E-function satisfies

$$(1-t^2) E_{rt} - t^{-1}(t^2+1) E_r + tr(E_{rr} + \frac{2}{r} E_r + FE) = 0 \quad .$$

The real solutions of (2.22) can be represented as

$$u = \mathrm{Re} \left\{ \int_{-1}^{1} E(r,t) f(z[1-t^2]) \frac{dt}{\sqrt{1-t^2}} \right\} \quad .$$

Now since E is real for real r and t one obtains

$$u = \int_{-1}^{1} E(r,t) \mathrm{Re}\{f(z[1-t^2])\} \frac{dt}{\sqrt{1-t^2}} \quad . \tag{2.23}$$

If one defines

$$h(\underset{\sim}{x}) = \int_{-1}^{1} \mathrm{Re}\{f(z\ [1-t^2])\} \frac{dt}{\sqrt{1-t^2}} \quad ,$$

and if $\mathrm{Re}\{f(z)\} = \sum a_n r^n Y_n(\theta)$ where the $Y_n(\theta)$ are circular harmonics, then by our previous computation one has that $h(\underset{\sim}{x}) = \sum \frac{\Gamma(1/2)\Gamma(n+1/2)}{\Gamma(n+1)} a_n r^n Y_n(\theta)$. If we integrate (2.23) by expanding the E-function first as

$$E(r,t) = 1+\sum_{n=1}^{\infty} e_n(r^2)t^{2n} ,$$

and then performing termwise integration, one obtains the following representation for the solution [G.]

$$(2.24) \qquad u(\underset{\sim}{x}) = h(\underset{\sim}{x})+\sum_{n=1}^{\infty} \frac{2e_n(r^2)}{\beta(n,1/2)} \int_0^1 \sigma(1-\sigma^2)^{n-1}h(\sigma^2\underset{\sim}{x})d\sigma$$

<u>Problem:</u> Verify the computation involved in (2.24).

Next, introducing $G(z) = \int_{-1}^{1} f(z[1-t^2]\frac{dt}{\sqrt{1-t^2}}$, then we may obtain the following representation for a complex solution of (2.2)

$$(2.25) \qquad V(z,\zeta) = G(z)+\sum_{n=1}^{\infty} \frac{e_n(z\zeta)z^{-n}}{2^{2n}\beta(n,n-1)} \int_0^z (z-t)^{n-1}G(t)dt .$$

Recalling the Riemann function representation (2.21) and noting from (2.17) that $R(z,\zeta;z,0) = 1$ for equations of the form (2.22) one obtains

$$\sum_{n=1}^{\infty} \frac{e_n(z\zeta)z^{-n}(z-s)^{n-1}}{2^{2n}\beta(n,n+1)} = -R_s(z,\zeta;s,0) .$$

Expanding $R(z,\zeta;s,0)$ as a Taylor series in s about center z , gives us the interesting relationship between the coefficients e_n and the Taylor coefficients of $R_s(z,\zeta;s,0)$, namely

$$e_n(z\zeta) = \frac{\sqrt{\pi}}{\Gamma(n+1/2)}(-z)^n \left[\frac{\partial^n}{\partial s^n} R(z,\zeta;s,0)\right]_{s=z} . \qquad (2.26)$$

If we put $e_n(z\bar{z})$ into the representation (2.24) one obtains

$$u(\underset{\sim}{x}) = h(\underset{\sim}{x}) + 2\sum_{n=1}^{\infty} \frac{(-1)^n z^n}{(n-1)!} \left[\frac{\partial^n}{\partial s^n} R(z,\bar{z};s,0)\right]_{s=z} \int_0^1 \sigma(1-\sigma)^{n-1} \cdot h(\underset{\sim}{x}\sigma^2)\,d\sigma$$

$$= h(\underset{\sim}{x}) - 2z \int_0^1 \sigma R_3(z,\bar{z};z\sigma^2,0)\,h(\underset{\sim}{x}\sigma^2)\,d\sigma \quad ,$$

which we obtain by regrouping the terms as the integral of a Taylor series. Since $h(\underset{\sim}{x})$ is real and we have a real solution, wherever a z appears in the function $zR_3(z,\bar{z},z\sigma^2,0)$ it must also combine with a $\bar{z}$ to yield an r^2 term. Hence, the above representation can also be written as a real integral, [Gi 2] pg. 102,

$$u(\underset{\sim}{x}) = h(\underset{\sim}{x}) - 2r \int_0^1 \sigma R_1(r,r;r\sigma^2,0)\,h(\underset{\sim}{x}\sigma^2)\,d\sigma \qquad (2.27)$$

Examples:

(A) $$\Delta u + \lambda^2 u = 0$$

For this case the E-function is already known to be $\cos \lambda rt$; hence, $e_n = (-1)^n \frac{(\lambda r)^{2n}}{(2n)!}$ and (2.27) gives us

$$u(\underset{\sim}{x}) = h(\underset{\sim}{x}) + \sum_{n=1}^{\infty} \frac{(-1)^n (r\lambda/2)^{2n}}{\Gamma(n)\Gamma(n+1)} \int_0^1 \sigma(1-\sigma^2)^{n-1} h(\underset{\sim}{x}\sigma^2)\,d\sigma$$

$$= h(\underset{\sim}{x}) - \lambda r \int_0^1 \sigma J_1(\lambda r\sqrt{1-\sigma^2})\, \frac{h(x\sigma^2)}{\sqrt{1-\sigma^2}}\, d\sigma \quad .$$

(B)

$$\Delta u + \frac{4\lambda(\lambda+1)}{(1+r^2)^2}\, u = 0 \quad .$$

In Vekna [Ve 1] the Riemann function for this equation is given by

$$R(z,\zeta;t,\tau) = P_\lambda\left(\frac{(1-z\zeta)(1-t\tau)+2z\tau+2\zeta t}{(1+z\zeta)(1+t\tau)}\right) \quad ,$$

where $P_\lambda(z) = F(\lambda+1,\ -\lambda;\ 1;\ \frac{1}{2}[1-z])$ is a Legendre function of degree λ.

It may be shown [Gi 2] pg. 104 that

$$e_n(r^2) = \sqrt{\pi}\left(\frac{r^2}{1+r^2}\right)^n \frac{\Gamma(\lambda+n+1)\Gamma(n-\lambda)}{\Gamma(\lambda+1)\Gamma(-\lambda)n!\Gamma(n+1/2)}$$

and hence

$$E(r,t) = \sqrt{\pi}\left(\frac{rt}{1+r^2-r^2t^2}\right)^{1/4} P_\lambda^{1/2}\left(\frac{1+r^2-2r^2t^2}{1+r^2}\right) \quad ,$$

where P_λ^μ is an associated Legendre function. One also has the general representation of the form

$$u(\underset{\sim}{x}) = h(\underset{\sim}{x}) - \frac{2\lambda(\lambda+1)r^2}{1+r^2} \int_0^1 \sigma F\left(\lambda+2,\ 1-\lambda;\ 2;\ \frac{r^2(1-\sigma^2)}{1+r^2}\right) \cdot h(\underset{\sim}{x}\sigma^2)\,d\sigma$$

The preceeding method can also be used the the elliptic equation with radially dependent analytic coefficients in n-dimensional space,

(2.28)
$$\Delta_n u + F(r^2)u = 0 \quad .$$

It can be shown [Gi 2] that if $E(r,t;n)$ is a solution of the partial differential equation

(2.29)
$$(1-t^2)E_{rt} + (n-3)(t^{-1}-t)E_r + rt(E_{rr} + \frac{n-2}{r} E_r + FE) = 0$$

which satisfies the boundary conditions

(2.30)
$$\lim_{t \to 0^+} (t^{n-3}E_r)r^{-1} = 0 \quad , \qquad \lim_{t \to 1^-} \left(\sqrt{1-t^2}\, E_r\right) r^{-1} = 0 \quad ,$$

$\lim_{r \to 0^+} E = 1$, and if $H(\underset{\sim}{x})$ is an arbitrary harmonic function defined in starlike with respect to the origin, the the function

(2.31)
$$u(\underset{\sim}{x}) = \int_0^1 E(r,t;n) H(\underset{\sim}{x}[1-t^2]) \frac{dt}{\sqrt{1-t^2}}$$

is a solution of (2.28).

Problem: Verify that (2.31) is a solution of (2.28).

If one seeks solutions of (2.29), (2.30) in the form

(2.32)
$$E(r,t;n) = 1 + \sum_{\ell=1}^{\infty} e_\ell(r;n) t^{2\ell} \quad ,$$

then the coefficients e_ℓ are defined recursively by

$$(n-1)e_1' = -rF$$

$$(2\ell+n-3)e_\ell' = (2\ell-3)e_{\ell-1}' - re_{\ell-1}'' - rFe_{\ell-1} \quad ,$$

$$\ell \geq 2 \ , \quad \text{with} \quad e_\ell(0,n) = 0 \ , \quad \ell \geq 1 \quad .$$

Assuming for the moment such solutions exist, and defining the harmonic function $h(\underset{\sim}{x})$ by

$$\text{(2.33)} \qquad h(\underset{\sim}{x}) = \int_0^1 t^{n-2} H(\underset{\sim}{x}[1-t^2]) \frac{dt}{\sqrt{1-t^2}} \quad ,$$

it may be shown by formal manipulation of series that the solution (2.31) also has the representation

$$\text{(2.34)} \qquad u(\underset{\sim}{x}) = h(\underset{\sim}{x}) + \sum_{=1}^{\infty} c_\ell(r;n) \int_0^1 \sigma^{n-1}(1-\sigma^2)^{\ell-1} h(\underset{\sim}{x}\sigma^2)\, d\sigma \quad ,$$

with

$$c_\ell(r;n) = 2e_\ell(r;n) \frac{\Gamma(\ell+n/2-1/2)}{\Gamma(n/2-1/2)\Gamma(\ell)} \quad .$$

<u>Problem:</u> Verify the formal computations involved in obtaining the representation (2.33), (2.34).

If one introduces the function $G(r,1-\sigma^2)$ by the series

(2.35) $$G(r,\tau) = \sum_{\ell=1}^{\infty} c_\ell(r;n)\tau^{\ell-1} \quad ,$$

then G formally satisfies the hyperbolic equation

(2.36) $$2(1-\tau)G_{r\tau}-G_r+r(G_{rr}+F(r^2)G) = 0 \quad .$$

Note that $G(r,\tau)$ must also satisfy the Goursat data

(2.37) $$G(0,\tau) = 0 \quad , \qquad G(r,0) = -\int_0^r F(r^2)rdr \quad ,$$

because $e_\ell(0,n) = c_\ell(0,n) = 0$, and $G(r,0) = c_1(r;n)$. The G-function defined by (2.35), (2.37) is clearly not a function of the dimension n . Hence, the dimension of the space appears in just a simple manner in the representation (2.34), which we may rewrite as

(2.38) $$u(\underset{\sim}{x}) = h(\underset{\sim}{x})+\int_0^1 \sigma^{n-1}G(r,1-\sigma^2)h(\underset{\sim}{x}\sigma^2)d\sigma \quad , \qquad n \geq 2 \quad .$$

When $n = 2$, the function $E(r,t;n)$ defined by (2.32) becomes Bergman's E-function, and hence one must have by (2.38) that

(2.39) $$G(r,1-\sigma^2) = -2rR_3(r,r;r\sigma^2,0) \quad .$$

Furthermore, the relationship between the coefficients e_ℓ and c_ℓ imply that the E-function can be expressed as an integral transformation of the G-function, namely

(2.40) $$E(r,t;n) = 1+\frac{t^2}{2}\int_0^1 (1-s)^{n/2-3/2}\, G(r,st^2)ds \quad , \qquad n \geq 2 \quad .$$

The integral transform (2.40) indicates that the E-function exists; furthermore, (2.27) indicates that the Taylor series (2.35) converges for $|\tau| \le 1$ with $|r| \le a$ (if the coefficient $F(r^2)$ is analytic in $|r| \le a$).

Problem: Verify directly that the series representation for the E-function converges in $\{|r| \le a\ ,\ |t| \le 1\}$.

Examples:

(A)
$$\Delta_n u + \lambda^2 u = 0$$

If we look for a solution in the form $f(\lambda rt)$ we find an E-function of the form

$$E(r,t;n) = \Gamma(n/2-1/2)\left(\frac{\lambda rt}{2}\right)^{-(n-3)/2} J_{(n-3)/2}(\lambda rt)\ ,$$

and since $e_\ell = (-1)^\ell \left(\frac{\lambda r}{2}\right)^{2\ell} \frac{1}{\ell!\Gamma(n/2-1/2+\ell)}$ one has

$$G(r,1-\sigma^2) = \frac{-\lambda r}{\sqrt{1-\sigma^2}} J_1\left(\lambda r\sqrt{1-\sigma^2}\right)\ .$$

(B)
$$\Delta_n u + \frac{\lambda(\lambda+1)}{(1+r^2)^2} u = 0$$

In [Gi 2] it is shown that if one seeks a solution of the form $f(X)$, with $X = r^2t^2(1+r^2)^{-1}$, then an E-function may be found of the form

$$E(r,t;n) = F(\lambda+1,-\lambda;n/2-1/2;r^2t^2(1+r^2)^{-1})$$

$$= \Gamma(n/2-1/2)\left[\frac{rt}{1+r^2(1-t^2)}\right]^{(3-n)/4} P_\lambda^{(3-n)/2}\left(\frac{1+r^2-2r^2t^2}{1+r^2}\right)$$

and a corresponding G-function

$$G(r,1-\sigma^2) = \frac{-2\lambda(\lambda+1)r^2}{1+r^2} F\left(\lambda+2;1-\lambda;2;\frac{r^2(1-\sigma^2)}{1+r^2}\right) .$$

Remark: The representations given in (A) were already known for $n > 2$ by Vekua [Ve 1] whereas, the representations in (B) were new [Gi 2].

Problem: Show that the function $K(r,t) \equiv t^{n-2}\frac{E(r,t;n)}{\sqrt{1-t^2}}$ satisfies

$$(1-t^2)K_{rt} - t^{-1}(t^2+1)K_r + rt(K_{rr} + \frac{n-1}{r} K_r + FK) = 0 ,$$

and that the transformed function

$$k(r,\alpha) \equiv \int_0^1 (1-t^2)^{\alpha} K(r,t)\,dt$$

satisfies

$$k'' + \frac{1}{r}(n+2\alpha-1)k' + Fk = 0 ,$$

with the two initial conditions $k(0,\alpha) = \frac{\Gamma(n/2-1/2)\Gamma(\alpha+1/2)}{\Gamma(n/2+\alpha)}$, $k'(0,\alpha) = 0$.

Problem: If $\mathcal{L}_u\{f\} \equiv \int_0^{\infty} e^{-u\alpha} f(u)\,du$ (the Laplace transform of f), show that

$$k(r,\alpha) = {}_u \frac{K(r,[1-e^{-u}]^{1/2}e^{-u/2}}{\sqrt{e^u-1}} .$$

For the case with $F(r^2) = \lambda^2$ show that

$$k(r,\alpha) = \Gamma(n/2-1/2)\Gamma(\alpha+1/2)\left(\frac{\lambda r}{2}\right)^{-(n/2+\alpha-1)} J_{n/2+\alpha-1}(\lambda r) ,$$

and hence

$$K(r,t) = t^{n-2} \frac{\Gamma(n/2-1/2)}{\sqrt{1-t^2}} \left(\frac{\lambda rt}{2}\right)^{-(n-3)/2} J_{(n-3)/2}(\lambda rt) \quad .$$

III. Boundary Value Problems Associated with $e[u] = 0$

In this section, we shall assume that the domains are bounded, star-like with respect to the origin, and Lyapunov. Such domains we shall refer to as being appropriate. The first case we shall consider is the equation with radially dependent coefficients, i.e., equation (2.22), or since no further difficulties are encountered, its n-dimensional analogue (2.28). If $a = \sup_{\underset{\sim}{x} \in D} r$, then we assume that $F(r^2)$ is analytic for $|r| \leq a$. The G-function representation for solutions of (2.28) it will be seen are effective in treating these boundary value problems. We adopt the notation

$$(3.1) \qquad u(\underset{\sim}{x}) = (\underset{\sim}{I}+\underset{\sim}{G})h(\underset{\sim}{x}) \equiv h(x)+\int_0^1 \sigma^{n-1}G(r,1-\sigma^2)h(\underset{\sim}{x}\sigma^2)d\sigma \quad .$$

We consider first the Dirichlet problem for (2.28) with continuous data, i.e., $u(\underset{\sim}{x}) = f(\underset{\sim}{x})$, $\underset{\sim}{x} \in \partial D$, where D is an appropriate domain, and $F(x) \leq 0$ in D. It is well known that the solution to this problem exists and is unique.

It is interesting that the solution of the above Dirichlet problem also has a representation in the form (3.1). To see this, we first show that the operator $\underset{\sim}{I}+\underset{\sim}{G}$ has an inverse on the class of functions that are in $\mathcal{C}\,(D+\partial D)$. By a simple change of integration parameter, the representation (3.1) may be rewritten as a Volterra integral equation (where we have suppressed all but the radial variables),

$$\hat{u}(r,\cdot) = \hat{h}(r,\cdot)+\int_0^r K(r,\rho)\hat{h}(\rho,\cdot)d\rho \quad ,$$

with $K(r,\rho) = \frac{1}{2}\left(\frac{\rho}{r}\right)^{n/2-1/2} \frac{G(r,1-\rho/r)}{\sqrt{r\rho}}$, $\hat{u}(r,\cdot) = ur^{n/2-1}$,

$\hat{h}(r,\cdot) = hr^{n/2-1}$. Since this Volterra integral equations may always be solved $(n > 2)$ we may write

$$h(r,\cdot) = u(r,\cdot)+r^{-n/2+1}\int_0^r \Gamma(r,\rho)\rho^{n/2-1}\, u(\rho,\cdot)d\rho$$

with $\Gamma(r,\rho) = \sum_{\ell=1}^{\infty} K^{(\ell)}(r,\rho)$,

$$K^{(\ell+1)}(r,\rho) = \int_\rho^r K(r,t)K^{(\ell)}(t,\rho)dt \ , \quad \ell \geq 1 \ .$$

Hence $(\underset{\sim}{I}+\underset{\sim}{G})^{-1}$ exists on $\mathbb{C}(D \cup \partial D)$. Next, we consider whether each solution of (2.28) which is in $\mathbb{C}(D \cup \partial D)$ has a representation of the form (3.1).

<u>Problem:</u> Show that the following identity holds for all functions $\phi(\underset{\sim}{x}) = (\underset{\sim}{I}+\underset{\sim}{G})\psi(\underset{\sim}{x})$, such that $\psi(\underset{\sim}{x}) \in \mathbb{C}^2(D)$:

(3.2) $$\Delta\phi+F\phi = (\underset{\sim}{I}+\underset{\sim}{G})\Delta\psi(\underset{\sim}{x}) \ , \quad \underset{\sim}{x} \in D \ .$$

Using the identity (3.2) we note that if ϕ is a solution of (2.28) in D then one has

(3.3) $$(\underset{\sim}{I}+\underset{\sim}{G})\Delta\psi = 0 \ , \quad \text{or} \quad r^{n/2-1}\Delta\psi+\int_0^r K(r,\rho)\rho^{n/2-1}\Delta\psi(\rho,\cdot)d\rho = 0 \ .$$

The only solution to this homogeneous equation is the trivial solution; hence, $\psi(\underset{\sim}{x})$ must be harmonic. We summarize the above discussion as:

<u>Theorem 3.1:</u> <u>Let</u> D <u>be appropriate and</u> $F(\underset{\sim}{x}) \leq 0$ <u>in</u> D , <u>then the correspondence</u> (3.1) <u>between harmonic functions in</u> $\mathbb{C}(D \cup \partial D)$ <u>and solutions</u>

of (2.28) in $\mathcal{C}$ (D+∂D) is one-to-one.

Because these correspond to each solution of (2.28) that is also in $\mathcal{C}$ (D+∂D) a harmonic function in $\mathcal{C}$ (D+∂D), and since each such harmonic function has a representation as a double-layer potential one may express $u(\underset{\sim}{x})$ in the form

$$u(\underset{\sim}{x}) = (\underset{\sim}{I}+\underset{\sim}{G})h(\underset{\sim}{x}) \quad ,$$

$$(3.4) \qquad h(\underset{\sim}{x}) = \begin{cases} \dfrac{\Gamma(n/2)}{(n-1)\pi^{n/2}} \displaystyle\int_{\partial D} \mu(\underset{\sim}{y}) \frac{\partial}{\partial \nu_y} \left(\frac{1}{|\underset{\sim}{x}-\underset{\sim}{y}|^{n-2}}\right) dw_y \ , & n > 2 \\[2ex] \dfrac{1}{\pi} \displaystyle\int_{\partial D} \mu(\underset{\sim}{y}) \frac{\partial}{\partial \nu_y} \log \left(\frac{1}{|\underset{\sim}{x}-\underset{\sim}{y}|}\right) dw_y \ , & n = 2 \ , \end{cases}$$

where $\mu(\underset{\sim}{y})$ is the double-layer density $\underset{\sim}{\nu}$ the normal direction to ∂D in the inner direction, and dw_y the surface measure. Using the Fubini theorem to interchange orders of integration, and letting $\underset{\sim}{x} \to \partial D$, yields a Fredholm integral equation for the density $\mu(\underset{\sim}{y})$, namely [Gi 2]

$$f(\underset{\sim}{x}) = \mu(\underset{\sim}{x}) + \int_{\partial D} K(\underset{\sim}{x},\underset{\sim}{y})\mu(\underset{\sim}{y})dw_y \ , \quad \underset{\sim}{x} \quad \partial D \ ,$$

with

$$(3.5) \qquad K(\underset{\sim}{x},\underset{\sim}{y}) = \frac{\Gamma(n/2)}{(n-1)\pi^{n/2}} \left\{ \frac{\partial}{\partial \nu_y} \left(\frac{1}{|\underset{\sim}{x}-\underset{\sim}{y}|^{n-2}}\right) + \int_0^1 \sigma^{n-1} G(r,1-\sigma^2) \frac{\partial}{\partial \nu_y} \left(\frac{1}{|\underset{\sim}{x}\sigma^2-\underset{\sim}{y}|^{n-2}}\right) d\sigma \right\}, \quad n > 2 \ .$$

<u>Remark</u>: The fact that the Fredholm integral equation (3.5) has a unique solution depends on the fact that a unique solution does exist to the Dirichlet problem for $\Delta u+Fu = 0$ in D, $u = f$ on ∂D, and that there corresponds to this solution a unique harmonic function in $\mathfrak{C}(D \cup \partial D)$. The density $\mu(\underset{\sim}{y})$ is the unique double-layer density which generates this harmonic function.

The Neumann problem $\Delta u+Fu = 0$, $\frac{\partial u}{\partial \nu} = g$ may be treated similarly. Here we introduce for the harmonic function the single-layer potential

$$h(\underset{\sim}{x}) = \frac{\Gamma(n/2)}{(n-1)\pi^{n/2}} \int_{\partial D} \rho(\underset{\sim}{y}) \frac{1}{|\underset{\sim}{x}-\underset{\sim}{y}|^{n-2}} dw_y ,$$

which leads to the Fredholm equation [Gi 3]

$$g(\underset{\sim}{x}) = -\rho(\underset{\sim}{x}) + \int_{\partial D} \rho(\underset{\sim}{y}) K(\underset{\sim}{x},\underset{\sim}{y}) dw_y ,$$

with

$$K(\underset{\sim}{x},\underset{\sim}{y}) = \frac{\Gamma(n/2)}{(n-1)\pi^{n/2}} \frac{\partial}{\partial \nu_x} \left\{ \left(\frac{1}{|\underset{\sim}{x}-\underset{\sim}{y}|^{n-2}} \right) \right.$$

(3.6)

$$\left. + \int_0^1 \sigma^{n-1} G(r,1-\sigma^2) \frac{1}{|\underset{\sim}{x}\sigma^2-\underset{\sim}{y}|^{n-2}} d\sigma \right\} , \quad n > 2 .$$

The Fredholm equations (3.5), and (3.6) suggest approximation methods for solving boundary value problems of the first and second kind. We restrict our attention to problems of the first kind; problems of the second kind may be treated analogously.

We introduce "approximations" of the operator $\underset{\sim}{G}$ as follows: let

$$(\underset{\sim}{G}_N h)(\underset{\sim}{x}) \equiv \int_0^1 \sigma^{n-1} G_N(r,1-\sigma^2) h(\underset{\sim}{x}\sigma^2)\, d\sigma \quad ,$$

(3.7) where

$$G_N(r,1-\sigma^2) = \sum_{\ell=1}^{N} c_\ell(r^2)(1-\sigma^2)^{\ell-1} \quad .$$

This suggest we study the "approximating" Fredholm equations,

$$f(\underset{\sim}{x}) = (\underset{\sim}{I}+\underset{\sim}{K}_N)\mu(\underset{\sim}{x}) \quad , \quad (N = 1,2,\ldots) \quad , \tag{3.8}$$

where $\underset{\sim}{K}_N$ is the integral operator which occurs when the term $G(r,1-\sigma^2)$ in the definition of the kernel $K(\underset{\sim}{x},\underset{\sim}{y})$ is replaced by $G_N(r,1-\sigma^2)$.

Using straight forward estimates involving the coefficients of $G(r,t)$ it is possible to show that the operators $\underset{\sim}{K}$ and $\underset{\sim}{K}_N$ are compact on $\mathcal{C}(\partial D)$. In particular, by showing that

$$\max_{\partial D} \left\{ \int_{\partial D} |K(\underset{\sim}{x},\underset{\sim}{y})|\, dw_y \right\} < \infty \quad , \tag{3.9}$$

and that

$$\lim_{(x_1-x_2)\to 0} \left\{ \int_{\partial D} |K(\underset{\sim}{x}_1,\underset{\sim}{y})-K(\underset{\sim}{x}_2,\underset{\sim}{y})|\, dw_y \right\} = 0 \tag{3.10}$$

uniformly in $\underset{\sim}{x}_1,\underset{\sim}{x}_2$, it follows from the Arzela-Ascoli theorem that $\underset{\sim}{K}$ is compact [At1]. We can show that (3.9) is true by recalling the integral identity

$$(3.11) \qquad \frac{\Gamma(n/2)}{(n-1)\pi^{n/2}} \int_{\partial D} \frac{\partial}{\partial \nu_y} \left(\frac{1}{|x-y|^{n-2}} \right) dw_y = \begin{cases} 1 , & x \in D \\ 1/2 , & x \in \partial D \\ 0 , & x \in D \cup \partial D \end{cases} .$$

The condition (3.10) follows also by using (3.11) plus the continuity of $G(r,1-\sigma^2)$.

If $\mu_N(x)$ denotes a solution of (3.8) we can show that $\mu_N(x) \to \mu(x)$ as $N \to \infty$, where $\mu(x)$ is the unique solution of (3.5). To this end we recall two inequalities from Taylor[Ta 1],pg. 164, namely

$$(3.12) \qquad ||(I+K_N)^{-1}|| \le \frac{||(I+K)^{-1}||}{1-||(I+K)^{-1}||\cdot||K-K_N||} ,$$

and

$$(3.13) \qquad ||(I+K_N)^{-1}-(I+K)^{-1}|| \le \frac{||(I+K)^{-1}||^2\cdot||K-K_N||}{1-||(I+K)^{-1}||\cdot||K-K_N||} .$$

Since we know that $(I+K)^{-1}$ exists, the inequality (3.12) implies that $(I+K_N)^{-1}$ exists; the inequality (3.13), furthermore, implies that $(I+K_N)^{-1} \to (I+K)^{-1}$.

These ideas have been used by Atkinson to obtain a computationally feasible approach to the Dirichlet problem for (2.28) [At 2],[At 3], [G 3].

There are other constructions of the solution of the Dirichlet problem which are suggested by the integral operator approach. For instance, in the case of dimension 2, if $h(x)$ is replaced in the representation $u(x) = (I+G)h(x)$ by $\frac{1}{2\pi} \log \frac{1}{|x-y|}$ then one obtains a fundamental singular solution for the differential equation,

$$S(\underset{\sim}{x},\underset{\sim}{y}) = \frac{1}{2\pi}\log\frac{1}{|x-y|}\left\{1+\int_0^1 \sigma\, G(r,1-\sigma^2)\right\}$$

$$(3.14) \qquad - \frac{1}{\pi}\int_0^1 \sigma\log\sigma\, G(r,1-\sigma^2)\,d\sigma$$

$$= \frac{1}{2\pi} R(r,r;0,0)\log\frac{1}{|\underset{\sim}{x}-\underset{\sim}{y}|} + \frac{1}{\pi}\int_0^1 \log\sigma\, R_\sigma(r,r;r\sigma^2,0)\,d\sigma \quad ,$$

$x = (x_1,x_2)$, $y = (y_1,y_2)$. The fundamental singular solutions are, as is well known, of use in obtaining representations for solutions of boundary value problems. One such method, given by Bergman and Schiffer [B S 1], is to construct the sequence of functions

$$V_{3k}(\underset{\sim}{x}) \equiv S(\underset{\sim}{x},\underset{\sim}{y}) \qquad , \quad \underset{\sim}{y} = \underset{\sim}{y}_k \ ,$$

$$(3.15) \qquad V_{3k+1}(\underset{\sim}{x}) \equiv \frac{\partial}{\partial y_1} S(\underset{\sim}{x},\underset{\sim}{y}) \ , \quad \underset{\sim}{y} = \underset{\sim}{y}_k \ ,$$

$$V_{3k+2}(\underset{\sim}{x}) \equiv \frac{\partial}{\partial y_2} S(\underset{\sim}{x},\underset{\sim}{y}) \ , \quad \underset{\sim}{y} = \underset{\sim}{y}_k \ , \quad (k = 1,2,\ldots) \ ,$$

where the points $\underset{\sim}{y}_k$ range over a denumerable set outside of D but which has every point on ∂D as an accumulation point. We note that every solution of (2.28) may be written in the form

$$u(\underset{\sim}{x}) = \int_C \left[u(\underset{\sim}{y})\frac{\partial}{\partial \nu_y} S(\underset{\sim}{x},\underset{\sim}{y}) - S(\underset{\sim}{x},\underset{\sim}{y})\frac{\partial u(\underset{\sim}{y})}{\partial \nu_y}\right] ds_y \ ,$$

where $C = \partial D$ and ds_y is the arc length differential. In any relatively compact subdomain D_o of D we may approximate this integral by a finite sum

$$u(\underset{\sim}{x}) = \sum_{\ell=1}^{N} \left[u(\underset{\sim}{\xi}_\ell) \frac{\partial}{\partial \nu_\ell} S(\underset{\sim}{x}, \underset{\sim}{\xi}_\ell) - S(\underset{\sim}{x}, \underset{\sim}{\xi}_\ell) \frac{\partial}{\partial \nu_\ell} u(\underset{\sim}{\xi}_\ell) \right] \Delta s_\ell$$

$$+ \eta(N, D_o),$$

where the points ξ_ℓ are points on C, chosen sufficiently close to each other, and $\eta(N,D_o)$ can be made arbitrarily small for each $D_o \subset\subset D$ by taking N sufficiently large. Now if $\underset{\sim}{x} \in D_o \subset\subset D$ then $S(\underset{\sim}{x}, \underset{\sim}{\xi})$ and $\text{grad}_\xi\ S(\underset{\sim}{x}, \xi)$ are uniformly continuous functions of $\xi \not\in D$. Thus the terms

$$S(\underset{\sim}{x}, \underset{\sim}{\xi}_\ell), \text{ and } \frac{\partial S}{\partial \nu_\ell}(\underset{\sim}{x}, \underset{\sim}{\xi}_\ell) = \frac{\partial S}{\partial \xi_{\ell,1}}(\underset{\sim}{x}, \xi_\ell) \cos(\nu_\ell, \xi_{\ell,1}) +$$

$$+ \frac{\partial S(\underset{\sim}{x}, \xi_\ell)}{\partial \xi_{\ell,2}} \cos(\nu_\ell, \xi_{\ell,2}),$$

$\xi_\ell = (\xi_{\ell,1}, \xi_{\ell,2})$, can be approximated arbitrarily closely by linear combinations of the functions (3.14).

Theorem 3.2: *Every function* $u(\underset{\sim}{x})$, *which is a solution of* (2.28), *may be approximated uniformly in any relatively compact subset of* D *by linear combinations of the functions* V_k *defined in* (3.15, (3.14).

We may introduce a metric in the linear space Σ of solutions of (2.28) and make it a Hilbert space; see for example, Garabedian [Ga 1]. To this end we introduce the inner product

$$(u,v) \equiv \int_D \left[\nabla u \cdot \nabla v - Fuv\right] d\underset{\sim}{x} = -\int u \frac{\partial v}{\partial \nu} ds. \tag{3.16}$$

We may show that the system of functions (3.15) is complete in the Hilbert space $\mathcal{H}$ of solutions possessing the Dirichlet inner product (3.15). By this we mean that every function $u \in \mathcal{H}$ can be approximated arbitrarily closely by a combination of one functions $v_k(\underset{\sim}{x})$ in one sense of the Dirichlet metric $\|u\| = (u,u)$, i.e. given an $\epsilon > 0$ $\exists$ a linear combination

$$\sum_{k=1}^{N} \alpha_{Nk} v_k(\underset{\sim}{x})$$

such that

$$\|u(\underset{\sim}{x}) - \sum_{k=1}^{N} \alpha_{Nk} v_k(\underset{\sim}{x})\| < \epsilon . \tag{3.17}$$

We show this as follows: let $u \in \Sigma$ and furthermore, be $C^2(D^*) \ni \bar{D} \subset D^* \subset D_1$. Then u can be approximated in the form (3.17) above. Applying our constructive method for obtaining the functions $\{\mathfrak{V}_k\}$ to the domain D_1 and D^*, we may approximate u and grad u uniformly by a linear combination of the $\{\mathfrak{V}_k\}$ and their gradients. It is easy to see then that the inequality (3.17) may be satisfied. Now let us consider the nested sequence of domains $D_{n+1} \subset D_n \ni D + \partial D \subset D_n$, which converge towards D. Let $K_n(\underset{\sim}{x},\underset{\sim}{y})$ be the kernel function of these domains for the corresponding classes of solutions Σ_n. Then as may be seen in Bergman-Schiffer [B S 1] one has for $u \in \Sigma$ that the functions

$$u_n(\underset{\sim}{x}) = (K_n(\underset{\sim}{x},\underset{\sim}{y}),\ u(\underset{\sim}{y})) \tag{3.18}$$

are of class $\Sigma \cap \Sigma_n$. Hence each $u_n(\underset{\sim}{x})$ can be approximated by functions $V_\nu(\underset{\sim}{x})$ in the sense of the Dirichlet metric. If the u_n converge to u in the Dirichlet metric, the completeness of the $\{\mathfrak{V}_k(\underset{\sim}{x})\}$ w.r.t. Σ would be established; i.e. given an $\epsilon > 0$ $\exists$ an n sufficiently large so that $\|u - u_n\| < \frac{\epsilon}{4}$, and $\exists$ a linear combination of the

$\mathcal{V}_k$, namely, $V_n \ni \|V_n - u_n\| < \frac{\varepsilon}{4}$

Hence, $$\|u - V_n\| = \|u - u_n + u_n - V_n\| < \frac{\varepsilon}{2}$$

which shows the completeness. Therefore it is necessary only that we show that $u_n \to u$ in the Dirichlet metric. It is known that the Green's $G(\underset{\sim}{x},\underset{\sim}{y})$ and Neumann's $N(\underset{\sim}{x},\underset{\sim}{y})$ functions depend continuously on the domain [B.S.1] Chapt. III, and hence, since $K(\underset{\sim}{x},\underset{\sim}{y}) \equiv N(\underset{\sim}{x},\underset{\sim}{y}) - G(\underset{\sim}{x},\underset{\sim}{y})$, that $u_n(\underset{\sim}{x})$ and grad $u_n(\underset{\sim}{x})$ converge uniformly to $u(\underset{\sim}{x})$ and grad $u(\underset{\sim}{x})$ respectively in each relatively compact subset of D.

Problem: Prove this.

It is furthermore shown in [B S 1] that $\|u_n\| \le \|u\|$. If the integrals $\|u\|_D$ are taken in the Lebesgue sense over D one then has that $\lim \|u_n\|_D = \|u\|_D$. Now if $\varepsilon > 0$ is arbitrarily prescribed one can determine a closed subdomain $\Delta \subset D \ni \|u\|_{D-\Delta} < \varepsilon$. Depending on ε choose an $N(\varepsilon) \ni$ for $n > N(\varepsilon)$ we have

$$| \|u\|_\Delta - \|u_n\|_\Delta | < \varepsilon \quad \text{and} \quad \|u - u_n\|_\Delta < \varepsilon$$

Then

$$\|u_n\|_{D-\Delta} = \|u_n\| - \|u_n\|_\Delta = \|u_n\| - \|u\| - \{\|u_n\|_\Delta - \|u\|_\Delta\} + \|u\|_{D-\Delta} < 3\varepsilon .$$

We then obtain that

$$\|u - u_n\| = \|u - u_n\|_\Delta + \|u - u_n\|_{D-\Delta} \leq \varepsilon + \varepsilon(1+\sqrt{3})^2$$

since

$$\text{(3.19)} \quad \|u - u_n\|_{D-\Delta} = (\|u\|^{\frac{1}{2}}_{D-\Delta} + \|u_n\|^{\frac{1}{2}}_{D-\Delta})^2 = (\varepsilon^{\frac{1}{2}} + \sqrt{3}\ \varepsilon^{\frac{1}{2}})^2 = \varepsilon(1+\sqrt{3})^2$$

We conclude that $u_n \to n$ in the Dirichlet metric.

Another method for obtaining a complete family of solutions can be obtained by mapping the harmonic functions onto solutions of (2.28). Let

$$H(p;\ m_k;\ \pm;\ \underset{\sim}{x}) = r^p Y(p;\ m_k;\ \theta_k,\ \pm\varphi),\ 0 \leq m_{n-2} \leq m_{n-3} \leq \ldots \leq m_1 \leq p,$$

$$p = 0,\ 1,\ 2, \ldots,$$

represent the homogeneous harmonic polynomials of degree p; see Endélyi (Higher Transcendental Functions, Vol II, Pg. 240) for a definition of these polynomials. Then if the domain D is appropriate, the functions $\psi(p;\, m_k;\, \pm;\, \underset{\sim}{x})$ defined by

$$\psi(p;\, m_k;\, \pm;\, \underset{\sim}{x}) = H(p;\, m_k;\, \pm;\, \underset{\sim}{x}) + \int_0^1 \sigma^{n-1}\, G(r, 1-\sigma^2)\, H(p;\, m_k;\, \pm;\, \underset{\sim}{x}\sigma^2)\, d\sigma \tag{3.20}$$

are complete w.r.t. with the Dirichlet inner product.

Proof: As in the previous development we may show if the functions ψ are complete with respect to uniform approximations in relatively compact subsets of D then they are complete in the Dirichlet metric. We turn to showing that they are complete w.r.t. uniform approximation. To this end we suppose that the set of functions $\psi(\underset{\sim}{x})$ are not complete. Then $\exists$ a function $u(\underset{\sim}{x})$ in the linear space of solutions to (2.28) which cannot be approximated arbitrarily closely by combinations of the functions $\psi(\underset{\sim}{x})$. However, to this function we may associate a function

$$h(\underset{\sim}{x}) = (\underset{\sim}{I} + \underset{\sim}{G})^{-1}\, u(\underset{\sim}{x}).$$

However, this leads us to a contradiction, since a result due to du Plessis [Pl 1] says that each harmonic function can be approximated arbitrarily closly on compact subsets.

Lemma: If D is appropriate and $F(r^2) \le 0$, then the operator $(\underset{\sim}{I} + \underset{\sim}{G})$ is monotone in the sense of Collatz on $C(D)$.

Proof: This follows immediately from the fact that $G(r,t) \ge 0$ in $[0,a] \times [0,1]$. To show that $G(r,t) \ge 0$ we prove instead the stronger claim: if $0 \le \lambda_1 \le F(r^2) \le \lambda_2$ for all $r \in [0,a]$ and $F(r^2)$ is analytic, then

(3.21) $$\frac{r\lambda_1}{\sqrt{t}} I_1(\lambda_1 rt) \le G(r,t) \le \frac{r\lambda_2}{\sqrt{t}} I_1(\lambda_2 r \sqrt{t}),$$

for $(r,t) \in [0,a] \times [0,1]$ where $I_1(Z)$ is a modified Bessel function of the first kind and of order one.

Proof. If $F = \lambda^2$ then $G(r,t) = \frac{r\lambda}{\sqrt{t}} I_1(\lambda rt)$. Furthermore, it is a consequence of the integral equation for the function $W(\rho,\tau) = (1-\tau)G$, $p = r\sqrt{1-t}$, $\tau = t$, namely

$$w(\rho,\tau) = \int_0^\rho \rho\, P(\rho^2)d\rho + \int_0^\rho d\rho \int_0^\tau d\tau \frac{\rho\, P\left(\frac{\rho^2}{1-\tau}\right) W}{(1-\tau)^2},$$

that if $F_2(r^2) \geq F_1(r^2)$ and W_1, W_2 are solutions corresponding to F_1, F_2 respectively that $W_2 \geq W_1$ in $[0,a] \times [0,1]$. The same is then true for the functions G_1 and G_2. That solutions exist for F_k $(k=1,2)$ analytic was already shown via the Riemann function representation.

Let $E^-(D)$ and $E^+(D)$ represent the class of sub (super) solutions of (2.28) which we define formally for our purposes as follows:

<u>Definition</u>: A function $w(\underset{\sim}{x})$ is said to be a sub (super) solution of (2.28) in D if the following conditions hold:

(i) $w(\underset{\sim}{x})$ is continuous in D

(ii) if $D_o \subset\subset D$ and $u(\underset{\sim}{x})$ is a solution of (2.28) in D_o, and $w(\underset{\sim}{x}) - u(\underset{\sim}{x}) \leq 0$ $(w(\underset{\sim}{x}) - u(\underset{\sim}{x}) \geq 0)$ on ∂D_o, then $w(\underset{\sim}{x}) < u(\underset{\sim}{x})$ $(w(\underset{\sim}{x}) > u(\underset{\sim}{x}))$ in D_o, unless $w(\underset{\sim}{x}) \equiv u(\underset{\sim}{x})$ in D_o.

<u>Remark</u>: If $F \equiv 0$ then E^- and E^+ become the classes of subharmonic $\mathfrak{H}^-$ and superharmonic functions $\mathfrak{H}^+$ respectively.

Since we have shown that $\Delta \mathfrak{b} - F\mathfrak{b} = (\underset{\sim}{I} + \underset{\sim}{G})\Delta\varphi$ if $\mathfrak{b} = (\underset{\sim}{I} + \underset{\sim}{G})\varphi$ and $\varphi \in C^2(D)$, then by G's monotoneness we have for $\varphi \in \mathcal{H}^-[D] \cap C^2[D]$ that $\Delta\varphi \geq 0$ and $\Delta\psi - P\psi \geq 0$.

It is well known (See for example Protter and Weinberger [PW 1]) that ψ cannot take on a positive maximum in D. Consequently, if $\psi - u \leq 0$ on ∂D_o for $D_o \subset\subset D$, then $\psi - u < 0$ in D_o unless $\psi \equiv u$ in D_o. $\therefore \psi \in E^-[D] \cap C^2[D]$.
In a similar way we may show if $\varphi \in \mathcal{H}^+[D] \cap C^2[D]$ then $\psi \in E^+[D] \cap C^2[D]$, if $\psi = (\underset{\sim}{I} + \underset{\sim}{G})\varphi$.

These results can be extended to the full classes of sub and super harmonic functions.

Theorem: The operator $\underset{\sim}{I} + \underset{\sim}{G}$ maps the classes $\mathcal{H}^-[D]$ into $E^-[D]$, and $\mathcal{H}^+[D]$ into $E^+[D]$.

Proof. For the cases where $\varphi(\underset{\sim}{x}) \in \mathcal{H}^-[D]$ (or $\mathcal{H}^+[D]$) we see that the properties (i) and (ii) are preserved by the operator $\underset{\sim}{I} + \underset{\sim}{G}$.

Remark: The previous theorem suggests approximating solutions $u(\underset{\sim}{x}) \in E[D]$ by sequences

$w_\ell^+(\underset{\sim}{x}) = (\underset{\sim}{I} + \underset{\sim}{G})\varphi_\ell^+(\underset{\sim}{x})$ $(\ell = 1,2,\ldots)$ of super functions such that $w_\ell^+(\underset{\sim}{x}) \downarrow u(\underset{\sim}{x})$ and also by subfunctions $w_\ell^-(\underset{\sim}{x}) \uparrow u(\underset{\sim}{x})$ with $w_\ell^-(\underset{\sim}{x}) = (\underset{\sim}{I} + \underset{\sim}{G})\varphi_\ell^-(\underset{\sim}{x})$, $(\ell = 1,2,\ldots)$.

We can show how the integral operator approach can be directly used for computation of solutions. (See the Appendix of [Gi 3] and also [At 3]. We consider the case of equation (2.28) for $n = 2$. In this case one obtains from (3.4), (3.5) the following Fredholm equation to solve

$$f(\underset{\sim}{x}) = -\pi\mu(\underset{\sim}{x}) - \int K(\underset{\sim}{x},\underset{\sim}{Y})\mu(\underset{\sim}{Y})\,ds_y, \quad \underset{\sim}{x} \in \Gamma \equiv \partial D,$$

(3.22)
with

$$K(\underset{\sim}{x},\underset{\sim}{Y}) = \frac{\partial}{\partial\nu_y} \ln|\underset{\sim}{x} - \underset{\sim}{y}| + \int_0^1 \sigma G(r, 1-\sigma^2)\frac{\partial}{\partial\nu_y} \ln|\underset{\sim}{Y} - \sigma^2\underset{\sim}{x}|\,d\sigma.$$

Let Γ be of class C^2, i.e. $\underset{\sim}{x}(t) = (x_1(t), x_2(t))$ has a C^2-parametrization for $t \in [0,A]$ with $|\underset{\sim}{x}^1(t)| \neq 0$. Then (3.22) becomes

$$f(t) = -\pi\mu(t) - \int_0^A \mu(s)k(t,s)\,ds, \quad t \in [0,A],$$

with

$$K(t,s) = \frac{x_2'(s)[x_1(s) - x_1(t)] - x_1'(s)[x_2(s) - x_2(t)]}{|\underset{\sim}{x}(s) - \underset{\sim}{x}(t)|^2}$$

$$(3.23) \qquad + \int_0^1 \sigma\, G(r(t),\, 1-\sigma^2)\, \frac{x_2'(s)[x_1(s) - \sigma^2 x_1(t)] - x_1'(s)[x_2(s) - \sigma^2 x_2(t)]}{|\underset{\sim}{x}(s) - \sigma^2 \underset{\sim}{x}(t)|^2}\, d\sigma$$

where $r(t) \equiv |\underset{\sim}{x}(t)|$. Let $K(t,s) = H(t,s) + L(t,s)$, where H and L denote the first and second terms on the right of (3.23). The first term has a removable singularity at $s = t$ since

$$\lim_{s \to t} H(t,s) = \tfrac{1}{2}\, \frac{x_1'(t)x_2''(t) - x_2'(t)x_1''(t)}{|\underset{\sim}{x}^1(t)|^{3/2}}$$

which is one-half the curvature of Γ at $\underset{\sim}{x}(t)$. The second term, however, can be shown to have a singularity of the form $\ln |\underset{\sim}{x}(s) - \underset{\sim}{x}(t)|$ as $s \to t$. The general approach taken below was developed by Atkinson in [At 1],[At 2], and is based on some work of Anselone [An 1]. Let $N > 0$ be an integer, $\delta = A/N$, and $t_j = j\delta$, $j = 0,1,\ldots,N$. We use, for simplicity, Simpson's rule on $[0,A]$. For the $H(t,s)$ part of the integral operator we use

$$\int_0^A H(t,s)\mu(s)\, ds \approx \sum_{j=0}^{N} w_j H(t_i,t_j)\mu(t_j).$$

For the singular part, let $t = t_i$, and break the integral up into smaller ones,

$$\int_0^A L(t_i, s)\mu(s)\,ds = \sum_{j=1}^{N/2} \int_{t_{2j-2}}^{t_{2j}} L(t_i, s)\mu(s)\,ds. \tag{3.24}$$

Let $\varepsilon > 0$ be some preassigned number, and if

$$\min_{k=0,1,2} \{|\underset{\sim}{x}(t_i) - \underset{\sim}{x}(t_{2j-k})|\} > \varepsilon \tag{3.25}$$

use the ordinary parabolic rule to get

$$\int_{t_{2j-2}}^{t_{2j}} L(t_i, s)\mu(s)\,ds \approx \frac{\delta}{3}\Big\{L(t_i, t_{2j-2})\mu(t_{2j-2}) \tag{3.26}$$

$$+ 4L(t_i, t_{2j-1})\mu(t_{2j-1}) + L(t_i, t_{2j})\mu(t_{2j})\Big\}$$

If (3.25) is not satisfied then let

$$\int_{t_{2j-2}}^{t_{2j}} L(t_i, s)\mu(s)\,ds \approx \int_{t_{2j-2}}^{t_{2j}} L(t_i, s)\hat{\mu}(s)\,ds, \tag{3.27}$$

where $\hat{\mu}(s)$ is the parabola interpolating $\mu(s)$ at t_{2j-2}, t_{2j-1}, and t_{2j}. Combining (3.24), (3.26), and (3.27) one obtains

$$\int_0^A L(t_i, s)\mu(s)\, ds \approx \sum_{j=0}^{N} V_{ij}\mu(t_j).$$

Most of the weights V_{ij} will be some integer multiple of $\frac{\delta}{3} L(t_i, t_j)$, provided $\epsilon > 0$ is reasonibly small. The other weights will involve integrals of the general form

$$\int_a^b L(t_i, s) g(s)\, ds$$

where $g(s)$ is independent of μ. We remark that in (3.27) only that part of the integral which is reasonibly well-behaved is approximated by a quadratic interpolating formula.

Using the above approximations replace (3.22) by the linear system

$$f(t_i) = -\pi\mu_N(t_i) - \sum_{j=0}^{N} \{W_j H(t_i, t_j) + V_{ij}\}\mu_N(t_j), i = 0, 1, \ldots, N.$$

The techniques of Anselone and Atkinson [An 1],[At 3] can now be used to show the convergence of μ_N to μ, uniformly on [0,A]. If H(t,s) is of class C^3 in S and $\mu \in C^3[0,A]$, then it can be shown that

$$\max_{i=0,1,\ldots,N} |\mu(t_i) - \mu_N(t_i)| \le O(\delta^3).$$

Further details and computational examples may be found in the Appendix of [Gi 3], and also in [At 3]. Work is currently in progress by the author with P. Linz, concerning other computational methods for solving boundary value problems. This will be reported on shortly.

Boundary Value Problems for Second Order Equations

We recall that for a simply connected domain Vekua [Ve 1] pg. 123, gave the following representation for solutions to second order elliptic equations

(3.28) $$u(x,y) = \text{Re}\left[H_o(Z)\varphi(Z) + \int_0^Z H(Z,t)\varphi(t)\ dt\right],$$

with $H_o(Z) \equiv R(Z,0;Z,\bar{Z})$,

and $$H(Z,t) = \frac{-\partial}{\partial t} R(t,0;Z,\bar{Z}) + B(t,0)\ R(t,0;Z,\bar{Z}).$$

Here $\varphi(Z)$ is taken to be a holomorphic function <u>in</u> D and is normalized by the condition $\varphi(0) = \overline{\varphi(0)}$. Vekua has shown [Ve 1] pg. 10 that one can assume, without loss of generality, that $A(0,f) = B(Z,0) = 0$. In this case one has

(3.29) $$H(Z,t) = -\frac{\partial}{\partial t} R(t,0,Z,\bar{Z}).$$

We consider the Dirichlet problem for the general linear, second order elliptic equation, namely we seek a solution $u(x,y)$ that takes on the values $f(t)$ for $t \in \partial D$. Furthermore, we shall require a certain amount of regularity for our data. In particular, we assume that ∂D is Lyapunov, and $f(t)$ satisfies a Hölder condition on ∂D. Under these conditions Vekna [Ve 1] showed that there exists a unique, real function $\mu(t)$, which is itself Hölder continuous, and such that we may represent $\varphi(Z)$ by

$$\varphi(Z) = \int_{\partial D} \frac{t\mu(t)\, ds}{t - Z}, \quad t \in \partial D, \ ds = ds(t),$$

where ds is the arc length differential. This leads to the following representation for the solution to the boundary value problem,

$$u(x,y) = \int_{\partial D} K(Z,t)\mu(t)\, ds, \qquad t \in \partial D, \quad Z \in D,$$

(3.30) where $$K(Z,t) = \mathrm{Re}\left[\frac{tH_o(Z)}{t - Z} + \int_0^Z \frac{tH(Z,t_1)}{t - t_1}\, dt_1\right].$$

We may also rewrite $K(Z,t)$ in the form

$$(3.31)\qquad K(Z,t) = \text{Re}\left[\frac{tH_o(Z)}{t-Z} - t\,H(Z,t)\,\log\left(1-\frac{Z}{t}\right) + H^*\,(Z,t)\right],$$

where

$$(3.32)\qquad H^*(Z,t) = \int_0^Z \frac{t[H(Z,t_1) - H(Z,t)]}{t-t_1}\,dt_1$$

If one allows Z to approach $t_o \in \partial D$ and computes the residue we obtain [Vel] pg. 124,

$$(3.33)\qquad A(t_o)\mu(t_o) + \int K(t_o,t)\mu(t)\,ds = f(t_o),$$

$$(3.34)\quad \text{where}\qquad A(t_o) = \text{Re}\,[i\pi t_o \bar{t}_o'\,H_o(t_o)],$$

$$(3.35)\qquad K(t_o,t) = \text{Re}\left[\frac{tH_o(t_o)}{t-t_o} - t\,H(t_o,t)\,\log\left(1-\frac{t_o}{t}\right) + H^*(t_o,t)\right]$$

and $t' = \frac{\partial t}{\partial s} = e^{i\theta(t)}$ (where $\theta(t)$ is the angle between the tangent to ∂D at t and to the x-axis.)

<u>Problem</u>: Verify the formulae (3.33) and (3.34). It is easy to see that (3.33), (334) may be rewritten as

$$(3.36)\qquad A(t_o)\mu(t_o) + \frac{B(t_o)}{i\pi}\int_{\partial D}\frac{\mu(t)}{t-t_o}\,dt + \int_{\partial D} K_o(t_o,t)\mu(t)\,ds = f(t_o),$$

with

(3.37) $$B(t_o) = i\pi \operatorname{Re}\left[t_o \bar{t}_o' H_o(t_o)\right],$$

(3.38) $$K_o(t_o,t) = K(t_o,t) - \frac{t' B(t_o)}{i\pi(t-t_o)} .$$

Problem: Show that $K_o(t_o,t)\,|t-t_o|^{\alpha}$ for some $0 \le \alpha < 1$, is Hölder continuous for the two points $t, t_0 \in \partial D$. From (3.37) and (3.38) it follows immediately that

(3.39) $$A(t_o) + B(t_o) = i\pi\, t_o \bar{t}_o'\, H_o(t_o)$$

$$A(t_o) - B(t_o) = -i\pi\, \bar{t}_o t_o'\, \overline{H_o(t_o)} .$$

From the theory of the Riemann function, it is clear that $H_o(t_o) \ne 0$ on ∂D. The index, $\varkappa$, of the singular integral equation (3.30) or (3.36) is therefore seen to be zero, [Mul] pg. 123, i.e.

(3.40) $$\varkappa = \frac{1}{2\pi}\left[\log \frac{A(t_o) - B(t_o)}{A(t_o) + B(t_o)}\right]_{\partial D} = 0 .$$

Consequently the Fredholm conditions hold for (3.36), i.e. equation (3.30) has a solution for any $f(t)$ iff the homogeneous case has only the trivial solution. Muskhelishvili has shown that

$$A(t_o)\mu(t_o) + \int_L K(t_o,t)\mu(t)ds = 0$$

has only the trivial solution if the homogeneous boudary value problem has only the trivial solution.

Since the singular integral equation has index zero it may be reduced directly to a Fredholm type equation by applying to it the singular operator

$$A(t_o)(\cdot) - \frac{B(t_o)}{\pi i} \int_{\partial D} \frac{(\cdot)\,dt}{t - t_o} \tag{3.41}$$

and applying the Poincaré-Bertrand theorem (see for example, Muskhelishvili [Mu 1], pg. 129, or Vekua [Ve 1].) One obtains an equation of one form

$$\mu(t_o) + \int_{\partial D} K^*(t_o,t)\mu(t)ds = f^*(t_o), \tag{3.42}$$

with

$$K^*(t_o,t) = \frac{1}{\pi^2 (t_o)^2 (H(t_o))^2} \Big[A(t_o)K(t_o,t) \quad - \tag{3.43}$$

$$\left. \frac{A(t_o)B(t_o)t'}{\pi i(t-t_o)} - \frac{B(t_o)}{\pi i} \int \frac{K(t_1,t)dt_1}{t_1-t_o} \right] ,$$

and

$$(3.44) \qquad f^*(t_o) = \frac{1}{\pi^2 \, |t_o|^2 \cdot |H_o(t_o)|^2} \left[A(t_o)f(t_o) - \frac{B(t_o)}{\pi i} \int_{\partial D} \frac{f(t)dt}{t-t_o} \right] .$$

We recall at this point that the representations of Velena and Bergman are the same. This suggests a method for obtaining approximate solutions to the Dirichlet problem. Instead of (3.42 - 3.44) we consider the sequence [Gi 2] pg. 100

$$(3.45) \qquad (\underset{\sim}{I} + \underset{\sim}{K}_\ell^*)\mu \equiv \mu(t_o) + \int_{\partial D} K_\ell^*(t_o,t)\mu(t)ds = f^*(t_o) ,$$

where $K^*_\ell(t_o,t)$ is obtained from $K^*(t_o,t)$ by replacing $H(z,t)$ by its uniform approximation

$$(3.46) \qquad H_\ell(z,t) \equiv H_o(z) \sum_{j=1}^{\ell} \frac{Q^{(j)}(z,\bar{z})(z-t)^j}{2^{2j}B(j,j+1)}$$

The singular kernel $K(t_o,t)$ of (3.35) may be approximated by the sequence of singular kernels,

$$(3.47)\quad K_\ell(t_o,t) = \mathrm{Re}\left\{\frac{tH_o(t_o)}{t-t_o} - tH_o(t_o)\ln\left(1-\frac{t_o}{t}\right)\cdot\sum_{n=1}^{\ell}\frac{Q^{(n)}(t_o,\bar{t}_o)(t_o-t)^{n-1}}{2^{2n}B(n,n+1)} + \right.$$

$$\left. + tH_o(t_o)\sum_{n=1}^{\ell}\frac{Q^{(n)}(t_o,\bar{t}_o)P_{n-1}(t_o,t)}{2^{2n}B(n,n+1)}\right\},$$

$$\text{where}\quad P_n(z,t) = \sum_{k=1}^{n}\frac{z^k(z-t)^{n-k}}{k} .$$

Using the same reasoning as before, in the case of the radially dependent coefficients, one can show that $\|(\underset{\sim}{I}+\underset{\sim}{K}{}^*_\ell)^{-1} - (\underset{\sim}{I}+\underset{\sim}{K})^{-1}\| \to 0$ as $\ell \to \infty$, and that the solutions $\mu_\ell(t)$ of (3.45) therefore approximate $\mu(t)$.

IV. Higher Order, Linear Elliptic Equations of Two Independent Variables

In this section we consider the fourth order equation,

$$\Delta\Delta u + au_{xx} + 2bu_{xy} + cu_{yy} + du_x + eu_y + fu = 0 \ , \tag{4.1}$$

where the coefficients are assumed to be holomorphic in x,y, and indeed have a holomorphic extension for $(z,\zeta) \in D \times D^*$, $z = x + iy$, $\zeta = x - iy$. Introducing, as before the complex form of (4.1) by making the substitution of the variables z,ζ one obtains

$$\underset{\sim}{L}[U] \equiv U_{zz\zeta\zeta} + MU_{zz} + LU_{z\zeta} + NU_{\zeta\zeta} + AU_z + BU_\zeta + CU = 0 \ . \tag{4.2}$$

For simplicity, we assume that the coefficients of (4.2) are holomorphic in the bicylinder $\Delta \equiv \{(z,\zeta) \,|\, |z| \le r_1 \ |\zeta| \le r_2\}$. If we assume the coefficients of (4.1) are real, then L and C are real, $M = \bar{N} = M_1 + iM_2$, $A = \bar{B} = A_1 + iA_2$ for real x and y. The coefficients of (4.2) are related to those of (4.1) (for real x and y) by the formulae $a = 4L + 8M_1$, $b = 16M_2$, $c = 4L - 8M_1$, $d = 16A_1$, $e = 16A_2$, $f = 16C$.

We want to show that it is possible to find solutions of (4.2) in the form [Be 1], [Be 2],

$$U(z,\zeta) = \sum_{k=1}^{2} \int_{-1}^{1} \Big[E^{(1,k)}(z,\zeta,t) f_k(\tfrac{z}{2}[1-t^2]) + E^{(2,k)}(z,\zeta,t) g_k(\tfrac{\zeta}{2}[1-t^2])\Big] \frac{dt}{\sqrt{1-t^2}} \ , \tag{4.3}$$

where f_k, g_k $(k = 1, 2)$ are arbitrary analytic functions. Substituting

$$\text{(4.4)} \qquad U(z, \zeta) = \int_{-1}^{1} E(z, \zeta, t) f\left(\frac{z}{2}[1 - t^2]\right) \frac{dt}{\sqrt{1 - t^2}} ,$$

into (4.2) shows that the E function must satisfy the differential equation

$$z^{-1}t^{-1}(1 - t^2)[E_{z\zeta\zeta t} + ME_{zt} + \tfrac{1}{2}LE_{t\zeta} + \tfrac{1}{2}AE_t]$$

$$+ \tfrac{1}{4}z^{-2}t^{-2}(1-t^2)^2[E_{\zeta\zeta tt} + ME_{tt}]$$

$$- z^{-1}t^{-2}[E_{z\zeta\zeta} + ME_z + \tfrac{1}{2}LE_\zeta + \tfrac{1}{2}AE]$$

$$\text{(4.5)} \qquad - 3/4\, z^{-2}t^{-3}(1 - t^4)[E_{t\zeta\zeta} + ME_t]$$

$$+ 3/4\, z^{-2}t^{-4}[E_{\zeta\zeta} + ME] + \underset{\sim}{L}[E] = 0 ,$$ and the regularity condition that

$$\underset{\sim}{D}_1(E)/zt ,$$

(4.6) and

$$[\underset{\sim}{D}_1(-2E_z + z^{-1}E - \tfrac{1}{2}z^{-1}t^{-1}E_t + \tfrac{1}{2}z^{-1}t\,E_t + \tfrac{1}{2}z^{-1}t^{-2}E)$$

$$- AE - LE_\zeta]/zt ,$$

where $\underset{\sim}{D}_1(H) \equiv H_{\zeta\zeta} + MH$, are continuous functions of t, z, ζ for $|t| \le 1$.

Problem: Prove that the E-functions must satisfy (4.5), (4.6) if (4.4) is a solution.

If one substitutes

$$U(z,\zeta) = \int_{-1}^{1} E(z,\zeta,t)\,f\left(\frac{\zeta}{2}[1-t^2]\right)\frac{dt}{\sqrt{1-t^2}} \tag{4.7}$$

into (4.2) then the E-function can be shown to satisfy an equation similar to (4.5) but with z and ζ interchanged, and with M and A replaced by N and D.

Problem: Prove the above statement.

It can also be shown that it is possible to determine the four E-functions such that they satisfy the initial conditions

$$\begin{aligned}
&E^{(1,1)}(z,0,t) = 1, \quad && E_{\zeta}^{(1,1)}(z,0,t) = 0,\\
&E^{(1,2)}(z,0,t) = 0, \quad && E_{\zeta}^{(1,2)}(z,0,t) = 1,\\
&E^{(2,1)}(0,\zeta,t) = 1, \quad && E_{z}^{(2,1)}(0,\zeta,t) = 0,\\
&E^{(2,2)}(0,\zeta,t) = 0, \quad && E_{z}^{(2,2)}(0,\zeta,t) = 1\,.
\end{aligned} \tag{4.8}$$

To this end one seeks a solution of (4.5) having the form

$$E(z,\zeta,t) = P^{(0)}(z,\zeta) + \sum_{\nu=1}^{\infty} t^{2\nu} z^{\nu} P^{(\nu)}(z,\zeta)\;, \tag{4.9}$$

where the $P^{(\nu)}$ are to be determined. Substituting (4.9) into (4.5) and equating to zero the coefficients of the powers of t yields the system of equations for the $P^{(\nu)}$,

$$\underset{\sim}{D}_1(P^{(0)}) = 0\;,$$

$$\underset{\sim}{D}_1(P^{(1)}) = -4\underset{\sim}{D}_1(P_z^{(0)}) - 2\underset{\sim}{D}_2(P^{(0)}) ,$$

$$\begin{aligned}\underset{\sim}{D}_1(P^{(n+2)}) = -\frac{1}{n^2+2n+3/4}\Big[&\underset{\sim}{D}_1(P_{zz}^{(n)}) + (2n+1)\underset{\sim}{D}_1(P_z^{(n+1)}) \\ &+ \underset{\sim}{D}_2(P_z^{(n)}) + (n+\tfrac{1}{2})\underset{\sim}{D}_2(P^{(n+1)}) \\ &+ NP_{\zeta\zeta}^{(n)} + DP_{\zeta}^{(n)} + CP^{(n)}\Big] , \quad n = 0, 1, \ldots ,\end{aligned} \tag{4.10}$$

where

$$\underset{\sim}{D}_1(H) \equiv H_{\zeta\zeta} + MH \quad \text{and} \quad \underset{\sim}{D}_2(H) = LH_{\zeta} + AH .$$

<u>Remark</u>: After $P^{(0)}, \ldots, P^{(n+1)}$ have been found $P^{(n+2)}$ can be obtained by solving the ordinary differential equation for this term.

<u>Problem</u>: Derive the recursive system (4.10).

It can be shown that $\exists$ two sequences of functions $P^{(1,1,n)}(z,\zeta)$ and $P^{(1,2,n)}(z,\zeta)$, $n = 0, 1, 2, \ldots$ satisfying the differential equations (4.10) $\ni$

$$P^{(1,1,0)}(z,0) = 0, \quad P_{\zeta}^{(1,1,0)}(z,0) = 0,$$

(4.11)

$$P^{(1,1,n)}(z,0) = 0, \quad P_{\zeta}^{(1,1,n)}(z,0) = 0, \quad (n = 1,2,\ldots)$$

$$P^{(1,2,0)}(a,0) = 0, \quad P_{\zeta}^{(1,2,0)}(z,0) = 1,$$

(4.12)

$$P^{(1,2,n)}(z,0) = 0, \quad P_{\zeta}^{(1,2,n)}(z,0) = 0, \quad (n = 1,2,\ldots)$$

and $\ni$ for $\dfrac{|z/r_1|}{|1 - z/r_1|^3 |1 - \zeta/r_2|} < \dfrac{1}{9}, \ |t| \le 1,$

the series

$$E^{(1,k)}(z,\zeta,t) = P^{(1,k,0)}(z,\zeta) + \sum_{n=1}^{\infty} t^{2n} z^n P^{(1,k,n)}(z,\zeta), \quad (k = 1,2) \tag{4.13}$$

converges uniformly and absolutely. The proof of this can be found in Bergman [Be2] Secs. 6,7, and 8.

We turn next to the case of coefficients which are radially dependent. As was the case for the second order equation, we can treat the situation of dimension $n \ge 2$ just as easily as dimension 2, i.e. the equation

(4.14) $$\Delta_n^2 u(\underset{\sim}{x}) + A(r^2)\Delta_n u(\underset{\sim}{x}) + B(r^2)u(\underset{\sim}{x}) = 0 .$$

Using the approach of Bergman, outlined above we attempt to find, first for $n = 2$, a solution of (4.14) in the form

(4.15) $$u(z,\zeta) = \mathrm{Re}\left\{\int_{-1}^{1}\left[\tilde{E}^{(1)}(z,\zeta,t)f^{(1)}(\tfrac{1}{2}z[1-t^2]) + \tilde{E}^{(2)}(z,\zeta,t)f^{(2)}(\tfrac{1}{2}z[1-t^2])\right]\cdot\frac{dt}{\sqrt{1-t^2}}\right\} .$$

It is easy to show that $\tilde{E}^{(1)}(z,\zeta,t)$, $\tilde{E}^{(2)}(z,\zeta,t)$ both satisfy the differential equation

$$z^{-1}t^{-1}(1-t^2)[\tilde{E}_{z\zeta\zeta t} + 1/8\ A(z\zeta)\tilde{E}_{t\zeta}]$$

$$+ 1/4\ z^{-2}t^{-2}(1-t^2)^2\tilde{E}_{\zeta\zeta tt} - z^{-1}t^{-2}[\tilde{E}_{z\zeta\zeta} + 1/8\ A(z\zeta)\tilde{E}_{\zeta}]$$

$$- 3/4\ z^{-2}t^{-3}(1-t^4)\tilde{E}_{t\zeta\zeta} + 3/4\ z^{-2}t^{-4}\tilde{E}_{\zeta\zeta} + \tilde{E}_{zz\zeta\zeta}$$

(4.16) $$+ 1/4\ A(z\zeta)\tilde{E}_{z\zeta} + 1/16\ B(z\zeta)\tilde{E} = 0 .$$

Bergman's two E-functions then must have the form

$$\tilde{E}^{(1)}(z,\zeta,t) = 1 + \sum_{\ell=1}^{\infty} t^{2\ell} z^{\ell} P^{(1,\ell)}(z,\zeta),$$

(4.17)

$$\tilde{E}^{(2)}(z,\zeta,t) = \zeta + \sum_{\ell=1}^{\infty} t^{2\ell} z^{\ell} P^{(2,\ell)}(z,\zeta) \;,$$

where $P^{(i,\ell)}$, $(i = 1,2)$ satisfy the recursive scheme

$$P^{(i,1)}_{\zeta\zeta} = -4P^{(i,0)}_{z\zeta\zeta} - 2A(z\zeta)P^{(i,0)}_{\zeta} \;,$$

$$P^{(i,\ell+2)}_{\zeta\zeta} = -\frac{1}{\ell^2 + 2\ell + 3/4} [P^{(i,\ell)}_{zz\zeta\zeta} + (2\ell+1)P^{(i,\ell+1)}_{z\zeta\zeta}$$

(4.18)

$$+ 1/4\; A(z\zeta)P^{(i,\ell)}_{z\zeta} + 1/4\; (\ell + 1/2)A(z\zeta)P^{(i,\ell+1)}_{\zeta} + 1/16\; B(z\zeta)P^{(i$$

with initial conditions

$$P^{(1,0)}(z,\zeta) = 1, \quad P^{(2,0)}(z,\zeta) = \zeta \;,$$

$$P^{(1,n)}(z,0) = P^{(1,n)}_{\zeta}(z,0) = 0 \;,$$

(4.19)

$$P^{(2,n)}(z,0) = P^{(2,n)}_{\zeta}(z,0) = 0 \;.$$

It is easy to see that the first Bergman E-function is a real function of $r^2 = z\bar{z}$ and t, whereas the second E-function is of the form $\tilde{E}^{(2)}(z,\bar{z},t) = \bar{z}E^{(2)}(r^2,t)$ where $E^{(2)}(r^2,t)$ is a real function of r^2 and t. To this end we first write $Q^{(1,n)}(z,\zeta) = z^n P^{(1,n)}(z,\zeta)$ and substitute into equations (4.17), (4.18). This yields

$$Q^{(1,0)}(z,\zeta) = 1, \quad Q^{(1,1)}(z,\zeta) = 0,$$

$$Q_{\zeta\zeta}^{(1,\ell+2)} = -\frac{1}{\ell^2 + 2\ell + 3/4}\,[z^2 Q_{zz\zeta\zeta}^{(1,\ell)} - \ell(\ell-3)Q_{\zeta\zeta}^{(1,\ell)} - 2\ell z Q_{\zeta\zeta z}^{(1,\ell)}$$

$$+ (2\ell+1)\{zQ_{\zeta\zeta z}^{(1,\ell+1)} - (\ell+1)Q_{\zeta\zeta}^{(1,\ell+1)}\} + 1/4\, A(z,\zeta)\{z^2 Q_{z\zeta}^{(1,\ell)}$$

$$- \ell z Q_{\zeta}^{(1,\ell)}\} + 1/4\,(\ell+1/2)A(z\zeta)\,zQ_{\zeta}^{(1,\ell+1)} + 1/16\, B(z\zeta)z^2 Q^{(1,\ell)}]\,,$$

for $\ell = 0,1,\ldots$ with initial conditions

$$Q^{(1,\ell)}(z,0) = Q_{\zeta}^{(1,\ell)}(z,0) = 0\,, \quad \ell = 1,2,\ldots,$$

which by a straight-forward computation is seen to reduce to

$$
\begin{aligned}
\tilde{Q}^{(1,\ell+2)}_{r^2r^2} = -\frac{1}{\ell^2+2\ell+3/4}\Big[& r^4\tilde{Q}^{(1,\ell)}_{r^2r^2r^2r^2} + 4r^2\tilde{Q}^{(1,\ell)}_{r^2r^2r^2} \\
& + 2\tilde{Q}^{(1,\ell)}_{r^2r^2} - \ell(\ell-3)\tilde{Q}^{(1,\ell)}_{r^2r^2} - 2\ell\{2\tilde{Q}^{(1,\ell)}_{r^2r^2} + r^2\tilde{Q}^{(1,\ell)}_{r^2r^2r^2}\} \\
& + (2\ell+1)\{2\tilde{Q}^{(1,\ell+1)}_{r^2r^2} + r^2\tilde{Q}^{(1,\ell+1)}_{r^2r^2r^2} - (\ell+1)\tilde{Q}^{(1,\ell+1)}_{r^2r^2}\} \\
& + 1/4\ A(r^2)\{\tilde{Q}^{(1,\ell)}_{r^2} + r^2\tilde{Q}^{(1,\ell)}_{r^2r^2} - \ell\tilde{Q}^{(1,\ell)}_{r^2}\} \\
& + 1/4(\ell+1/2)A(r^2)\tilde{Q}^{(1,\ell+1)}_{r^2} + 1/16\ B(r^2)\tilde{Q}^{(1,\ell)}\Big],
\end{aligned}
\tag{4.20}
$$

where $\tilde{Q}^{(1,\ell)}(r^2) \equiv Q^{(1,\ell)}(z,\bar{z})$.

<u>Problem</u>: Verify equation (4.19).

The initial conditions will be satisfied if we set

$$
\tilde{Q}^{(1,\ell)}(0) = \tilde{Q}^{(1,\ell)}_{r^2}(0) = 0 . \tag{4.21}
$$

In a similar manner, if we define $Q^{(2,\ell)}(z,\zeta)$ by $\zeta Q^{(2,\ell)}(z,\zeta) = z^{\ell}P^{(2,\ell)}(z,\zeta)$, one obtains the following system of equations for $\tilde{Q}^{(2,\ell)}(r^2) = Q^{(2,\ell)}(z,\zeta)$:

$$\tilde{Q}^{(2,0)}(r^2) = 1, \quad \frac{1}{r^2}\left(r^4 \tilde{Q}^{(2,1)}_{r^2}\right)_{r^2} = -2A(r^2) \;,$$

$$\frac{1}{r^2}\left(r^4 \tilde{Q}^{(2,\ell+2)}_{r^2}\right)_{r^2} = -\frac{1}{\ell^2 + 2\ell + 3/4}\Big[r^2\{3\tilde{Q}^{(2,\ell)}_{r^2r^2} + 5r^2\tilde{Q}^{(2,\ell)}_{r^2r^2r^2}$$

$$+ r^4\tilde{Q}^{(2,\ell)}_{r^2r^2r^2r^2} + 2\tilde{Q}^{(2,\ell)}_{r^2}\} - \ell(\ell-3)\{r^2\tilde{Q}^{(2,\ell)}_{r^2r^2} + 2\tilde{Q}^{(2,\ell)}_{r^2}\}$$

$$-2\ell\{4r^2\tilde{Q}^{(2,\ell)}_{r^2r^2} + r^4\tilde{Q}^{(2,\ell)}_{r^2r^2r^2} + 2\tilde{Q}^{(2,\ell)}_{r^2}\}$$

$$+ (2\ell+1)\{r^4\tilde{Q}^{(2,\ell+1)}_{r^2r^2r^2} - (\ell-3)r^2\tilde{Q}^{(2,\ell+1)}_{r^2r^2} - 2\ell\tilde{Q}^{(2,\ell+1)}_{r^2}\}$$

$$+ 1/4\, A(r^2)\{r^4\tilde{Q}^{(2,\ell)}_{r^2r^2} - (\ell-2)r^2\tilde{Q}^{(2,\ell)}_{r^2} - \ell\tilde{Q}^{(2,\ell)}\}$$

.22)

$$+ 1/8\,(2\ell+1)A(r^2)\{r^2\tilde{Q}^{(2,\ell+1)}_{r^2} + \tilde{Q}^{(2,\ell+1)}\} + 1/16\, r^2 B(r^2)\tilde{Q}^{(\ell)}\Big]$$

$$\ell = 0,1,2,\ldots,$$

with $\tilde{Q}^{(2,\ell)}(0) = \tilde{Q}^{(2,\ell)}_{r^2}(0) = 0, \quad \ell = 1,2,\ldots$.

These equations imply $Q^{(2,\ell)}(z,\bar{z})$ depends only on r^2, [CG 1].

Convergence of $\bar{z}(1 + \sum_{\ell=1}^{\infty} t^{2\ell}\tilde{Q}^{(2,\ell)}(r^2))$ follows directly from Bergman's work.

Bergman showed that in the representation (4.15) we could take $f^{(2)}(0) = 0$. Hence if we set $H^{(1)}(x_1,x_2) = \mathrm{Re}\{f^{(1)}(z/2)\}$ and $r^2H^{(2)}(x_1,x_2) = \mathrm{Re}\{\bar{z}f(z/2)\}$, the representation (4.15) may be rewritten as

$$(4.23)\quad \begin{aligned} u(x_1,x_2) = & \int_{-1}^{1} E^{(1)}(r^2,t)H^{(1)}\left(x_1\sqrt{1-t^2},x_2\sqrt{1-t^2}\right)\frac{dt}{\sqrt{1-t^2}} \\ & + \int_{-1}^{1} E^{(2)}(r^2,t)H^{(2)}\left(x_1\sqrt{1-t^2},x_2\sqrt{1-t^2}\right)\frac{dt}{\sqrt{1-t^2}} \end{aligned}$$

with

$$(4.24)\quad \begin{aligned} E^{(1)}(r^2,t) &= 1+\sum_{\ell=1}^{\infty} e_\ell^{(1)}(r^2)t^{2\ell} \equiv 1+\sum_{\ell=1}^{\infty}\tilde{Q}^{(1,\ell)}(r^2)t^{2\ell}\ , \\ E^{(2)}(r^2,t) &= r^2+\sum_{\ell=1}^{\infty} e_\ell^{(2)}(r^2)t^{2\ell} \equiv r^2(i-t^2)[1+\sum_{\ell=1}^{\infty}\tilde{Q}^{(2,\ell)}(r^2)t^{2\ell}]. \end{aligned}$$

It can be shown [CG 1] that the two E-functions in (4.24) satisfy the differential equation

$$\frac{(1-t^2)}{rt}\left[2E_{rrrt}+\frac{2}{r}E_{rrt}-\frac{2}{r^2}E_{rt}+A(r^2)E_{rt}\right]$$

$$+\frac{(1-t^2)^2}{r^2t^2}\left[E_{rrtt}-\frac{1}{r}E_{rtt}\right]$$

$$-\frac{1}{rt^2}\left[2E_{rrr}+\frac{2}{r}E_{rr}-\frac{2}{r^2}E_r+A(r^2)E_r\right] \tag{4.25}$$

$$-\frac{3(1-t^4)}{r^2t^3}\left[E_{rrt}-\frac{1}{r}E_{rt}\right]+\frac{3}{r^2t^4}\left[E_{rr}-\frac{1}{r}E_r\right]$$

$$+E_{rrrr}+\frac{2}{r}E_{rrr}-\frac{1}{r^2}E_{rr}+\frac{1}{r^3}E_r$$

$$+A(r^2)\left[E_{rr}+\frac{1}{r}E_r\right]+B(r^2)E=0\ ,$$

with the initial conditions

$$E^{(1)}(0,t)=1\ ,\qquad E^{(2)}(0,t)=0\ ,$$

$$E^{(1)}_{r^2}(0,t)=0\ ,\qquad E^{(2)}_{r^2}(0,t)=1-t^2\ . \tag{4.26}$$

<u>Problem</u>: Verify the equation (4.25) and initial conditions (4.26).

If we define the harmonic functions

$$h^{(k)}(x_1,x_2) = \int_{-1}^{1} H^{(k)}(x_1[1-t^2],x_2[1-t^2])\frac{dt}{\sqrt{1-t^2}}, \tag{4.27}$$

$(k=1,2)$, then the representation (4.23) may be put into the form [CG 1] pg. 66.

$$u(x_1,x_2) = h^{(1)}(x_1,x_2) + r^2h^{(2)}(x_1,x_2) + \int_0^1 \sigma G^{(1)}(r^2,1-\sigma^2)h^{(1)}(x_1\sigma^2,x_2\sigma^2)\,d\sigma$$

$$+ \int_0^1 \sigma G^{(2)}(r^2,1-\sigma^2)h^{(2)}(x_1\sigma^2,x_2\sigma^2)\,d\sigma, \tag{4.28}$$

where

$$G^{(k)}(r^2,\tau) = \sum_{\ell=1}^{\infty} 2e_\ell^{(k)}(r^2)\,\frac{\Gamma(\ell+\frac{1}{2})}{\Gamma(\frac{1}{2})\Gamma(\ell)}\,\tau^{\ell-1}, \quad (k=1,2). \tag{4.29}$$

Problem: Verify the representation formula (4.28),(4.29).

Remark: Since Bergman's operator is invertible [Be 2], one may use the fact that the harmonic polymonials are complete, to generate a complete family of solutions

$$\varphi_{\ell m}(x_1,x_2) = h_m^{(1)}(x_1,x_2) + r^2 h_n^{(2)}(x_1,x_2)$$

$$+ \int_0^1 \sigma G^{(1)}(r^2, 1-\sigma^2) h_m^{(1)}(x_1\sigma^2, x_2\sigma^2)\, d\sigma$$

$$+ \int_0^1 \sigma G^{(2)}(r^2, 1-\sigma^2) h_\ell^{(2)}(x_1\sigma^2, x_2\sigma^2)\, d\sigma \,, \tag{4.30}$$

where $h_m(x_1,x_2)$ is an arbitrary harmonic polynomial.

We now turn to the case of dimension $n > 2$. It may be shown [CG 1] that one may represent solutions in the form

$$u(\underset{\sim}{x}) = \int_0^1 t^{n-2} E^{(1)}(r^2,t;n) H^{(1)}(\underset{\sim}{x}[1-t^2]) \frac{dt}{\sqrt{1-t^2}}$$

$$+ \int_0^1 t^{n-2} E^{(2)}(r^2,t;n) H^{(2)}(\underset{\sim}{x}[1-t^2]) \frac{dt}{\sqrt{1-t^2}} \,, \tag{4.31}$$

where the $E^{(k)}(r^2,t;n)$ are solutions of

$$\frac{(1-t^2)}{rt} \left[2E_{rrrt} + \frac{(6-2n)}{r} E_{rrt} + A(r^2) E_{rt}\right]$$

$$+ \frac{(1-t^2)^2}{r^2 t^2} \left[E_{rrtt} - \frac{1}{r} E_{rtt}\right]$$

$$+ \frac{(n-3)}{rt^2}\left[2E_{rrr} + \frac{2}{r}E_{rr} - \frac{2}{r^2}E_r + A(r^2)E_r\right]$$

$$+ \frac{(2n-7)(1-t^4)}{r^2t^3}\left[E_{rrt} - \frac{1}{r}E_{rt}\right] + \frac{(n-5)(n-3)}{r^2t^4}\left[E_{rr} - \frac{1}{r}E_r\right]$$

(4.32)

$$+ E_{rrrr} + \frac{2}{r}E_{rrr} - \frac{1}{r^2}E_{rr} + \frac{1}{r^3}E_r + A(r^2)\left(E_{rr} + \frac{1}{r}E_r\right) + B(r^2)E =$$

and satisfy the following initial conditions

$$E^{(1)}(0,t) = 1, \quad E^{(1)}_{r^2}(0,t) = 0 ,$$

(4.33)

$$E^{(2)}(0,t) = 0 , \quad E^{(2)}_{r^2}(0,t) = 1 - \frac{t^2}{n-1} .$$

Problem: Verify that (4.31) is a solution of (4.14), where the E-functions are defined by (4.32), (4.33).

The initial conditions (4.33) suggest that we seek E-functions having the form

$$E^{(1)}(r^2,t;n) = 1 + \sum_{\ell=1}^{\infty} e_\ell^{(1)}(r^2;n)t^{2\ell} ,$$

(4.34)

$$E^{(2)}(r^2,t;n) = r^2 + \sum_{\ell=1}^{\infty} e_\ell^{(2)}(r^2;n)t^{2\ell} .$$

Formal substitution of these series into (4.31) yields the following recursion formula for the $e_\ell^{(k)}(r^2;n)$, $(k=1,2)$,

$$[(2(\ell+1)+n-1)(2\ell+n-1)][e''_{\ell+2} - \frac{1}{r} e'_{\ell+2}] + 2(2\ell+n-1)re'''_{\ell+1}$$

$$-(4\ell+2)(2\ell+n-1)[e''_{\ell+1} - \frac{1}{r} e'_{\ell+1}] + (2\ell+n-1)rA(r^2)e'_{\ell+1}$$

$$+ r^2 e'''_\ell + r(2-4\ell)e'''_\ell - (4\ell^2-1)[e''_\ell - \frac{1}{r} e'_\ell]$$

.35) $$+ r^2 A(r^2)[e''_\ell + \frac{1-2\ell}{r} e'_\ell] + r^2 B(r^2) e_\ell = 0 ,$$

$$e_\ell(0;n) = \frac{d}{d(r^2)} e_\ell(0;n) = 0 , \quad \ell = 2,3,\ldots,$$ where $'$ denotes differentiation with respect to r. The functions $e_1^{(k)}(r^2;n)$ are defined by

$$e_1^{(1)''} - \frac{1}{r} e_1^{(1)'} = 0 ,$$

.36) $$e_1^{(1)}(0;n) = \frac{de_1^{(1)}}{d(r^2)}(0;n) = 0 ,$$

and

$$e_1^{(2)''} - \frac{1}{r} e_1^{(2)'} = -\frac{2r^2}{n-1} A(r^2)$$

$$e_1^{(2)}(0;n) = 0\ , \quad \frac{de_1^{(2)}}{d(r^2)}(0;n) = -\frac{1}{n-1}\ .$$

The equations (4.36) imply that

$$e_1^{(1)}(r^2;n) \equiv 0, \text{ and } e_1^{(2)}(r^2;n) = \frac{1}{n-1}\left(-r^2 + \int_0^r \xi^3 A(\xi^2)d\xi - r^2\int_0^r \xi A(\xi^2)d\xi\right)$$

Remark: When $n=2$ the represnetation (4.31) with E-functions defined by (4.34),(4.35)(4.36) coincide with Bergman's representation.

If we now define the new functions $c_\ell^{(i)}(r^2;n)$ $(i=1,2)$, $(\ell = 0,1,2,...)$ by the formulas

$$c_0^{(1)}(r^2;n) = 1, \quad c_0^{(2)}(r^2;n) = r^2$$

$$c_\ell^{(i)}(r^2;n) = 2e_\ell^{(i)}(r^2;n)\frac{\Gamma(\ell + n/2 - 1/2)}{\Gamma(n/2 - 1/2)\Gamma(\ell)}\ ,$$

then one obtains the following recursion formulae for the $c_\ell^{(i)}(r^2;n)$:

$$c_1^{(1)}(r^2;n) \equiv 0\ ,$$

$$c_1^{(2)}(r^2;n) = -r^2 + \int_0^r \xi^3 A(\xi^2)d\xi - r^2\int_0^r \xi A(\xi^2)d\xi\ ,$$

$$c_2^{(1)''} - \frac{1}{r} c_2^{(1)'} = - \frac{r^2}{2} B(r^2)$$

$$(c_2^{(2)''} - \frac{1}{r} c_2^{(2)'}) + rc_1^{(2)'''} - (c_1^{(2)''} - \frac{1}{r} c_1^{(2)'}) + \frac{1}{2} rA(r^2) c_1^{(2)'}$$

$$= -2r^2 A(r^2) - \frac{1}{2} r^4 B(r^2) ,$$

and for $\ell = 1,2,3,\ldots,$ $c_\ell^{(1)}(r^2;n), c_\ell^{(2)}(r^2;n)$ both satisfy

$$4\ell(\ell+1)[c''_{\ell+2} - \frac{1}{r} c'_{\ell+2}] + 4\ell r c'''_{\ell+1}$$

$$-4\ell(2\ell+1)(c''_{\ell+1} - \frac{1}{r} c'_{\ell+1}) + 2\ell r A(r^2) c'_{\ell+1}$$

$$+ r^2 c''''_\ell + r(2-4\ell) c'''_\ell + (4\ell^2 - 1)(c''_\ell - \frac{1}{r} c'_\ell) \tag{4.37}$$

$$+ r^2 A(r^2)(c''_\ell + \frac{1-2\ell}{r} c'_\ell) + r^2 B(r^2) c_\ell = 0 ,$$

with $c_\ell(0;n) = \frac{\partial c_\ell}{\partial (r^2)}(0;n) = 0$, $n = 2,3,\ldots$

<u>Theorem</u>: Let $h^{(k)}(\underset{\sim}{x})$ be arbitrary harmonic functions defined in a star-like region $\mathfrak{S}$ about the origin. Then the function defined by

$$u(\underset{\sim}{x}) = h^{(1)}(\underset{\sim}{x}) + r^2 h^{(2)}(\underset{\sim}{x}) + \int_0^1 \sigma^{n-1} G^{(1)}(r^2, 1-\sigma^2) h^{(1)}(\underset{\sim}{x}\sigma^2)\, d\sigma$$

$$+ \int_0^1 \sigma^{n-1} G^{(2)}(r^2, 1-\sigma^2) h^{(2)}(\underset{\sim}{x}\sigma^2)\, d\sigma \tag{4.38}$$

where $G^{(k)}(r^2,t)$ are solutions of the partial differential

$$G_{rrrr} + \frac{4(1-t)}{r} G_{rrrt} + \frac{4(1-t)^2}{r^2} G_{rrtt} - \frac{12(1-t)}{r^2} G_{rrt}$$

$$- \frac{4(1-t)^2}{r^3} G_{rtt} + \frac{2(1-t)}{r}\left(\frac{6}{r^2} + A(r^2)\right) G_{rt} - \frac{2}{r} G_{rrr}$$

$$+ \left(\frac{3}{r^2} + A(r^2)\right) G_{rr} - \frac{1}{r}\left(\frac{3}{r^2} + A(r^2)\right) G_r + B(r^2) G = 0 \,, \tag{4.39}$$

and satisfy the initial conditions

$$\left\{\begin{array}{ll} G^{(1)}(0,t) = 0 \,, & \lim\limits_{r\to 0} \frac{1}{r} G_r^{(1)}(r,t) = 0 \\ G^{(1)}(r,0) = 0 \,, & G_{rrt}^{(1)}(r,0) - \frac{1}{r} G_{rt}^{(1)}(r,0) + \frac{r^2}{2} B(r^2) = 0 \\ G^{(2)}(0,t) = 0 \,, & \lim\limits_{r\to 0} \frac{1}{r} G_r^{(2)}(r,t) = -2 \end{array}\right. \tag{4.40}$$

$$(4.41)\quad \begin{cases} G^{(2)}(0,t) = 0\,, \quad \lim\limits_{r\to 0} \frac{1}{r} G_r^{(2)}(r,t) = -2 \\[2ex] G_{rr}^{(2)}(r,0) - \frac{1}{r} G_r^{(2)}(r,0) + 2r^2 A(r^2) = 0\,, \\[2ex] G_{rrt}^{(2)}(r,0) - \frac{1}{r} G_{rt}^{(2)}(r,0) + rG_{rrr}^{(2)}(r,0) - (G_{rr}^{(2)}(r,0) - \frac{1}{r} G_r^{(2)}(r,0)) \\[2ex] + 2r^2 A(r^2) + \frac{r}{2} A(r^2) G_r^{(2)}(r,0) + \frac{r^4}{2} B(r^2) = 0 \end{cases}$$

is a solution of the differential equation (4.14).

<u>Problem</u>: Verify this by direct substitution of (4.38) into (4.14) and then integrating by parts.

If the region $\mathfrak{S}$ is a sphere, then it is clear that the functions $G^{(k)}(r^2,t)$ are given by

$$(4.42)\qquad G^{(k)}(r^2,t) \equiv \sum_{\ell=1}^{\infty} c_\ell^{(k)}(r^2;n)\,t^{\ell-1}$$

<u>Remark</u>: The representation (4.38) defines the same solution as (4.31) if one associates the harmonic functions $h^{(k)}(\underset{\sim}{x})$ and $H^{(k)}(\underset{\sim}{x})$ by the following formula:

$$h^{(k)}(\underset{\sim}{x}) = \int_0^1 t^{n-2} H^{(k)}(\underset{\sim}{x}[1-t^2]) \frac{dt}{\sqrt{1-t^2}}\,, \quad (k=1,2)\ .$$

Problem: Find an integral representation for the $H^{(k)}(\underset{\sim}{x})$ in terms of the $h^{(k)}(\underset{\sim}{x})$.

Problem: Show that the E-functions and G-functions are related by the formula

$$E^{(i)}(r^2,t;n) = 1 + t^2 \int_0^1 \sigma^{n-1} G^{(i)}(r,[1-\sigma^2]t^2)\,d\sigma \ , \quad (i=1,2) \ .$$

Definition: A domain D is said to be firmly regular if $\exists$ an open covering $\{N_k\}$ of the boundary $\ni$ for each k $\exists$ a positive constant h_k and a $\underset{\sim}{\xi}_k \in E^n$ with $|\underset{\sim}{\xi}_k| = 1$ for which the punctured cone

$$C^1_{\underset{\sim}{x}} \equiv \{\underset{\sim}{y} \,|\, \underset{\sim}{y} = \underset{\sim}{x} + r\underset{\sim}{\xi} \ , \ 0 < r < h_k \ , \ |\underset{\sim}{\xi} - \underset{\sim}{\xi}_k| < h_k \ , \ |\underset{\sim}{\xi}| = 1\}$$

lies in D for any point $\underset{\sim}{x} \in N_k \cap D$.

Theorem: Let D be a simply connected, firmly regular, starlike domain containing the origin, and let $\{h_n(\underset{\sim}{x})\}$ denote the set of harmonic polynomials. Assume further that for some sphere S containing D in its interior, the

Dirichlet problem for (4.14) is well-posed. Then the set of solutions $\{\varphi_{nm}\}$ defined by

$$\varphi_{nm}(\underset{\sim}{x}) = h_m^{(1)}(\underset{\sim}{x}) + r^2 h_n^{(2)}(\underset{\sim}{x}) + \int_0^1 \sigma^{n-1} G^{(1)}(r^2, 1-\sigma^2) h_m^{(1)}(\underset{\sim}{x}\sigma^2)\, d\sigma$$

(4.43)

$$+ \int_0^1 \sigma^{n-1} G^{(2)}(r^2, 1-\sigma^2) h_n^{(2)}(\underset{\sim}{x}\sigma^2)\, d\sigma$$

is a complete family of solutions in D.

<u>Remark</u>: Sufficient conditions for the Dirichlet problem to be well-posed may be found in Agmon [Ag 1]. For the special case were $A(r^2) \equiv 0$, uniqueness is assured for $B(r^2) \geq 0$.

<u>Proof</u>: Since $A(r^2)$ and $B(r^2)$ are analytic the adjoint equation to (4.14) also has analytic solutions. It is shown in Garabedian [Ga 1] pg. 188, that the solutions of an elliptic equation with analytic coefficients has a unique continuation property. In [Br 2] Browder showed that the unique continuation property for the adjoint equation is equivalent to the Runge approximation property for the original equation in firmly regular domains D. Therefore, if the $\{\varphi_{nm}\}$ form a complete family of solutions in a sphere $S \supset\supset D$, then $\{\varphi_{nm}\}$ are also complete in D.

<u>Claim</u>: The set $\{\varphi_{nm}\}$ are complete for C^4 solutions of (4.14) in a sphere of radius a, $S_a \supset D$.

<u>Proof</u>: To this end we note that since $u(\underset{\sim}{x})$ depends continuously on its boundary data, and since the boundary data can be expanded in a uniformly convergent series of sperical harmonics, it suffices to show that $\exists$ harmonic functions $h^{(1)}(\underset{\sim}{x})$ and $h^{(2)}(\underset{\sim}{x}) \ni$ on $|\underset{\sim}{x}| = a$

(4.44)
$$A_n Y_n(m_k;\theta;\varphi) = \Big[h^{(1)}(\underset{\sim}{x}) + r^2 h^{(2)}(\underset{\sim}{x}) + \int_0^1 \sigma^{n-1} G^{(1)}(r^2;1-\sigma^2) \cdot h^{(1)}(\underset{\sim}{x}\sigma^2)\,d\sigma + \int_0^1 \sigma^{n-1} G^{(2)}(r^2;1-\sigma^2) h^{(2)}(\underset{\sim}{x}\sigma^2)\,d\sigma\Big]\Big|_{|\underset{\sim}{x}|=a}$$

$$B_n Y_n(m_k;\theta;\varphi) = \frac{\partial}{\partial r}\Big[h^{(1)}(\underset{\sim}{x}) + r^2 h^{(2)}(\underset{\sim}{x}) + \int_0^1 \sigma^{n-1} G^{(1)}(r^2,1-\sigma^2) \cdot h^{(1)}(\underset{\sim}{x}\sigma^2)\,d\sigma + \int_0^1 \sigma^{n-1} G^{(2)}(r^2,1-\sigma^2) h^{(2)}(\underset{\sim}{x}\sigma^2)\,d\sigma\Big]\Big|_{|\underset{\sim}{x}|=a},$$

where the $Y_n(m_k;\theta;\varphi)$ denote arbitrary spherical harmonics, and A_n, B_n are arbitrary constants. However, these harmonic functions can be immediately determined by setting

$$h^{(1)}(\underset{\sim}{x}) + a_n r^n Y_n(m_k; \theta; \varphi), \text{ and } h^{(2)}(\underset{\sim}{x}) = b_n r^n Y_n(m_k; \theta; \varphi) ,$$

substituting these in (4.44), and solving the resulting system of algebraic equations for a_n and b_n. That this is always possible follows from the assumption, regarding $A(r^2)$ and $B(r^2)$, that the solution of the Dirichlet problem is unique.

<u>Remark</u>: In the preceding approach we have shown, under the assumption of the Dirichlet problem being well-posed, that to each solution of (4.14) in D there corresponds a unique pair of harmonic functions $\{h^{(1)}, h^{(2)}\}$, or what is the same thing a unique biharmonic function $b(\underset{\sim}{x}) = h^{(1)}(\underset{\sim}{x}) + r^2 h^{(2)}(\underset{\sim}{x})$. We must consider the possibility of finding solutions of (4.14) in the form

$$u(\underset{\sim}{x}) = (\underset{\sim}{I} + \underset{\sim}{G})b(\underset{\sim}{x}) \equiv b(\underset{\sim}{x}) + \int_0^1 \sigma^{n-1} G(r, 1-\sigma^2) b(\underset{\sim}{x}\sigma^2) d\sigma , \quad n \geq 2 , \tag{4.45}$$

$\Delta_n^2 b(\underset{\sim}{x}) = 0$ in D. It can be shown, by the usual methods, that the G-function must satisfy the equation

$$G_{rrrr} + \frac{4(1-t)}{r} G_{rrrt} + \frac{6(1-t^2)}{r^2} G_{rrtt} + \frac{4(1-t)^3}{r^3} G_{rttt}$$

$$- \frac{2}{r} G_{rrr} - \frac{16(1-t)}{r^2} G_{rrt} - \frac{28(1-t)^2}{r^3} G_{rtt}$$

$$+ [A - \frac{(n-1)(n-2)}{2r^2}] G_{rr} + [\frac{2(1-t)}{r} A - \frac{(1-t)}{r^3}(n^2 - 6n - 38)]G_{rt}$$

$$+ \frac{A(1-t)^2}{r^2} G_{tt} + (\frac{3n^2 - 18n - 6}{2r^3} - \frac{A}{r}) G_r - \frac{3(1-t)}{r^2} AG_t$$

$$+ (B + A\frac{(-n^2 + 5n)}{4r^2}) G = 0 ,$$

(4.46)

where the G-function satisfies the following initial conditions:

$$G(0,t) = 0 , \quad G(r,0) = -\frac{1}{2}\int_0^r A(r^2)rdr$$

$$G_t(r,0) = \frac{1}{8}\int_0^r A(r^2)[\int_0^\rho A(\rho^2)\rho d\rho - 19]\, rdr + \frac{3}{4} r^2 A(r^2)$$

$$G_{tt}(r,0) = \frac{1}{4}\int_0^r rA(r^2)[-\frac{169}{4} + \frac{43}{8} n) + nrtA(r^2)\frac{r^2}{4}$$

$$+ (\frac{45 + A}{8})\int_0^r A(\rho^2)\rho d\rho - \frac{1}{8}\int_0^r A(\rho^2)\rho \int_0^\rho \eta A(\eta^2) d\eta d\rho]dr$$

$$+ \frac{1}{4}A(r^2)r^2\left[\frac{69+9n}{4} - \frac{3}{4}\int_0^r A(r^2)rdr - r\right]$$

(4.47)

$$- \frac{5}{8} r^3 A^1(r^2) + (\frac{n-5}{16}) \int_0^r r \int_0^r \rho A(\rho^2) d\rho dr - \frac{1}{2} \int^r r^3 B(r^2) dr .$$

Problem: Verify that the equation (4.46) must hold for the G-function, and furthermore, show that the initial conditions (4.47) must be satisfied.

Remark: The integral representation may be rewritten, using a change of integration parameter, as a Volterra integral equation. Consequently, just as in the second order case, it can be shown that there is a one-to-one correspondence on the class of functions that are $C[\bar{D}]$, where D is appropriate. Furthermore, since the folowing identity holds,

$$\Delta_n^2 \varphi + A(r^2) \Delta \varphi + B(r^2) \varphi = \int_0^1 \sigma^{n-1} G(r, 1-\sigma^2) (\Delta_n^2 \psi)(\underset{\sim}{x}\sigma^2) d\sigma,$$

with $\varphi = (\underset{\sim}{I} + \underset{\sim}{G}) \psi(\underset{\sim}{x})$, there is a unique correspondence between the solutions of (4.14) that are in $C(\bar{D})$ and the class of biharmonic functions in $C(\bar{D})$. This permits us to reduce the study of the solutions of (4.14) to the study of biharmonic functions in D. In the next chapter we shall discuss in some detail, therefore, the analytic methods for treating the biharmonic functions.

V: Further Representations for Solutions of Higher Order Elliptic Differential Equations with Analytic Coefficients

In this chapter we shall follow the approach of I. N. Vekua [Ve 1] for treating equations of the form

$$(5.1) \qquad \Delta^{(n)} u + \sum_{k=1}^{n} \underset{\sim}{L}_k (\Delta^{n-k} u) = f(x,y), \ n \geq 1,$$

where $\Delta^{o} \equiv 1$, and Δ^{m} stands here for the m th iterated, two-dimensional Laplacian, and

$$\underset{\sim}{L}_k = \sum_{p,q=0}^{p+q \leq k} a_k^{pq}(x,y) \frac{\partial^{p+q}}{\partial x^p \partial y^q} .$$

We assume as usual that the coefficients are analytic functions of x and y and have a holomorphic extension to $D \times D^*$.

First we consider the special case of $\Delta^n u = 0$, which we may rewrite upon substituting $z = x + iy$, $\zeta = x - iy$, as

$$(5.2) \qquad \frac{\partial^{2n} U}{\partial z^n \partial \zeta^n} = 0 , \quad U(z,\zeta) \equiv u\left(\frac{z+\zeta}{z}, \frac{z-\zeta}{zi}\right).$$

If u is a solution of (5.2) then

$$(5.3) \qquad u_o(x,y) = \Delta^{n-1} u$$

is harmonic, and we may write

(5.4) $$u_o(x,y) = 4^{n-1}(\varphi_o(z) + \varphi_o^*(\zeta)),$$

where φ_o, φ_o^* are holomorphic functions in D and D^*. Without any loss of generality we may take $\varphi_o(o) = \varphi_o^*(o)$, and in this case the functions φ_o, φ_o^* are uniquely defined in terms of u. From (5.4) one may then write

$$\frac{\partial^{n-2}U}{\partial z^{n-1}\partial\zeta^{n-1}} = \varphi_o(z) + \varphi_o^*(\zeta) ,$$

which has a solution of the form

$$U(z,\zeta) = V(z,\zeta) + \frac{z^{n-1}}{(n-1)!}\int_o^\zeta \frac{(\zeta-z)^{n-2}}{(n-2)!}\varphi_o^*(\tau)d\tau$$

(5.5) $$+\frac{\zeta^{n-1}}{(n-1)!}\int_o^z \frac{(z-t)^{n-2}}{(n-2)!}\varphi_o(t)dt ,$$

where $V(z,\zeta)$ is a solution of $\Delta^{n-1}v = 0$. Applying the same reasoning to $\frac{\partial^{2n-2}V}{\partial z^{n-1}\partial\zeta^{n-1}} = 0$ one obtains a representation

$$V(z,\zeta) = W(z,\zeta) + \frac{z^{n-2}}{(n-2)!}\int_o^\zeta \frac{(\zeta-\tau)^{n-3}}{(n-3)!}\varphi_1^*(\tau)d\tau$$

(5.6) $$+\frac{\zeta^{n-2}}{(n-2)!}\int_o^z \frac{(z-t)^{n-3}}{(n-3)!}\varphi_1(t)dt ,$$

where $W(z,\zeta)$ is a solution of $\Delta^{n-2}w=0$, and φ_1 and φ_1^* are arbitrary holomorphic functions which may be normalized by $\varphi_1(0) = \varphi_1^*(0)$. Preceeding in this way we obtain

$$U(z,\zeta) = \varphi_{n-1}(z)+\overset{*}{\varphi}_{n-1}(\zeta) + \sum_{k=1}^{n-1}\Big[\frac{z^k}{k!}\int_0^{\zeta}\frac{(\zeta-\tau)^{k-1}}{(k-1)!}\overset{*}{\varphi}_{n-k-1}(\tau)d\tau$$

$$(5.7) \quad + \frac{\zeta^k}{k!}\int_0^z \frac{(z-t)^{k-1}}{(k-1)!}\varphi_{n-k-1}(t)dt\Big] ,$$

where the $\varphi_k(z)$, $\overset{*}{\varphi}_k(\zeta)$, $(k = 0, 1, 2,\ldots,n-1)$ are holomorphic functions in D and $\overset{*}{D}$ respectively, and which are subjected to the conditions

$$\varphi_k(0) = \overset{*}{\varphi}_k(0) \quad , \quad (k = 0, 1, \ldots, n-1) ,$$

This latter condition permits us to write

$$\varphi_{n-k-1}(z) = \tfrac{1}{2}a_k + \int_0^z \chi_k(t)dt ,$$

$$\overset{*}{\varphi}_{n-k-1}(\zeta) = \tfrac{1}{2}a_k + \int_0^{\zeta}\overset{*}{\chi}_k(\tau)d\tau , \quad (k = 0,1,\ldots,n-1) ,$$

where the a_k are constants and χ_k, $\overset{*}{\chi}_k$ are holomorphic in D, $\overset{*}{D}$ respectively. Consequently, the representation (5.7) may be rewritten by interchanging orders of integration as

$$(5.8) \quad u=\sum_{k=0}^{n-1} a_k \frac{z^k\zeta^k}{k!k!} +\sum_{k=0}^{n-1}\Big[\frac{z^k}{k!}\int_0^{\zeta}\frac{(\zeta-\tau)^k}{k!}\overset{*}{\chi}_k(\tau)d\tau + \frac{\zeta^k}{k!}\int_0^z\frac{(z-t)^k}{k!}\chi_k(t)dt\Big]$$

which implies that the terms

$$(5.9) \qquad a_k = \left(\frac{\partial^{2k} u}{\partial z^k \partial \zeta^k}\right)_{z=\zeta=0}, \quad \chi_k(z) = \left(\frac{\partial^{2k+1} u}{\partial z^{k+1} \partial \zeta^k}\right)_{\zeta=0}$$

$$\chi^*_k(\zeta) = \left(\frac{\partial^{2k+1} u}{\partial z^k \partial \zeta^{k+1}}\right)_{z=0},$$

are uniquely determined. This latter observance means that the Goursat problem for $\Delta^n u = 0$, i.e. the solution of this equation which satisfies the initial data

$$(5.10) \qquad \left(\frac{\partial^k u}{\partial \zeta^k}\right)_{\zeta=0} = f_k(z), \quad \left(\frac{\partial^k u}{\partial z^k}\right)_{z=0} = f_k^*(\zeta), \quad (k=0,\ldots,n-1)$$

with holomorphic f_k, f_k^* in D, D^* respectively, is solvable in terms of the representation (5.8). The solution satisfying the Goursat data (5.10) is given by setting

$$a_k = f_k^{(k)}(0), \quad \chi_k(z) = f_k^{(k+1)}(z), \quad \chi_k^*(\zeta) = f_k^{*(k+1)}(\zeta).$$

The real solutions are obtained by setting $\zeta=\bar{z}$ and taking the real part; hence, we have the following representation for real solutions,

$$(5.11) \qquad u(x,u) = \sum_{k=0}^{n-1} \alpha_k r^{2k} + \sum_{k=0}^{n-1} \frac{\bar{z}^k}{k!} \int_0^z \frac{(z-t)^k}{k!} \chi_k(t)\,dt,$$

with real constants α_k.

An alternate representation for solutions is obtained in the form

$$u = \sum_{k=0}^{n-1} \left[z^k \psi_k^*(\zeta) + \zeta^k \psi_k(z) \right] ,$$

with
$$\psi_k(t) = \tfrac{1}{2} \frac{a_k z^k}{k!k!} + \frac{1}{k!k!} \int_0^z (z-t)^k \chi_k(t)\,dt ,$$

(5.12)
$$\psi_k^*(\zeta) = \tfrac{1}{2} \frac{a_k \zeta^k}{k!k!} + \frac{1}{k!k!} \int_0^\zeta (\zeta-\tau)^k \chi_k^*(\tau)\,d\tau .$$

From (5.12) it is clear that $\psi_k^{(m)}(0) = \psi_k^{*(m)}(0) = 0$, $(m=1,\ldots,k-1; k \geq 1)$ $\psi_k^{(k)}(0) = \psi_k^{*(k)}(0)$, $(k=0,1,\ldots,n-1)$. A real solution, therefore, may be put into the form

(5.13) $$u = \mathrm{Re}\left\{ \sum_{k=0}^{n-1} z^k \overline{\psi_k(z)} \right\}$$

where $\psi_k^{(m)}(0)=0$, $(m=1,\ldots,k-1;\ k \geq 1)$, $\psi_k^{(k)}(0) = \overline{\psi_k^{(k)}(0)}$, $(k=0,1,\ldots,n-1)$. From (5.12) and (5.13) it is clear that a real solution may also be represented as

(5.14) $$u = \sum_{k=0}^{n-1} \omega_k(x,y) r^{2k} , \quad r = |z| ,$$

where the ω_k are harmonic in D and uniquely determined from $u(x,y)$.

If one substitutes into (5.14) $\omega_0 = \omega_1 = \ldots = \omega_{n-2} = 0$, $\omega_{n-1} = C \log \frac{1}{r}$, where $r = |z-z_0|$, C a real constant, the one obtains the particular solution

$$\text{(5.15)} \qquad \omega(x,y;\, x_0,y_0) = C\, r^{2n-2} \log \frac{1}{r}\,.$$

We normalize this solution by setting $C = \dfrac{1}{2\pi 4^{n-1}[(n-1)!]^2}$.

Using successively the Green's formula

$$\text{(5.16)} \qquad \iint\limits_{D} (v\Delta^k u - \Delta v \Delta^{k-1} u)\, dxdy = - \int\limits_{\partial D} \left(v \frac{\partial \Delta^{k-1} u}{\partial \nu} - \frac{\partial v}{\partial \nu} \Delta^{k-1} u\right) ds\,,$$

one obtains by addition, that

$$\text{(5.17)} \qquad \iint (v\Delta^n u - u\Delta^n v)\, dxdy = - \sum_{k=0}^{n-1} \int\limits_{\partial D} \left(\Delta^k v \frac{\partial \Delta^{n-k-1} u}{\partial \nu} - \frac{\partial \Delta^k v}{\partial \nu} \Delta^{n-k-1} u \right) ds.$$

In the special case where $\Delta u = \Delta v = 0$ in D, this becomes

$$\text{(5.18)} \qquad \sum_{k=0}^{n-1} \int\limits_{\partial D} \left(\Delta^k v \frac{\partial \Delta^{n-k-1} u}{\partial \nu} - \frac{\partial \Delta^k v}{\partial \nu} \Delta^{n-k-1} u \right) ds = 0.$$

The solution $\omega(x,y;\, x_0,y_0)$ given by (5.15) is an elementary solution, and if v_0 is a regular solution in D, so is the function

$$\text{(5.19)} \qquad v = \omega(x,y;\, x_0,y_0) + v_0 = \frac{r^{2n-2}}{2\pi 4^{n-1}[(n-1)!]^2} \log \frac{1}{r} + v_0\,.$$

Using the elementary solution given above, and the residue theorem, one obtains by substituting v into the following representation formula

$$(5.20)\quad u(x_o,y_o) = \sum_{k=0}^{n-1} \int_{\partial D} \left(\Delta^k u \frac{\partial \Delta^{n-k-1} v}{\partial \nu} - \frac{\partial \Delta^k u}{\partial \nu} \Delta^{n-k-1} v \right) ds \ .$$

The function v, defined above, is called the Green's function for the equation $\Delta^n u = 0$, if it satisfies in addition the boundary conditions

$$(5.21)\quad v_o = -\omega,\ \frac{\partial v_o}{\partial \nu} = -\frac{\partial \omega}{\partial \nu}, \ldots, \frac{\partial^{n-1} v_o}{\partial \nu^{n-1}} = -\frac{\partial^{n-1} \omega}{\partial \nu^{n-1}}, \text{ on } \partial D \ .$$

From (5.21) it follows that all derivatives of order $\leq n-1$ of $Z \equiv \upsilon$ vanish on ∂D. Consequently, (5.20) reduces to

$$(5.22)\quad u(x_o,y_o) = \sum_{k=0}^{[\frac{1}{2}(n-1)]} \int_{\partial D} \left(\Delta^k u \frac{\partial \Delta^{n-k-1} Z}{\partial \nu} - \frac{\partial \Delta^k u}{\partial \nu} \Delta^{n-k-1} Z \right) ds \ ,$$

where $[p/2]$ is the greatest integer $\leq p/2$. It is clear from (5.22) that having the Green's function permits one to solve the Dirichlet problem with continuous data, i.e. to find a solution of $\Delta^n u = 0$ satisfying the boundary conditions

$$u = f_o, \quad \frac{\partial u}{\partial \nu} = f_1, \ldots, \frac{\partial^{n-1} u}{\partial \nu^{n-1}} = f_{n-1} \quad \text{on} \quad \partial D . \tag{5.23}$$

We turn next to the homogeneous equation

$$\underset{\sim}{M}[u] \equiv \Delta^n u + \sum_{k=1}^{n} \underset{\sim k}{L}(\Delta^{n-k} u) = 0 \tag{5.24}$$

where the coefficients of the operators

$$\underset{\sim k}{L}[u] \equiv \sum_{p,q=0}^{p+q \leq k} a_k^{pq}(x,y) \frac{\partial^{p+q} u}{\partial x^p \partial y^q} \tag{5.25}$$

are holomorphic functions of $z = x + iy$, $\zeta = x - iy$ in $D \times D^*$. It can be shown (See Vekua [Ve 1], pg. 184.) that every regular solution of (5.24) is analytic. This is known as Picard's Theorem.

Following Vekua's approach we note that (5.24) may be written in the form

$$\sum_{k=0}^{n} \sum_{m=0}^{n} A_{km} \frac{\partial^{k+m} u}{\partial z^k \partial \zeta^m} = 0 , \quad \text{or} \tag{5.26}$$

$$\sum_{k=0}^{n} \sum_{m=0}^{n} \frac{\partial^{k+m} B_{km} u}{\partial z^k \partial \zeta^m} = 0 , \qquad B_{km} = B_{km}(z,\zeta),$$

with $A_{nn} = B_{nn} = 1$. If $U(z,\zeta)$ is a solution of (5.26) in $D \times D^*$ then it is immediate that

$$\frac{\partial^{2n}}{\partial z^n \partial \zeta^n}\Big[U(z,\zeta) + \sum_{k=0}^{n-1} \int_{z_o}^{z} \frac{(z-t)^{n-k-1}}{(n-k-1)!} B_{kn}(t,\zeta)U(t,\zeta)dt$$

$$+ \sum_{k=0}^{n-1} \int_{\zeta_o}^{\zeta} \frac{(\zeta-\tau)^{n-k-1}}{(n-k-1)!} B_{nk}(z,\tau)U(z,\tau)d\tau$$

$$+ \sum_{k=0}^{n-1} \sum_{m=0}^{n-1} \int_{z_o}^{z} dt \int_{\zeta_o}^{\zeta} \frac{(z-t)^{n-k-1}}{(n-k-1)!} \frac{(\zeta-\tau)^{n-m-1}}{(n-m-1)!} B_{km}(t,\tau)U(t,\tau)d\tau\Big] = 0,$$

$(z_o, \zeta_o) \in D \times D^*$, and therefore that

$$U(Z,\zeta) - \int_{z_o}^{z} K_1(z,\zeta,t)U(t,\zeta)dt - \int_{\zeta_o}^{\zeta} K_2(\zeta,z,\tau)U(z,\tau)$$

(5.27) $$- \int_{z_o}^{z} dt \int_{\zeta_o}^{\zeta} K(z,\zeta,t,\tau)U(t,\tau)d\tau = U_o(z,\zeta) \quad ,$$

where $U_o(z,\zeta)$ is a solution of $\Delta^n u = 0$, and

$$K_1(z,\zeta,t) = - \sum_{k=0}^{n-1} \frac{(z-t)^{n-k-1}}{(n-k-1)!} B_{kn}(t,\zeta) \quad ,$$

$$K_2(\zeta,z,\tau) = - \sum_{k=0}^{n-1} \frac{(\zeta-\tau)^{n-k-1}}{(n-k-1)!} B_{nk}(z,\tau),$$

$$(5.28)\quad K(z,\zeta,t,\tau) = -\sum_{k=0}^{n-1}\sum_{m=0}^{n-1} \frac{(z-t)^{n-k-1}(\zeta-\tau)^{n-m-1}}{(n-k-1)!(n-m-1)!} B_{km}(t,\tau).$$

It is shown in Vekua (see also Gilbert [Gi 1] pp. 134-140.) that a solution of (5.27), (5.28) may be found in the form

$$U(z,\zeta) = U_o(z,\zeta) + \int_{z_o}^{z} \Gamma_1(z,\zeta,t)U_o(t,\zeta)dt$$

$$(5.29)\quad + \int_{\zeta_o}^{\zeta} \Gamma_2(\zeta,z,\tau)U_o(z,\tau)d\tau + \int_{z_o}^{z} dt \int_{\zeta_o}^{\zeta} \Gamma(z,\zeta,t,\tau)U_o(t,\tau)d\tau\ ,$$

where the kernels Γ_k are holomorphic functions of their arguments for $z,t \in D$, and $\zeta,\tau \in D^*$. Consequently, $U(z,\zeta)$ is holomorphic in $D \times D^*$.

If one replaces $U_o(z,\zeta)$ in the above by

$$(5.30)\quad U_o(z,\zeta) = \frac{1}{4^n} \int_{z_o}^{z} dt \int_{\zeta_o}^{\zeta} \frac{(z-t)^{n-1}(\zeta-\tau)^{n-1}}{(n-1)!(n-1)!} f\left(\frac{t+\tau}{2}, \frac{t-\tau}{2i}\right)d\tau\ ,$$

which is a solution of $\Delta^n U = f(x,y)$, we obtain

$$U^*(z,\zeta) = \frac{1}{4^n}\int_{z_o}^{z} dt \int_{\zeta_o}^{\zeta} R(t,\tau;z,\zeta) f\left(\frac{t+\tau}{2}, \frac{t-\tau}{2i}\right)d\tau\ ,$$

where

$$R(t,\tau;z,\zeta) = \frac{(z-t)^{n-1}(\zeta-\tau)^{n-1}}{(n-1)!(n-1)!} + \int_{t}^{z} \frac{(t_1-t)^{n-1}(\zeta-\tau)^{n-1}}{(n-1)!\ (n-1)!} \Gamma_1(z,\zeta,t_1)dt_1$$

$$+\int_{\tau}^{\zeta}\frac{(z-t)^{n-1}(\tau_1-\tau)^{n-1}}{(n-1)!\,(n-1)!}\Gamma_2(\zeta,z,\tau_1)d\tau_1$$

$$(5.31)\qquad +\int_{t}^{z}dt_1\int_{\tau}^{\zeta}\frac{(t_1-t)^{n-1}(\tau_1-\tau)^{n-1}}{(n-1)!\,(n-1)!}\Gamma(z,\zeta;t_1,\tau_1)d\tau_1\,,$$

which satisfies $\underset{\sim}{M}[u] = f(x,y)$.

Problem: Verify that U^* as given above satisfies the non-homogeneous equation.

Remark: $R(t,\tau;z,\zeta)$ is a solution of $\underset{\sim}{M}[u] = 0$ in the last two variables, since it is obtained from (5.29) by replacing U_o by $\frac{(z-t)^{n-1}(\zeta-\tau)^{n-1}}{(n-1)!\,(n-1)!}$.

Problem: Verify the following ordinary differential equations for g and g^*:

$$\frac{d^n g}{d\zeta^n}+\sum_{m=0}^{n-1}A_{nm}(z,\zeta)\frac{d^m g}{d\zeta^m}=0\,,$$

$$\frac{d^n g^*}{dz^n}+\sum_{m=0}^{n-1}A_{mn}(z,\zeta)\frac{d^m g^*}{dz^m}=0\,,$$

where $g=\left(\frac{\partial^{n-1}R}{\partial z^{n-1}}\right)_{t=z}$, $g^*=\left(\frac{\partial^{n-1}R}{\partial \zeta^{n-1}}\right)_{\tau=\zeta}$, where g and g^* are seen to satisfy

.32) $$g = \frac{dg}{d\zeta} = \dots = \frac{d^{n-2}g}{d\zeta^{n-2}} = 0, \quad \frac{d^{n-1}g}{d\zeta^{n-1}} = 1 \quad \text{for} \quad \zeta = \tau,$$

$$g^* = \frac{dg^*}{dz} = \dots = \frac{d^{n-2}g^*}{dz^{n-2}} = 0, \quad \frac{d^{n-1}g^*}{dz^{n-1}} = 1 \quad \text{for} \quad z = t.$$

The functions g and g^*, therefore, can be assumed to be determined by using the theory of ordinary differential equations. From (5.32) the function $R(t,\tau;z,\zeta)$ is seen to satisfy the initial conditions

$$\frac{\partial^k R}{\partial z^k} = 0 \, (k = 0,1,\dots,n-2), \quad \frac{\partial^{n-1} R}{\partial z^{n-1}} = g, \quad \text{for} \quad z = t,$$

and

.33) $$\frac{\partial^k R}{\partial \zeta^k} = 0 \quad (k = 0,1,\dots,n-2), \quad \frac{\partial^{n-1} R}{\partial \zeta^{n-1}} = g^*, \quad \text{for} \quad \zeta = \tau.$$

The function, $R(t,\tau;z,\zeta)$, is the solution of a Goursat problem and can be seen to be uniquely determined by these functions, and we shall refer to it as the Riemann function. From (5.31) the Riemann function was seen to be expressible in terms of the resolvent kernels Γ_1, Γ_2, and Γ. Contrarily one also has

$$\Gamma_1(z,\zeta,t) = - \left.\frac{\partial^{2n-1} R(t,\tau;z,\zeta)}{\partial t^n \partial \tau^{n-1}}\right|_{\tau=\zeta},$$

$$(5.34)\quad \Gamma_2(\zeta,z,\tau) = - \left.\frac{\partial^{2n-1} R(t,\tau;z,\zeta)}{\partial t^{n-1} \partial \tau^{n}}\right|_{t=z},$$

$$\text{and}\quad \Gamma(z,\zeta;t,\tau) = \frac{\partial^{2n} R(t,\tau;z,\zeta)}{\partial t^n \, \partial \tau^n}.$$

Using the general representation for $U_o(z,\zeta)$,

$$U_o(z,\zeta) = \sum_{k=0}^{n-1} \left\{ a_k \frac{(z-z_o)^k (\zeta-\zeta_o)^k}{k!\, k!} + \frac{(z-z_o)^k}{k!} \int_{\zeta_o}^{\zeta} \frac{(\zeta-\tau)^k}{k!} \chi_k^*(\tau)\, d\tau \right.$$

$$\left. + \frac{(\zeta-\zeta_o)^k}{k!} \int_{z_o}^{z} \frac{(z-t)^k}{k!} \chi_k(t)\, dt \right\},$$

where $a_o,\ldots,a_{n-1}$ are arbitrary constants, and $\chi_k(z)$, $\chi_k^*(\zeta)$, $(k=0,1,\ldots,n-1)$ are arbitrary holomorphic functions one obtains the following general expression for solutions of (5.24),

$$U(z,\zeta) = \sum_{k=0}^{n-1} a_k R_k(z_o,\zeta_o;z,\zeta) + \sum_{k=0}^{n-1} \left\{ \int_{z_o}^{z} R_k(t,\zeta_o;z,\zeta)\, \chi_k(t)\, dt \right.$$

.35) $$+ \int_{\zeta_o}^{\zeta} R_k(z_o, \tau; z, \zeta) \overset{*}{\chi_k}(\tau)\, d\tau \Big\} ,$$

where

$$R_k(t,\tau;z,\zeta) = \frac{(z-t)^k(\zeta-\tau)^k}{k!k!} + \frac{(\zeta-\tau)^k}{k!}\int_t^z \frac{(\xi-t)^k}{k!}\Gamma_1(z,\zeta,\xi)\,d\xi$$

$$+ \frac{(z-t)^k}{k!}\int_\tau^\zeta \frac{(\eta-\tau)^k}{k!}\,\Gamma_2(\zeta,z,\eta)\,d\eta$$

.36) $$+ \int_t^z d\xi \int_\tau^\zeta \frac{(\xi-t)^k(n-\tau)^k}{k!k!}\,\Gamma(z,\zeta;\xi,\eta)\,d\eta$$

$$(k = 0,1,\ldots,n-1) .$$

The function $R_k(t,\tau;z,\zeta)$ is seen to be a solution of the integral equation [Ve 1, pg. 188]

$$R_k(t,\tau;z,\zeta) - \int_t^z K_1(\zeta,z,\xi)R_k(t,\tau;\xi,\zeta)\,d\xi - \int_\tau^\zeta K_2(z,\zeta,\eta)$$

$$R_k(t,\tau;z,\eta)\,d\eta - \int_t^z d\xi \int_\tau^\zeta K(z,\zeta;\xi,\eta)R_k(t,\tau;\xi,\eta)\,d\eta$$

.37) $$= \frac{(t-\tau)^k(\zeta-\tau)^k}{k!k!} , \quad (k=0,1,\ldots,n-1)$$

From (5.31) it is clear that the functions R_k are also defined by $R_k(t,\tau;z,\zeta) = \dfrac{\partial^{2(n-k-1)} R(t,\tau;z,\zeta)}{\partial t^{n-k-1}\partial\tau^{n-k-1}}$.

It may be shown that the functions χ_k, χ_k^*, and the constants a_k may be uniquely determined from the solution $U(z,\zeta)$ [Ve 1, pg 189], furthermore, this fact permits one to solve uniquely the corresponding Goursat problem. Since the Riemann function satisfies the Goursat data (5.33) this indicates it is uniquely determined by this data.

When the coefficients of (5.24) are real, one may express the general formula for real solutions as

$$u(x,y) = \sum_{k=0}^{n-1}\left[a_k R_k(z_0,\bar{z}_0,z,\bar{z}) + \mathrm{Re}\int_{z_0}^{z} R_k(t,\bar{z}_0,z,\bar{z})\chi_k(t)\,dt\right] \tag{5.38}$$

where the coefficients a_k are now taken to be real.

If one introduces the functions

$$\varphi_k(z) = \tfrac{1}{2}a_k + \int_{z_0}^{z}\chi_k(t)\,dt,\quad \varphi_k^*(\zeta) = \tfrac{1}{2}a_k + \int_{\zeta_0}^{\zeta}\chi_k^*(\tau)\,d\tau,$$

$(k=0,1,\ldots,n-1)$, (5.35) may be rewritten as

$$u(x,y) = R_o(z,\zeta_o;z,\zeta)\varphi_o(z) + R_o(z_o,\zeta;z,\zeta)\varphi_o^*(\zeta)$$

$$(5.38)\qquad - \sum\left\{\int_{z_o}^{z} \varphi_k(t)\frac{\partial}{\partial t} R_k(t,\zeta_o;z,\zeta)\,dt + \int_{\zeta_o}^{\zeta} \varphi_k^*(\tau)\frac{\partial}{\partial \tau} R_k(z_o,\tau;z,\zeta)\,d\tau\right\}.$$

When we seek real solutions for the case where the differential equation in question has real coefficients then (5.38) takes the form

$$u(x,y) = \mathrm{Re}\{R_o(z,\bar{z}_o;\ z,\bar{z})\varphi_o(z)$$

$$(5.39)\qquad - \sum_{k=0}^{n-1} \int_{z_o}^{z} \varphi_k(t)\frac{\partial}{\partial t} R_k(t,\overline{z_o},z,\bar{z})\,dt\},$$

with $\varphi_k(z_o) = \overline{\varphi_k(z_o)}$, $(k=0,1,\ldots,n-1)$.

If we reverse the coordinates (z,ζ) and (t,τ) then it may be shown (see Ve 1, pp. 190-192), that $R(z,\zeta;t,\tau)$ is the Riemann function for the adjoint equation

$$M^*[v] \equiv \sum_{k,m=0}^{n-1} (-1)^{k+m} \frac{\partial^{k+m}}{\partial z^k \partial \zeta^m}(A_{km}(z,\zeta)v) = 0.$$

For equation (5.24)we define a fundamental solution to have the form

$$\omega = g(x,y)r^{2n-2}\log\frac{1}{r} + g_o(x,y) \;, \quad r = |z-z_o|$$

where $g,\ g_o \in C^{2n}$, and $g(x_o,y_o) \neq 0$. If in (5.38) we set

$\zeta_o = \bar{z}_o$, $\varphi_k(z) = 0$, $\varphi_k^*(z) = 0$ $(k=0,1,\ldots,n-2)$

and

$$\varphi_{n-1}(z) = C\log(z-z_o),\ \varphi_{n-1}^*(\zeta) = C\log(\zeta-\zeta_o).$$

<u>Problem</u>: Show that the following expression is a fundamental solution for equation (5.24)

$$\omega_1(x,y;x_o,y_o) = -C\left[2R(z_o,\bar{z}_o;z,\bar{z})\log\frac{1}{r} + R^*(z_o,\bar{z}_o;z,\bar{z})\right],$$

where $r = |z-z_o|$, and

$$R^*(z_o,\bar{z}_o;z,\bar{z}) = \int_0^1 \log\sigma \ \frac{\partial}{\partial\sigma}\Big[R(z_o+(z-z_o)\sigma,\bar{z}_o;z,\bar{z})$$

$$+ R(z_o,\overline{z_o} + (\bar{z}-\overline{z_o})\sigma;z,\bar{z})\Big]d\sigma\ .$$

From (5.31) it is clear that one may factor a term $|z-z_o|^{2n-2}$ from $R(z_o,\overline{z_o};z,\bar{z})$; hence, we have

$$R(z_o,\overline{z_o};z,\bar{z}) = r^{2n-2}g(x,y;x_o,y_o),$$

where $g(x,y;x_o,y_o)$ is an analytic function of its arguments satisfying $g(x_o,y_o;x_o,y_o) = \dfrac{1}{[(n-1)!]^2}$.

To normalize the fundamental solution we assume that $C = -\frac{1}{4^n\pi}$.

We consider next the boundary value problems associated with

$$(5.24) \qquad \underset{\sim}{M}[u] \equiv \Delta^n u + \sum_{k=1}^{n} L_k(\Delta^{n-k}u) = 0 .$$

We will assume that D is bounded by a simple, smooth contour and may be parametrized by $x - x(s)$, $y = y(s)$, such that $x(s+\ell) = x(s)$, $y(s+\ell) = y(s)$, where s is the arc length parameter and ℓ the length of the boundary ∂D. We seek a solution of the Dirichlet problem, namely to find a solution of (5.24) which satisfies the boundary conditions

$$u = f_o, \ \frac{\partial u}{\partial \nu} = f_1, \ldots, \frac{\partial^{n-1} u}{\partial \nu^{n-1}} = f_{n-1} ,$$

where ν is the outward normal to ∂D.

<u>Problem</u>: Verify that on ∂D that

$$.40) \qquad \frac{\partial^k u}{\partial \nu^k} = i^k \sum_{\ell=0}^{k} (-1)^{k-\ell} \frac{k!}{\ell!(k-\ell)!} \left(\frac{\partial^k u}{\partial z^{k-\ell} \partial \zeta^{\ell}}\right)\left(\frac{\partial z}{\partial s}\right)^{k-2\ell}$$

and

$$\left(\frac{\partial^{m+1} u}{\partial z^{m+1}}\right)^{+}\left(\frac{dz}{ds}\right)^{m+1} + (-1)^{m}\left(\frac{\partial^{m+1} u}{\partial \zeta^{m+1}}\right)^{+}\left(\frac{d\zeta}{ds}\right)^{m+1}$$

$$(5.41) \quad = \sum_{\ell=0}^{m} (-1)^{\ell}\left(\frac{dz}{ds}\right)^{2m-2\ell} \frac{d}{ds}\left(\frac{\partial^{m} u}{\partial z^{m-\ell}\partial \zeta^{\ell}}\right)^{+}, \quad (m=0,1,\ldots;\ \zeta=\bar{z}).$$

From this identity it is clear that the derivatives $\left.\frac{\partial^{k+m}}{\partial z^{k}\partial \zeta^{m}}\right|_{\partial D}$ may be expressed in terms of the Dirichlet data and their derivatives. (see Vekua [Ve 1] pg. 214).

Introducing the notation $g_{km}(s) = \frac{\partial^{k+m} u^{+}}{\partial z^{k}\partial \zeta^{m}}$, $k+m \le n-1$, we note that by the above discussion the Dirichlet data may be rewritten in the form

$$(5.42) \qquad u = g_{oo},\ \frac{\partial u}{\partial z} = g_{1o}, \ldots, \frac{\partial^{n-1}}{\partial z^{n-1}} = g_{n-1,o}\ .$$

It can be shown that the conditions (5.40) may be rewritten as

$$\mathrm{Re}\left[\frac{1}{\pi i}\int_{\partial D} \frac{u-g_{oo}dt}{t-z} - \frac{1}{\pi i}\int_{\partial D}\frac{u-g_{oo}}{t}\, dt\right] = 0$$

$$(5.43) \quad \mathrm{Re}\left[\frac{1}{\pi i}\int_{\partial D} t^{\ell}\left(\frac{\partial^{\ell} u}{\partial t^{\ell}} - g_{\ell o}\right)\frac{dt}{t-z} - \frac{1}{\pi i}\int_{\partial D} t^{\ell}\left(\frac{\partial^{\ell} u}{\partial t^{\ell}} - g_{\ell o}\right)\frac{dt}{t}\right] = 0,$$

where z is any point $\ni$ $z \notin D + \partial D$, and where it is assumed that $0 \in D$. That the conditions (5.41) follow from (5.40) is self evident. To show the converse we note from the first equation of (5.41) we have

$$(5.44)\quad \frac{1}{\pi i}\int_{\partial D} \frac{u-g_{oo}}{t-z}\,dt = 0\ , \text{ and } \frac{1}{\pi i}\int_{\partial D} \frac{u-g_{oo}}{t}\,dt = i\,C$$

where C is a real constant. Because $u-g_{oo}$ is real we have (see Chapter III of Vekua [V.1]) from the first of these equations that $u-g_{oo}=C_o$, where C_o is a real constant. This in turn implies from the second equation that $C = C_o = o$.

From the second of equations (5.41) one has that

$$(5.45)\quad t^{\ell}\left(\frac{\partial^{\ell} u}{\partial t^{\ell}} - g_{\ell o}\right) = \chi_{\ell}^{+}(t)$$ on ∂D [where χ_{ℓ}^{+} is the boundary value of a function $\chi_{\ell}(z)$ which is holomorphic in D], and $\chi_{\ell}(0)^{\ell} = i\,C_{\ell}$, C_{ℓ} being a real constant. Now using (5.41) we see that for $\zeta = \bar{z}$

$$(5.46)\quad \mathrm{Re}\left\{\frac{1}{t}\,\chi_1^{+}\,\frac{dt}{ds}\right\} = \tfrac{1}{2}\left[\frac{dt}{ds}\left(\frac{\partial u}{\partial t} - g_{1o}\right) + \frac{d\bar{t}}{ds}\left(\frac{\partial u}{\partial \bar{t}} - g_{1o}\right)\right]$$

$$= \tfrac{1}{2}\,\frac{d}{ds}\left(u^{+} - g_{oo}\right) = 0.$$ $\chi_1(z)$ satisfies therefore, a certain homogeneous boundary value problem. Vekua [Ve 1]

Sec. 27 shows that χ_1 must therefore be of the form

$$\chi_1(z) = iA_1 e^{-ip(z)}$$

where A_1 is a real constant, and $p(z)$ holomorphic in D and satisfying the boundary condition $\operatorname{Re} p^+(t) = \theta - \varphi$, where $\theta = \arg \frac{dt}{ds}$, $\varphi = \arg t$, $p(0) = \overline{p(0)}$. This implies $\operatorname{Re}\left[2\pi i\, \chi_1(0)\right] = 0$ or $\cos p(0) = 0$, which implies that $0 = iA_1(\cos p(0) + i \sin p(0)) = i\, C_1$ or $A_1 = C_1 = 0$, i.e. $\chi_1(z) = 0$ in D. Therefore we have obtained the first of the two equations (5.42) from the first of the two equations (5.43). A similar argument may be used to show that the other equations of (5.42) are equivalent to the corresponding equations of (5.43). For details see Vekua [V.1] pp. 215-216.

The equations (5.43) may be rewritten, using the residue theorem, in the form

$$\operatorname{Re}\left[-t_o^{\ell} \frac{\partial^{\ell} u}{\partial t_o^{\ell}} + \frac{1}{\pi i}\int_{\partial D} t^{\ell} \frac{\partial^{\ell} u}{\partial t^{\ell}} \frac{dt}{t-t_o} - \frac{1}{\pi i}\int_{\partial D} t^{\ell-1} \frac{\partial^{\ell} u}{\partial t^{\ell}}\, dt\right] = F_{\ell}(t_o)$$

(5.47) $(\ell = 0,1,\ldots,n-1;\ t_o \in \partial D)$,

with

$$F_\ell(t_o) = \mathrm{Re}\left[-t_o^\ell g_{\ell o}(t_o) + \frac{1}{\pi i}\int_{\partial D}\frac{t^\ell g_{\ell o}(t)\,dt}{t-t_o} - \frac{1}{\pi i}\int_{\partial D} t^{\ell-1} g_{\ell o}(t)\,dt\right]$$

$(\ell=0,1,\ldots,n-1;\ t_o \in \partial D)$.

Following Vekua [V.1] pg. 216, we now seek a solution to the Dirichlet problem in the form

$$u(x,y) = \mathrm{Re}\sum_{k=0}^{n-1}\left[a_k R_k(0,0;z,\bar{z}) + \int_0^z R_k(t,0;z,\bar{z})\chi_k(t)\,dt\right]$$

An alternate representation is obtained by introducing $\varphi_k(z) = a_k \dfrac{z^k}{k!} + \displaystyle\int_0^z \frac{(z-t)^k}{k!}\chi_k(t)\,dt$ and using the equations $\dfrac{\partial^\ell}{\partial t^\ell} R_k(t,\tau;z,\bar{z})\Big|_{t=z} = 0$, $(\ell=0,1,\ldots,k-1;\ k=1,\ldots,n-1)$.

<u>Problem</u>: Verify by using integration by parts that

.48) $$u(x,y) = \mathrm{Re}\sum_{k=0}^{n-1}\left[A_k(z)\varphi_k(z) - \int_0^z B_k(z,t)\varphi_k(t)\,dt\right],$$

where

$$A_k(z) = (-1)^k\left[\frac{\partial^k}{\partial t^k} R_k(t,0;z,\bar{z})\right]_{t=z},$$

.49) $$B_k(z,t) = (-1)^k \frac{\partial^{k+1}}{\partial t^{k+1}} R_k(t,0;z,\bar{z})\ (k=0,1,\ldots,n-1)$$

Furthermore, introducing the holomorphic functions ψ_k, by $z^k\psi_k(z) = \varphi_k(z)$, $(k=0,1,\ldots,n-1)$, we may rewrite (5.48) by

$$(5.48)\quad u(x,y) = \mathrm{Re}\left\{\left[\underset{\sim}{A}(z)\psi(z) - \int_0^z \underset{\sim}{B}(z,t)\underset{\sim}{\psi}(t)dt\right]\right\},$$

where the vectors $\underset{\sim}{A}$, $\underset{\sim}{B}$, $\underset{\sim}{\psi}$ have the components $A_k z^k$, $B_k t^k$, and ψ_k respectively. We assume, in what follows, that $\psi_k(z)$ and its derivatives of order $\le n-1$ are continuous in $D + \partial D$, and hence one has that the derivatives $\frac{\partial^{k+m} u}{\partial z^k \partial \zeta^m}$, $(k \le n,\ m \le n)$ are also continuous in $D + \partial D$. It has also been shown that the converse is true, which is the reason for our assumption. From (5.48) one has that

$$(5.49)\quad z^\ell \frac{\partial^\ell u}{\partial z^\ell} = \sum_{k=0}^{\ell} \underset{\sim}{A}_{\ell k}(z)\psi^{(k)}(z) + \underset{\sim}{A}_\ell^*(z)\overline{\underset{\sim}{\psi}(t)}$$

$$+ \int_0^z \underset{\sim}{B}_\ell(z,t)\underset{\sim}{\psi}(t)dt + \int_0^{\bar z} B_\ell^*(z,\bar t)\overline{\psi(t)}\,\overline{dt},$$

where

$$\underset{\sim}{A}_{\ell k}(z) = \tfrac{1}{2}\frac{\ell ! z^{\ell}}{k!(\ell-k)!}\frac{\partial^{\ell-k}\underset{\sim}{A}(z)}{\partial z^{\ell-k}} + \underset{\sim}{A}^{o}_{\ell k}, \quad \underset{\sim}{A}^{*}_{\ell}(z) = \tfrac{1}{2} z^{\ell}\frac{\partial^{\ell}\bar{A}}{\partial z^{\ell}},$$

$$\underset{\sim}{B}_{\ell}(z,t) = -\tfrac{1}{2} z^{\ell}\frac{\partial^{\ell}\underset{\sim}{B}(z,t)}{\partial z^{\ell}}, \quad B^{*}_{\ell}(z,\bar{t}) = -\tfrac{1}{2} z^{\ell}\frac{\partial^{\ell}\overline{B(z,t)}}{\partial z^{\zeta}}$$

$(k = 0,1,\ldots,\ell;\ \ell = 0,1,\ldots,n-1)$

The equation (5.49) may also be put in a vector form. For instance, let $\underset{\sim}{V}$ be the vector with components u, $z\frac{\partial u}{\partial z}$, $\ldots$, $z^{n-1}\frac{\partial^{n-1}u}{\partial z^{n-1}}$, and let α_k, α^*, γ, γ^* be matrices whose rows are respectively the vectors $\underset{\sim}{A}_{\ell k}$, $\underset{\sim}{A}^{*}_{\ell}$, $\underset{\sim}{B}_{\ell}$, $\underset{\sim}{B}^{*}_{\ell}$ $(\ell=0,1,\ldots,n-1)$ respectively; then (5.49) may be rewritten as

$$\underset{\sim}{V} = \sum_{k=0}^{n-1}\alpha_k(z)\underset{\sim}{\psi}^{(k)}(z) + \alpha^{*}(z)\overline{\underset{\sim}{\psi}(z)}$$

$$+ \int_0^z \gamma(z,t)\underset{\sim}{\psi}(t)\,dt + \int_0^{\bar{z}}\gamma^{*}(z,\bar{t})\overline{\psi(t)}\,\overline{dt}\ . \tag{5.50}$$

Putting (5.50) into the integral equation for the boundary conditions (5.43) yields a variety of terms; some of which are

$$\underset{\sim}{P}_k(t_o) = -\alpha_k(t_o)\underset{\sim}{\psi}^{(k)}(t_o) + \frac{1}{\pi i}\int_{\partial D}\frac{\alpha_k(t)\underset{\sim}{\psi}^{(k)}(t)}{t-t_o}\,dt\ ,$$

$$\underset{\sim}{q}_k = \frac{1}{\pi i}\int_{\partial D}\alpha_k(t)\underset{\sim}{\psi}^{(k)}(t)\frac{dt}{t}\ ,\quad (k=0,1,\ldots,n-1);\ t_o \in \partial D.$$

Since the $\underset{\sim}{\psi}^{(k)}(t)$ are holomorphic vectors in D and in $C(D+\partial D)$, we have

$$\underset{\sim}{P}_k(t_o) = \frac{1}{\pi i}\int_{\partial D}\frac{\alpha_k(t)-\alpha_k(t_o)}{t-t_o}\,\psi^{(k)}(t)dt.$$

Following Vekua [Ve1] pg. 219, we notice that integration by parts leads to

$$\mathrm{Re}\Bigg\{-\alpha^*(t_o)\overline{\psi(t_o)} + \frac{1}{\pi i}\int_{\partial D}\frac{\alpha^*(t)t_o\overline{\psi(t)}}{t(t-t_o)}\,dt + \int_{\partial D}Q(t_o,t)\psi(t)dt$$

$$-\int_0^{t_o}\gamma(t_o,t)\underset{\sim}{\psi}(t)dt + \frac{1}{\pi i}\int_{\partial D}\frac{t_o\,dt}{t(t-t_o)}\int_0^{t}\gamma(t,t_1)\underset{\sim}{\psi}(t_1)dt_1$$

(5.51) $$-\int_0^{\overline{t_o}}\gamma^*(t_o,\bar{t})\overline{\psi(t)}\,\overline{dt} + \frac{1}{\pi i}\int_{\partial D}\frac{t_o\,dt}{t(t-t_o)}\int_0^{\bar{t}}\gamma^*(t,\overline{t_1})\overline{\psi(t_1)}\,\overline{dt_1}\Bigg\} = F(t_o),$$

where $Q(t_o,t) = \sum_{k=0}^{n-1}\left[P_k(t_o,t) + Q_k(t)\right]$, with $Q_k(t) = \frac{(-1)^k}{\pi i} \frac{d^k \alpha_k(t)}{dt^k t}$,

$$P_k(t_o,t) = \frac{(-1)^k}{\pi i} \frac{d^k}{dt^k}\left(\frac{\alpha_k(t) - \alpha_k(t_o)}{t-t_o}\right) .$$

Muskhelishvili shows [Mu 1] pg. 180 that if $\psi(0) = \overline{\psi(0)}$ that one may write $\psi(z)$ in terms of a real (vector) density which has $(n-1)$ Hölder continuous derivatives; i.e.

$$\psi(z) = \frac{1}{\pi i}\int_{\partial D} \frac{\mu(t)\,dt}{t-z} , \quad z \in D .$$

It follows that

$$\int_0^{t_o} \gamma(t_o,t)\underset{\sim}{\psi}(t)\,dt = \lim_{z\to t_o} \left[\int_0^z \gamma(z,t)\,dt \frac{1}{\pi i}\int \frac{\underset{\sim}{\mu}(t_1)\,dt_1}{t_1-t}\right]$$

$$= \lim_{z\to t_o} \frac{1}{\pi i}\int_{\partial D}\left(\int_0^z \frac{\gamma(z,t)\,dt}{t_1-t}\right)\underset{\sim}{\mu}(t_1)\,dt_1$$

$$= \frac{1}{\pi i}\int_{\partial D}\left[\int_0^{t_o} \frac{\gamma(t_o,t)-\gamma(t_o,t_1)}{t_1-t}\,dt - \gamma(t_o-t_1)\log\left(1 - \frac{t_o}{t_1}\right)\right]\mu(t_1)\,dt_1 ,$$

and similarly that

$$\int_0^{\overline{t_o}} \gamma^*(t_o,\bar t)\overline{\underset{\sim}{\psi}(t)}\,\overline{dt} = -\frac{1}{\pi i}\int_{\partial D}\left[\int_0^{\overline{t_o}} \gamma^* \frac{(t_o,\bar t)-\gamma^*(t_o,\overline{t_1})}{\bar t_1 - \bar t}\,d\bar t\right.$$

$$- \gamma^{*}(t_o,\bar{t}_1)\log\left(1-\frac{\bar{t}_o}{\bar{t}_1}\right)\Big]\, \underset{\sim}{\mu}(t_1)dt_1$$

Putting all these together, and interchanging some orders of integration leads to an integral equation of the form

$$\text{(5.52)} \qquad a(t_o)\underset{\sim}{\mu}(t_o) + \int_{\partial D} K(t_o,t)\underset{\sim}{\mu}(t)ds = F(t_o) \,,$$

where $a(t_o) = \tfrac{1}{2}\left[\alpha^{*}(t_o) + \overline{\alpha^{*}(t_o)}\right]$,

$$K(t_o,t)=\mathrm{Re}\Big\{\frac{\alpha^*(t_o)t'}{\pi i(\bar{t}-\bar{t}_o)} + \frac{\alpha^*(t)t't_o}{\pi i t(t-t_o)} + \frac{\overline{t'}}{\pi^2}\int_{\partial D}\frac{t_o\alpha^*(t_1)-t_1\alpha^*(t_o)}{(t_1-t_o)(\bar{t}-\bar{t}_1)t_1}\,dt_1$$

$$+ \frac{\alpha^*(t)}{\pi^2}\int_{\partial D}\left(\frac{d}{ds}\log\frac{\bar{t}-\bar{t}_1}{t-t_1}\right)\frac{dt_1}{t_1-t_o} + t'Q(t_o,t) + \frac{t'}{\pi i}\int_{\partial D} Q\,\frac{(t_o,t_1)}{t-t_1}\,dt$$

$$- \frac{t'}{\pi i}\int_0^{t_o}\frac{\gamma(t_o,t_1)-\gamma(t_o,t)}{t-t_1}\,dt_1 + \frac{t'\gamma(t_o,t)}{\pi i}\log\left(1-\frac{t_o}{t}\right)$$

$$- \frac{t'}{\pi^2}\int_{\partial D}\frac{t_o dt_1}{t_1(t_1-t_o)}\left[\int_0^{t_1}\frac{\gamma(t_1,t_2)-\gamma(t_1,t)}{t-t_2}\,dt_2 - \gamma(t_1,t)\log\left(1-\frac{t_1}{t}\right)\right]$$

$$+ \frac{\overline{t'}}{\pi i}\int_0^{\overline{t_o}}\frac{\gamma^*(t_o,\overline{t_1}) - \gamma^*(t_o,\bar{t})}{\bar{t}-\bar{t}_1}\,d\bar{t}_1 - \frac{\overline{t'}\gamma^*(t_o,\bar{t})}{\pi i}\log\left(1-\frac{\overline{t_o}}{t}\right)$$

$$+ \frac{\overline{t'}}{\pi^2}\int_{\partial D}\frac{t_o dt_1}{t_1(t_1-t_o)}\left[\int_o^{\overline{t_1}}\frac{\gamma^*(t_1,\overline{t_2}) - \gamma^*(t,\bar{t})}{\bar{t}-\bar{t}_2}\,d\bar{t}_2\right.$$

$$- \gamma^*(t_1,\bar{t})\log\left(1-\frac{\bar{t}_1}{\bar{t}}\right)\Big]\Big\} . \tag{5.52}$$

Problem: Verify the expression for the kernel in (5.53).

The matrix integral equation (5.52) may be rewritten as

$$a(t_o)\underset{\sim}{\mu}(t_o) + \frac{b(t_o)}{\pi i}\int_{\partial D}\frac{\underset{\sim}{\mu}(t)dt}{t-t_o} + \int_{\partial D}K_o\ (t_o,t)\underset{\sim}{\mu}(t)dt = \underset{\sim}{F}(t_o)\ , \tag{5.54}$$

with $b(t_o) = \frac{1}{2}\left[\alpha^*(t_o - \overline{\alpha^*(t_o)}\right]$,

$$K_o(t_o,t) = K(t_o,t) - \frac{t'b(t_o)}{\pi i(t-t_o)} .$$

It is shown in Vekua [Ve 1] pp. 222-223 that the system of equations (5.54) is of normal type and that the theory of systems of singular integral equations may be applied. In particular, it is shown that the index is zero and the Fredholm theory may be applied. Indeed, the system of equations may be regularized; i.e. the system can actually be put into the form of a system of Fredholm equations. For more details the reader is referred to Chapter V of Vekua, [Ve 1]

VI. Elliptic Equations in Dimension $n \geq 3$.

The simplest example we shall consider is that of Laplace's equation,

$$\Delta_n u \equiv \frac{\partial^2 u}{\partial x_1^2} + \ldots + \frac{\partial^2 u}{\partial x_n^2} = 0 \ , \quad n \geq 2 \ . \tag{6.1}$$

For $n = 3$ one has an integral operator due to Bergman and Whittaker which maps functions of two complex variables onto solutions of (6.1). See in this regard [Gi 1] pp. 44-62. This operator is given by

$$(\underset{\sim}{B}_3 f)(\underset{\sim}{x}) = \frac{1}{2\pi i} \int_{\mathcal{L}} f(\mu, \zeta) \frac{d\zeta}{\zeta} \ , \quad H(\underset{\sim}{x}) = (\underset{\sim}{B}_3 f)(\underset{\sim}{x}) \ , \tag{6.2}$$

$$\mu = Z\zeta + X + Z^* \zeta^{-1} \ , \quad X = x_3, \ 2Z = (x_1 + ix_2), \ 2Z^* = -x_1 + ix_2 .$$

The integral powers of the auxillary variable μ may be seen to be harmonic. Indeed, they may be used as a generating function for the harmonic polynomials,

$$\mu^n = \sum_{m=-n}^{n} h_{nm}(\underset{\sim}{x}) \zeta^{-m} \ . \tag{6.3}$$

From (6.3) it is seen that if $f(\mu, \zeta)$ has a power series

6.4) $$f(\mu, \zeta) = \sum_{n=0}^{\infty} \sum_{m=-n}^{n} a_{nm} \mu^n \zeta^m$$

then $H(\underset{\sim}{x}) = (\underset{\sim}{B}_3 f)(\underset{\sim}{x})$ has the expansion

6.5) $$H(\underset{\sim}{x}) = \sum_{n=0}^{\infty} \sum_{m=-n}^{n} a_{n,m} h_{n,m}(\underset{\sim}{x}).$$

By the residue theorem one may also show that for $|m| > n$, $\underset{\sim}{B}_3(\mu^n \zeta^m) \equiv 0$. Consequently, modulo a function of the type

$$f^*(\mu, \zeta) = \sum_{n=0}^{\infty} \sum_{|m|>n} a_{nm} \mu^n \zeta^m ,$$

the function f which generates a particular harmonic (regular about the origin) function is unique. We refer to f (as given by (6.4)) as the B_3 - associate of $H(\underset{\sim}{x})$ (as given by (6.5)).

Bergman has given a representation for an inverse operator $\underset{\sim}{B}_3^{-1}$. Other representations have been given by Gilbert [Gi 1] pg 60 (for harmonic functions regular at infinity), and [GL 1] (for harmonic functions regular at the origin). Colton [Co 3], and Gilbert-Kukral [GK 1]. We present below Bergman's representation; some of the others will be listed later

(6.6) $$(\underset{\sim}{B}_3^{-1} H)(\mu,\zeta) \equiv 2 \int_0^1 \mu^{\frac{1}{2}} \frac{\partial}{\partial\mu} \{\chi(\zeta\mu t^2, \mu\zeta^{-1}(1-t^2))\}dt ,$$

$f(\mu,\zeta) = (\underset{\sim}{B}_3^{-1} H)(\mu,\zeta)$, where $\chi(z,z^*)$ is the restriction of $\tilde{H}(x,z,z^*) \equiv H(\underset{\sim}{x})$ to the characteristic space $X = 2\sqrt{zz^*}$.

An operator which generates solutions to (6.1) for $N = 4$ was given by Gilbert [Gi 1] pp. 75-82.

(6.7) $$(\underset{\sim}{G}_4 f)(\underset{\sim}{x}) = -\frac{1}{4\pi^2} \int_{|\zeta|=1} \frac{d\zeta}{\zeta} \int_{|\eta|=1} \frac{d\eta}{\eta} f(\mu,\zeta,\eta), \quad H(\underset{\sim}{x}) \equiv (\underset{\sim}{G}_4 f)(\underset{\sim}{x}) ,$$

here the auxillary variable μ is defined in this case to be

$$\mu = \frac{1}{\eta\zeta} (\eta\zeta Y + \eta z + \zeta z^* + Y^*) ,$$

where $Y = x_1 + ix_2$, $Y^* = x_1 - x_2$, $Z = x_3 + ix_4$, $Z^* = -x_3 + ix_4$, and f is a function of the type

$$(6.8)\qquad f(\mu,\zeta,\eta) = \sum_{n=0}^{\infty} \sum_{m,p=0}^{n} a_{nmp}\mu^{n}\zeta^{m}\eta^{p} .$$

<u>Remark</u>: It may be shown that μ^n is a generating function for the hyperspherical harmonics; se [Gi 1] pg. 76 for further details.

A representation was given for $\underset{\sim}{G}_4^{-1}$ by Gilbert [Gi 1] using the hyperspherical harmonics, a Bergman-type inverse was subsequently given by Kreyszig [Kr 1], and recently Colton [Co 4] has added another representation. Kreyszig's representation is given below; the development of this may be found in [Kr 1] or in [Gi 1] pg. 78. Let $H(\underset{\sim}{x}) = \tilde{H}(Y,Y^*,Z,Z^*)$ be a harmonic function regular about the origin, and $\chi(Y,Z,Z^*)$ the restriction of $\tilde{H}$ to the characteristic space, $YY^* = ZZ^*$. Then $f(\mu,\zeta,\eta) = (\underset{\sim}{G}_4^{-1}\chi)(\mu,\zeta,\eta)$, with

$$(6.9)\qquad (\underset{\sim}{G}_4^{-1}\chi)(\mu,\zeta,\eta) \equiv \int_0^1 \int_0^1 \frac{\partial}{\partial\mu}\{\mu \frac{\partial}{\partial\mu}[\mu\chi(\mu[1-\alpha][1-\beta],\mu\eta\beta[1-\alpha],$$
$$\mu\zeta\alpha[1-\beta])]\}d\alpha d\beta$$

There are several ways to introduce operators for harmonic functions in dimensions greater than 4, [Gi 1] pp. 82-86. The most obvious one was given by Bergman [Be 1], [Gi 1] pg.82. This operator has the form

$$(6.10)\qquad (\underset{\sim}{B}_n f)(\underset{\sim}{x}) = \left(\frac{1}{2\pi i}\right)^{n-2} \int_a f(\mu;\zeta)\,\frac{d\zeta}{\zeta},$$

where a is a product of $n-2$ unit circles $\zeta = (\zeta_1, \ldots, \zeta_{n-2})$, and $\frac{d\zeta}{\zeta} = \frac{d\zeta_1}{\zeta_1} \cdots \frac{d\zeta_{n-2}}{\zeta_{n-2}}$. The auxillary variable is chosen so that it may be written in the form $\mu = \underset{\sim}{N}(\zeta) \cdot \underset{\sim}{x}$ with $\underset{\sim}{N} \cdot \underset{\sim}{N} = 0$. Bergman's choice for $\underset{\sim}{N}$ is

$$N_1 = \frac{i}{2}\left(\zeta_1 + \frac{1}{\zeta_1}\right),\quad N_2 = \frac{1}{4}\left(\zeta_1 - \frac{1}{\zeta_1}\right)\left(\zeta_2 + \frac{1}{\zeta_2}\right),\quad \ldots,$$

$$\vdots$$

$$N_{n-1} = \left(\frac{-i}{2}\right)^{n-2}\left(\zeta_1 - \frac{1}{\zeta_1}\right)\left(\zeta_2 - \frac{1}{\zeta_2}\right) \cdots \left(\zeta_{n-3} - \frac{1}{\zeta_{n-3}}\right)\left(\zeta_{n-2} + \frac{1}{\zeta_{n-2}}\right)$$

$$(6.11)\qquad N_n = \left(\frac{-i}{2}\right)^{n-1}\left(\zeta_1 - \frac{1}{\zeta_1}\right)\left(\zeta_2 - \frac{1}{\zeta_2}\right) \cdots \left(\zeta_{n-3} - \frac{1}{\zeta_{n-3}}\right)\left(\zeta_{n-2} - \frac{1}{\zeta_{n-2}}\right)$$

The difficulty with this choice is that if we try and use powers of μ as a generating function for the harmonic polynomials we obtain too many. Actually, Kukral [Ku 1][Ku 2] has recently shown, in his thesis, that it is impossible to find a Bergman type operator for $n \geq 5$. We will present his argument somewhat later on.

We next turn to the equation

$$\Delta_3 u + F(\underset{\sim}{x})u = 0, \quad \underset{\sim}{x} \equiv (x_1, x_2, x_3), \quad F \text{ analytic.} \tag{6.12}$$

The special case where $F(\underset{\sim}{x}) \equiv F(x_1, x_2)$ was investigated by Bergman [Be 1] pg. 74. Tjong and Maric [Tj 1], [MT 1] were the first to succeed in obtaining an integral operator which generated solutions to (6.12). However, they did not succeed in showing this was a map onto the space of solutions. Gilbert and Lo [G L 1] showed for the case where $F \leq 0$ that the Tjong operator was onto. Colton and Gilbert [CG 2] showed that for $F(\underset{\sim}{x}) = F(x_1, x_2)$ the Tjong operator was onto, and obtained an analogous operator for

$$\Delta_4 u + F(\underset{\sim}{x})u = 0, \quad \underset{\sim}{x} = (x_1, x_2, x_3, x_4), \quad F \text{ analytic.} \tag{6.13}$$

For $F(\underset{\sim}{x}) = F(x_1, x_2, x_3)$ this latter operator was also seen to be onto. Colton [Co 5] then modified these operators and obtained a proof, for general analytic F, that these operators were onto. We shall present his results subsequent to Tjong's. Tjong introduces the new variables

$$\xi_1 = X, \quad \xi_2 = X + 2\zeta Z, \quad \xi_3 = X + 2\zeta^{-1} Z^* , \tag{6.14}$$

in which we may write $\mu = \frac{1}{2}(\xi_2 + \xi_1)$. (This is an important feature of these variables.) The equation (6.12) may be written as

$$\frac{\partial^2 U}{\partial X^2} - \frac{\partial^2 U}{\partial Z \partial Z^*} + Q(X,Z,Z^*)U = 0 . \tag{6.15}$$

Tjong then seeks a solution to (6.15) in the form

$$\begin{aligned} U(X,Z,Z^*) &= (\underset{\sim}{T}_3 f)(X,Z,Z^*) \\ &\equiv \frac{1}{2\pi i} \int_{|\zeta|=1} \int_{\gamma} E(X,Z,Z^*,\zeta,t) f(\nu,\zeta) \frac{dt}{\sqrt{1-t^2}} \frac{d\zeta}{\zeta} , \end{aligned} \tag{6.16}$$

where γ is a rectifiable curve joining $t = -1$ to 1, $\nu = (1-t^2)\mu$, and where $\hat{E}(\xi_1, \xi_2, \xi_3, \zeta, t) \equiv E(X,Z,Z^*,\zeta,t)$

satisfies the partial differential equation

$$\mu t\left(\frac{\partial^2\hat{E}}{\partial\xi_1^2} + \frac{\partial^2\hat{E}}{\partial\xi_2^2} + \frac{\partial^2\hat{E}}{\partial\xi_3^2} + 2\frac{\partial^2\hat{E}}{\partial\xi_1\partial\xi_2} + 2\frac{\partial^2\hat{E}}{\partial\xi_1\partial\xi_3} - 2\frac{\partial^2\hat{E}}{\partial\xi_2\partial\xi_3} + \hat{Q}\hat{E}\right.$$

(6.17)
$$+ (1-t^2)\frac{\partial^2\hat{E}}{\partial\xi_1\partial t} - \frac{1}{t}\frac{\partial\hat{E}}{\partial\xi_1} = 0$$

Here $f(\mu, \zeta)$ is a holomorphic function of the type given in (6.4).

Tjong goes on to show that it is possible to find a solution of (6.17) in the form

(6.18)
$$\hat{E}(\xi_1, \xi_2, \xi_3, \zeta, t) = 1 + \sum_{n=1}^{\infty} t^{2n}\mu^n p^{(n)}(\xi_1, \xi_2, \xi_3, \zeta) ,$$

where the $p^{(n)}$ are defined recursively by

(6.19)
$$p_1^{(n+1)} = -\frac{1}{2n+1}\{p_{11}^{(n)} + p_{22}^{(n)} + p_{33}^{(n)} + 2p_{12}^{(n)} + 2p_{13}^{(n)} - 2p_{23}^{(n)} + \hat{F}p^{(n)}\}$$

and

$$p^{(n+1)}(0, \xi_2, \xi_3, \zeta) = 0 .$$

An alternate representation for the solution is obtained in the form

$$U = \underset{\sim}{T}_3 f \equiv \frac{1}{2\pi i} \int_{|\zeta| = 1} g(\mu, \zeta) \frac{d\zeta}{\zeta}$$

$$(6.20) \qquad + \sum_{n=1}^{\infty} \frac{1}{2\pi i \beta(n, \frac{1}{2})} \int_{|\zeta| = 1} \{p^{(n)} (\xi_1, \xi_2, \xi_3, \zeta) \int_0^{\mu} (\mu - s)^{n-1} g(s, \xi) ds\} \cdot$$

where $\quad g(\mu, \zeta) = \int_x f(\nu, \zeta) \frac{dt}{\sqrt{1 - t^2}}$

It was first shown that this operator is, for $F \leq 0$, invertible in a regular, bounded domain D, by making use of the following representation for $\underset{\sim}{B}_3^{-1}$, namely

$$(6.21) \qquad (B_3^{-1} H)(\mu, \zeta) \equiv \frac{1}{\pi} \int_{\partial D} \rho(\underset{\sim}{y}) \frac{1}{\underset{\sim}{N} \cdot (\underset{\sim}{x} - \underset{\sim}{Y})} d\omega_y ,$$

which is arrived at by noting that the parametrix for Laplace's equation may be written as

(6.22)
$$\frac{1}{|\underset{\sim}{x} - \underset{\sim}{y}|} = \frac{1}{2\pi i} \int_{|\zeta|=1} \frac{1}{\underset{\sim}{N} \cdot (\underset{\sim}{x} - \underset{\sim}{Y})} \frac{d\zeta}{\zeta},$$

(where $\underset{\sim}{N}(\zeta)$ is the usual isotropic vector used in the auxilary variable $\mu = \underset{\sim}{N} \cdot \underset{\sim}{x})$, and representing $H(\underset{\sim}{x})$ as a single layer potential

(6.23)
$$H(\underset{\sim}{x}) = \frac{1}{\pi} \int_{\partial D} \rho(\underset{\sim}{Y}) \frac{d\omega_Y}{|\underset{\sim}{x} - \underset{\sim}{Y}|} .$$

If one substitutes (6.21) into (6.20) one obtains that

(6.24)
$$u(\underset{\sim}{x}) = \int_{\partial D} \{ \frac{1}{\pi} \frac{1}{|\underset{\sim}{x} - \underset{\sim}{Y}|} + K(\underset{\sim}{x}, \underset{\sim}{Y}) \} \rho(\underset{\sim}{Y}) \, d\omega_Y,$$

with

(6.25)
$$K(\underset{\sim}{x}, \underset{\sim}{Y}) \equiv \frac{1}{2\pi i} \int_{|\zeta|=1} P(\underset{\sim}{x}, \underset{\sim}{Y}; \zeta) \frac{d\zeta}{\zeta},$$

$$P(\underset{\sim}{x}, \underset{\sim}{Y}; \zeta) \equiv \sum_{n=1}^{\infty} \frac{1}{\beta(n, \frac{1}{2})} \phi^{(n)}(\underset{\sim}{\xi}; \zeta) \Phi_n(\mu; \underset{\sim}{N} \cdot \underset{\sim}{Y}),$$

$$\Phi_n(\mu; \underset{\sim}{N} \cdot \underset{\sim}{Y}) = \int_0^{\mu} \frac{(\mu - s)^{n-1}}{s - \underset{\sim}{N} \cdot \underset{\sim}{Y}} \, ds .$$

We note that the Φ_n are universal functions; whereas $P(\underset{\sim}{x},\underset{\sim}{Y};\zeta)$ depends on $F(\underset{\sim}{x})$.

If $u(\underset{\sim}{x}) \in C^1(\bar{D})$ then one may formulate the Neumann problem, $\frac{\partial u}{\partial \nu} = f$ on ∂D, for (6.12) as

$$f(\underset{\sim}{x}) = -\rho(\underset{\sim}{x}) + \frac{1}{\pi} \int_{\partial D} \rho(\underset{\sim}{Y}) \frac{\partial}{\partial \nu_x} \left(\frac{1}{|\underset{\sim}{x} - \underset{\sim}{Y}|}\right) d\omega_Y \tag{6.26}$$

$$+ \int_{\partial D} \rho(\underset{\sim}{Y}) \frac{\partial}{\partial \nu_x} K(\underset{\sim}{x},\underset{\sim}{Y}) d\omega_Y \; .$$

It can be shown [GL 1] that the second integral corresponds to a compact operator. To show that the integral equation (6.26) has a unique solution we consider the homogeneous transposed equation,

$$\mu(\underset{\sim}{Y}) = \frac{1}{\pi} \int_{\partial D} \mu(\underset{\sim}{x}) \frac{\partial}{\partial \nu_x} \left(\frac{1}{|\underset{\sim}{x} - \underset{\sim}{Y}|}\right) d\omega_x + \int_{\partial D} \mu(\underset{\sim}{x}) \frac{\partial}{\partial \nu_x} K(\underset{\sim}{x},\underset{\sim}{Y}) d\omega_x \tag{6.27}$$

This implies that the "double-layer solution"†

---

† This is what one would obtain by formal computation if one represented $H(\underset{\sim}{x})$ as a double layer potential. The integrations in (6.20) make this possible.

$$v(\underset{\sim}{y}) = \frac{1}{2\pi} \int_{\partial D} \mu(\underset{\sim}{x}) \frac{\partial}{\partial \nu_x} \left(\frac{1}{|\underset{\sim}{x} - \underset{\sim}{Y}|}\right) d\omega_x + \int_{\partial D} \mu(\underset{\sim}{x}) \frac{\partial}{\partial \nu_x} K(\underset{\sim}{x}, \underset{\sim}{Y}) d\omega_x \tag{6.28}$$

has the boundary value zero. If $F(\underset{\sim}{x})$ is extended to D^1 such that the corresponding exterior problem is unique, then $v(\underset{\sim}{x}) \equiv 0$. Since $\frac{\partial}{\partial \nu_x} K$ has a weak singularity, the normal derivitative $\frac{\partial v}{\partial \nu}$ is continuous across ∂D. Hence for the interior problem $v(\underset{\sim}{x})$ obeys the boundary condition $\frac{\partial v}{\partial \nu} = 0$, $\underset{\sim}{x} \in \partial D$. Since $F \le 0$ in D it follows immediately that $v(\underset{\sim}{x}) \equiv 0$ in D, and hence the only solution to the transposed, homogeneous equation is $\mu(\underset{\sim}{x}) \equiv 0$. We conclude that Neumann data $f(\underset{\sim}{x})$ is orthogonal to each solution of the homogeneous transposed equation, and hence (6.26) has a unique solution. This implies there exists a unique harmonic function in D, and by the fact that the Bergman-Whittaker operator will map functions $g(\mu, \zeta)$ onto these harmonic functions, we have that the mapping $\underset{\sim}{T}f$ is onto, with

$$f(\mu, \zeta) = -\frac{1}{2\pi} \int_{\mathfrak{L}} g(\mu[1 - t^2], \zeta) \frac{dt}{t^2}, \tag{6.29}$$

where $\mathfrak{L}$ is an arc from -1 to +1 but not passing through the origin.

We next turn to the equation

$$\Delta_4 u + F(\underset{\sim}{x})u = 0 \ , \quad \underset{\sim}{x} = (x_1, \ldots, x_4) \ , \tag{6.30}$$

where $F(\underset{\sim}{x})$ is an entire function of its arguments. By using the variables Y, Y^*, Z, Z^* introduced earlier, this may be rewritten as

$$\frac{\partial^2 U}{\partial Y \partial Y^*} - \frac{\partial^2 U}{\partial Z \partial Z^*} + Q(Y, Y^*, Z, Z^*)U = 0 \ . \tag{6.31}$$

Colton and Gilbert [CG 2] developed an integral operator for this equation which is analogous to the three-variable operator of Marić - Tjong [MT 1]. In order to present their work we must first introduce some notation,

$$\xi_1 = Z - Z^*, \ \xi_2 = \zeta^{-1}Z + Y, \ \xi_3 = \eta^{-1}Z^* + Y, \ \xi_4 = \eta^{-1}Z^* + \eta^{-1}\xi^{-1}Y^* \ ,$$

$$\underset{\sim}{\xi} = (\xi_1, \xi_2, \xi_3, \xi_4), \ \mu = \xi_2 + \xi_4 = Y + \zeta^{-1}Z + \eta^{-1}Z^* + \zeta^{-1}\eta^{-1}Y^* \ ,$$

$$\nu = (1-t^2)\mu \ , \ Q(\underset{\sim}{\xi};\zeta,\eta) \equiv Q(Y,Y^*,Z,Z^*) \ ,$$

32) $$E(\underset{\sim}{\xi};\zeta,\eta,t) = E(Y,Y^*,Z,Z^*,\zeta,\eta,t) \ ,$$

$$E_i \equiv \frac{\partial E}{\partial \xi_i} \ , \quad E_{ij} = \frac{\partial^2 E}{\partial \xi_i \partial \xi_j} \ , \quad E_{it} = \frac{\partial^2 E}{\partial \xi_i \partial t} \ .$$

They seek a solution of (6.31) in the form

$$U(Y,Y^*,Z,Z^*) = (\underset{\sim}{T}_4 f)(Y,Y^*,Z,Z^*)$$

$$\equiv \frac{-1}{4\pi^2} \int_{|\zeta|=\frac{1}{2}} \int_{|\eta|=1} \int_{\gamma} E(Y,Y^*,Z,Z^*,\zeta,\eta,t) f(\nu,\zeta,\eta) \frac{dt}{\sqrt{1-t^2}} \frac{d\eta}{\eta} \frac{d\zeta}{\zeta} \ ,$$

where γ is a path joining -1 to $+1$, and $f(\nu,\zeta,\eta)$ is a function of the type (6.8). By substituting (6.33) into (6.31), using $f_\nu = -\frac{1}{2t\mu} f_t$, and integrating by parts we obtain

$$U_{ZZ^*} - U_{YY^*} - QU = -\frac{1}{4\pi 2} \int_{|\zeta|=\frac{1}{2}} \int_{|\eta|=1} \int_{\gamma} [E_{ZZ^*} - E_{YY^*} - EQ)$$

$$+ \frac{1}{2\mu}\left(\frac{E_{z^*t}}{t\zeta}(1-t^2) - \frac{E_{z^*}}{t^2\zeta} + \frac{E_{zt}}{t\eta}(1-t^2) - \frac{E_z}{t^2\eta} - \frac{E_{Y^*t}}{t}(1-t^2)\right.$$

$$\left. + \frac{E_{Y^*}}{t^2} - \frac{E_{Yt}}{t\eta\zeta}(1-t^2) + \frac{E_Y}{t^2\eta\zeta}\right)\Big]f \frac{dt}{\sqrt{1-t^2}}\frac{d\eta}{\eta}\frac{d\zeta}{\zeta} .$$

In order for (6.33) to be a solution the bracketed term in the integrand must vanish. Changing to the $\xi_1, \xi_2, \xi_3, \xi_4$ variables this yields the following partial differential equation that E must satisfy,

$$2\mu t[-\eta\zeta E_{11} - \eta E_{12} + \zeta E_{13} + E_{23} + \zeta E_{14} - E_{34} - \eta\zeta EQ]$$
$$+ (1-t^2)(\zeta - \eta)E_{1t} - \frac{1}{t}(\zeta - \eta)E_1 = 0 .$$

(6.34)

Colton and Gilbert then go onto show, using the method of dominants, that a solution may be found to (6.34) in the form

(6.35) $$E(\underset{\sim}{\xi};\zeta,\eta,t) = 1 + \sum_{n=1}^{\infty} t^{2n}\mu^n p^{(n)}(\underset{\sim}{\xi};\zeta,\eta) .$$

The $p^{(n)}$ are defined recursively by the equations

$$p_1^{(n+1)} = \frac{2}{(\zeta-\eta)(2n+1)} \{\eta\zeta p_{11}^{(n)} + \eta p_{12}^{(n)} - \zeta p_{13}^{(n)} - p_{23}^{(n)} - \zeta p_{23}^{(n)}$$

$$-\zeta p_{14}^{(n)} + p_{34}^{(n)} + \eta\zeta p^{(n)} Q\} ,$$

where $p^{(n+1)}(0,\xi_2,\xi_3,\xi_4,\zeta,\eta) = 0, \quad (n = 0,1,2,\ldots)$

$$p^{(0)}(\underset{\sim}{\xi};\zeta,\eta) \equiv 1,$$

and $p_i^{(n)} = \dfrac{\partial p^{(n)}}{\partial \xi_i}$, $p_{ij}^{(n)} = \dfrac{\partial p^{(n)}}{\partial \xi_i \partial \xi_j}$.

The reader is directed to [CG 2] for further details. Colton and Gilbert were able to prove that this mapping was onto in the case where $F(\underset{\sim}{x})$ depended on three variables rather than four. (They also accomplished this for the Marić-Tjong operator.) The method they use is to consider Cauchy data for the solution, where the data is given on the coordinate plane which corresponds to the missing variable. A variation on this idea, which leads to a more tractible inversion formula for $\underset{\sim}{T}_k^{-1}$, $(k=3,4)$, was obtained by Gilbert and Kukral [GK 1], and we present this below. However, we first make a short digression on the Cauchy problem for Laplace's equation,

$$u_{x_n x_n} = -u_{x_1 x_1} - u_{x_2 x_2} - \dots - u_{x_{n-1} x_{n-1}} ,$$

$$(6.37) \qquad u(x_1, \dots, x_{n-1}, 0) = f(x_1, \dots, x_{n-1}) ,$$

$$\frac{\partial}{\partial x_n} u(x_1, \dots, x_{n-1}, 0) = g(x_1, \dots, x_{n-1})$$

We shall assume in what follows that the data f, g is analytic. We first note that if u is a solution of (6.37) with $f \equiv 0$, then $V(\underset{\sim}{x}) \equiv \frac{\partial u}{\partial x_n}(\underset{\sim}{x})$ is a harmonic function satisfying on $x_n = 0$ the Cauchy data $V = g$, $\frac{\partial V}{\partial x_n} = 0$. Hence $U = u + \tilde{u}_1$ is a solution of the complete Cauchy problem if $\tilde{u}$ is a harmonic function satisfying the Cauchy data $\tilde{u} = 0$, $\frac{\partial \tilde{u}}{\partial x_n} = f$. Without loss of generality then we consider the case with $f = 0$. From the Cauchy-Kowelewski theorem we know there exists a local solution. Using an idea of Garabedian [Ga 1] we introduce complex variables in the following way: $z_j = x_j + iy_j$, $(j = 1, \ldots, n-1)$, x_n = real. Since U is an analytic function one must have

$\frac{\partial U}{\partial \bar{z}_j} = 0$, $(J = 1, \ldots, n-1)$, and hence Laplace's equation,

$\frac{\partial^2 U}{\partial x_j^2} + \frac{\partial^2 U}{\partial y_j^2} = 0$, is satisfied for $j = 1, \ldots, n-1$. We may rewrite our differential equation (6.37) therefore as

$$U_{x_n x_n} = \sum_{j=1}^{n-1} 2(U_{x_j x_j} + U_{y_j y_j}) - U_{x_1 x_1} - \ldots - U_{x_{n-1} x_{n-1}}$$

(6.38) $$= U_{x_1 x_1} + \ldots + U_{x_{n-1} x_{n-1}} + 2U_{y_1 y_1} + \ldots + 2U_{y_{n-1} y_{n-1}} .$$

If we now introduce the new independent variables $\tilde{y}_j = \frac{1}{\sqrt{2}} y_j$, we obtain a wave-equation in (2n - 2)-space variables,

$$v_{x_n x_n} = v_{x_1 x_1} + \ldots + v_{x_{n-1} x_{n-1}} + v_{\tilde{y}_1 \tilde{y}_1} + \ldots + v_{\tilde{y}_{n-1}, \tilde{y}_{n-1}} ,$$

the initial data is now

(6.40) $$v(x_1, \ldots, x_{n-1}, 0, \tilde{y}_1, \ldots, \tilde{y}_{n-1}) = 0$$

$$v_{x_n}(x_1, \ldots, x_{n-1}, 0, \tilde{y}_1, \ldots, \tilde{y}_{n-1}) = g(x_1 + i\sqrt{2}\,\tilde{y}_1, \ldots, x_{n-1} + i\sqrt{2}\tilde{y}$$

The equations (6.39), (6.40) constitute a well-posed Cauchy problem for the wave equation and one has a known solution expressed in terms of spherical means [Ga 1]

$$v(x_1, \ldots, x_n, \tilde{y}_1, \ldots, \tilde{y}_{n-1})$$

$$\frac{1}{(2n-4)!} \frac{\partial^{2n-4}}{\partial x_n^{2n-4}} \int_0^{x_n} (x_n^2 - \rho^2)^{n-3/2} \rho \Omega_n(\underset{\sim}{x}; \rho)\, d\rho ,$$

with $\Omega_n(\underset{\sim}{x}; \rho) = \frac{1}{\omega_{2n-2}} \int \cdots \int \tilde{g}(\underset{\sim}{x} + \underset{\sim}{\alpha}\rho)\, d\omega_{2n-2}$,

$\underset{\sim}{x} = (x_1, \ldots, x_{n-1}, \tilde{y}_1, \ldots, \tilde{y}_{n-1})$, $\underset{\sim}{\alpha} = (\alpha_1, \ldots, \alpha_{2n-2})$,

$|\underset{\sim}{\alpha}| = 1$, and where ω_{2n-2} is the surface area of the unit sphere in $(2n-2)$ - dimensional space.

We next turn to consider the problem of inverting Tjong's operator for

$$\Delta_3 u + F(x_1, x_2) u = 0 \ .$$

To this end we consider the Cauchy data

$$u(x_1, x_2, 0) = \hat{\varphi}_1(x_1, x_2) = \varphi_1(z, z^*),$$

$$u_{x_3}(x_1, x_2, 0) = \hat{\varphi}_2(x_1, x_2) = \varphi_2(z, z^*),$$

where F and the data are assumed be holomorphic in some neighborhood of the origin. The solution can be expressed, we recall, in the form

$$u(\underset{\sim}{x}) = u_1(\underset{\sim}{x}) + \int_0^{x_3} u_2(x_1, x_2, s)\, ds \ ,$$

where u_1 is a solution of the previous Cauchy problem with $\varphi_2 \equiv 0$ and u_2 is a solution with φ_1 replaced by φ_2 and φ_2 replaced by zero. We are thereby led to consider, without loss of generality, the reduced Cauchy problem,

$$\frac{\partial^2 U}{\partial x^2} - \frac{\partial^2 U}{\partial z \partial z^*} + \hat{F}(z,z^*)U = 0 ,$$

(6.41)

$$U(0,z,z^*) = \varphi(z,z^*) ,$$

$$U_x(0,z,z^*) = 0 .$$

Now since F is independent of $x_3 = X$, we observe that [GK 1]

$$U(X,Z,Z^*) = \frac{1}{2\pi i} \int_{|\zeta|=1} \int_{\gamma} \tfrac{1}{2}\{E(X,Z,Z^*,\zeta,t)f(\nu,\zeta)$$

(6.42) $$+ E(-X,Z,Z^*,\zeta,t)f(\nu^1,\zeta)\} \frac{dt}{\sqrt{1-t^2}} \frac{d\zeta}{\zeta} ,$$

where $\nu^1 = (-X + \zeta Z + \zeta^{-1}Z^*)(1-t^2)$, is a solution of (6.41). Furthermore, $U(X,Z,Z^*)$ is an even function of X, so that $U_X(0,Z,Z^*) = 0$. We now try to determine $f(\mu,\zeta)$ from $\varphi(Z,Z^*)$. Along $X = 0$, (6.42) becomes

(6.43) $$U(0,Z,Z^*) = \frac{1}{2\pi i} \int_{|\zeta|=1} \int_{\gamma} f(\nu_1,\zeta) \frac{dt}{\sqrt{1-t^2}} \frac{d\zeta}{\zeta} ,$$

$$\nu_1 = (\zeta Z + \zeta^{-1} Z^*)(1 - t^2) = \mu_1 (1 - t^2) ,$$

since $E(0,Z,Z^*,\zeta,t) = 1$. Furthermore, if one defines

44) $$g(\mu_1, \zeta) = \int_\gamma f(\nu_1, \zeta) \frac{dt}{\sqrt{1 - t^2}} ,$$

then the reciprocal integral representation is given by [G.]

45) $$f(\mu, \zeta) = -\frac{1}{2\pi} \int_{\gamma^1} g(\mu(1 - t^2), \zeta) \frac{dt}{t^2} ,$$

where γ^1 is an arc from -1 to 1 which does not pass through the origin. As before, we assume that $g(\mu, \zeta)$ has the expansion

$$g(\mu, \zeta) = \sum_{n=0}^{\infty} \sum_{m=-n}^{+n} a_{n,m} \mu^n \zeta^m .$$

Hence, in order to invert Tjong's operator it is sufficient for us to find coefficients $a_{n,m}$ such that

$$U(0,Z,Z^*) = \varphi(Z,Z^*) = \frac{1}{2\pi i}\int_{|\zeta|=1} \sum_{n=0}^{\infty} \sum_{m=-n}^{n} a_{n,m}\mu_1^{\ n}\zeta^m \frac{d\zeta}{\zeta}$$

$$= \frac{1}{2\pi i}\int_{|\zeta|=1} \{\sum_{n=0}^{\infty} \sum_{m=-n}^{n} a_{n,m} \sum_{j=0}^{n} \binom{n}{j} Z^{*\,n-j} Z^j \zeta^{2j-n+m}\} \frac{d\zeta}{\zeta}$$

$$(6.46) \quad = \sum_{n=0}^{\infty} \sum_{m=-n}^{n} a_{n,m} \binom{n}{\frac{1}{2}[n-m]} Z^{\frac{1}{2}(n-m)} Z^{*\,\frac{1}{2}(n+m)} \quad .$$

Let us assume $a_{n,m} = 0$ if $n+m =$ odd, and then set

$k = \dfrac{n-m}{2}$, $\ell = \dfrac{n+m}{2}$ (i.e. $n = k+\ell$, $m = \ell - k$). Then (6.46)

becomes

$$(6.47) \qquad \varphi(Z,Z^*) = \sum_{k=0}^{\infty} \sum_{\ell=0}^{\infty} b_{k,1} Z^k Z^{*\ell} ,$$

with $b_{k,\ell} = \binom{k+\ell}{k} a_{k+\ell,\ell-k}$. Using the beta function

representation (6.47) can be rewritten as

$$\int_0^1 \varphi(Z^*, Z^*(1-t))\,dt = \sum_{n=0}^{\infty} \sum_{m=-n}^{n} \frac{a_{n,m}}{n+1} Z^{\frac{1}{2}(n-m)} Z^{*\,\frac{1}{2}(n+m)} .$$

Upon substituting $\mu\zeta^{-1}$ for Z and $\mu\zeta$ for Z^* and differentiating with respect to μ, one obtains the following representation for T_3^{-1}, [G K 1]

$$g(\mu,\zeta) = (\underset{\sim}{T}_3^{-1}U)(\mu,\zeta) \equiv \frac{\partial}{\partial\mu}\{\mu\int_0^1 \varphi(\frac{\mu t}{\zeta}, \mu\zeta(1-t))dt\}$$

(6.48)
$$= \sum_{n=0}^{\infty} \sum_{m=-n}^{n} a_{n,m}\mu^n\zeta^m .$$

Remark: It is interesting to note that when $F(\underset{\sim}{x}) \equiv 0$, $E \equiv 1$, then (6.48) yields another representation for $\underset{\sim}{B}_3^{-1}$.

Colton [Co 5] modified Tjong's operator $\underset{\sim}{T}_3$ in an interesting and very useful manner. He replaces Tjong's variable $\xi_1 = X$ by $\xi_1 = 2\zeta Z$. This still preserves the two essential properties of Tjong's change of variables, namely $\mu = \frac{1}{2}(\xi_2 + \xi_3)$ and $J_{\underset{\sim}{\xi}}(\underset{\sim}{x}) \neq 0$ (i.e. The Jacobian of the mapping $\underset{\sim}{x} \to \underset{\sim}{\xi}$ is not zero). The essential feature of Colton's variation is that he has $E \equiv 1$ on the characteristic plane $\xi = 2\zeta Z$, (of equation (6.15)). Now, since he considers real solutions of (6.12) (for real x_1, x_2, x_3), it follows that U is uniquely determined by its Goursat data on $Z = 0$. To verify this latter statement we prove the following lemma which generalizes Colton's observation, and which will be of further use when we later consider higher order equations.

Lemma (6.1): Let $X = x_3$, $Z = \frac{1}{2}(x_1 + ix_2)$, $Z^* = \frac{1}{2}(-x_1 + ix_2)$, and let $u(\underset{\sim}{x})$ be a real-valued C^{2n} solution in $N(\underset{\sim}{0}) \subset \mathbb{R}^3$, of the following differential equation with entire coefficients,

(6.49) $$\Delta^n u + \sum_{k=1}^{n} \underset{\sim}{L}'_k(\Delta^{n-k} u) = 0 ,$$

where $$\underset{\sim}{L}'_k \equiv \sum_{m,p,q=0}^{m+p+q \le k} a^{(k)}_{m,p,q}(\underset{\sim}{x}) \frac{\partial^{m+p+q}}{\partial x_1^m \partial x_2^p \partial x_3^q} ,$$

Then $U(X,Z,Z^*) \equiv u(x_1,x_2,x_3)$ is an analytic function of X,Z,Z^* in some neighborhood of the origin in $\mathbb{C}^3$, and is uniquely determined by the Goursat data

(6.50) $$U(X,0,Z^*) = F_0(X,Z^*), \quad \frac{\partial U}{\partial Z}(X,0,Z^*) = F_1(X,Z^*), \quad \ldots, \quad \frac{\partial^{n-1} U}{\partial Z^{n-1}}(X,0,Z^*) = F_{n-1}(X,Z^*) .$$

<u>Proof</u>: Since the coefficients are entire in $\mathbb{C}^3$ the solution is analytic; see for example Fritz John [Jo 1] pg. 142; hence, the function $U(X,Z,Z^*)$ is analytic. Hence, locally one has

$$U(X,Z,Z^*) = \sum_{\ell,m,n=0}^{\infty} \alpha_{m,n,\ell} X^{\ell} Z^{n} Z^{*m} ,$$

which implies

$$U(X,0,Z^*) = \sum_{\ell,m=0}^{\infty} \alpha_{m,0,\ell} X^{\ell} Z^{*m} ,$$

and

1) $$\frac{\partial^k}{\partial Z^k} U(X,0,Z^*) = \sum_{\ell,m=0}^{\infty} k!\, \alpha_{m,k,\ell} X^{p} Z^{*m} ,$$

$(k = 0,1, \ldots, n-1)$. Also, one has that

$$U(X,Z,0) = \sum_{\ell,n=0}^{\infty} \alpha_{0,n,\ell} X^{\ell} Z^{n}$$

and

(6.52) $$\frac{\partial^k U(X,Z,0)}{\partial Z^{*k}} = \sum_{\ell,n=0}^{\infty} k!\, \alpha_{k,n,\ell} X^{\ell} Z^{n} ,$$

$(k = 0, 1, \ldots, n-1)$.

It is an immediate conclusion since $u(\underset{\sim}{x})$ is real-valued, that for real $\underset{\sim}{x}$

$$\overline{U(X,Z,Z^*)} = U(X,Z,Z^*) .$$

Hence, for real $\underset{\sim}{x}$

$$\sum_{\ell,m,n=0}^{\infty} \alpha_{m,n,\ell} X^{\ell} Z^{n} Z^{*m} = \sum_{\ell,m,n=0}^{\infty} \overline{\alpha_{m,n,\ell}}\; X^{\ell} (-Z^*)^{n} (-Z)^{m} ,$$

which implies $\alpha_{m,n,\ell} = (-1)^{n+m}\, \overline{\alpha_{n,m,\ell}}$. This shows in the case of real solutions that the functions (6.52) are completely determined by the functions given in (6.51).

The equation (6.49), now, can be put into the form

$$\sum_{j=0}^{n} (-1)^{n-j} \binom{n}{j} \frac{\partial^{2n} U}{\partial Z^{j}\, \partial Z^{*j}\, \partial X^{2n-2j}}$$

(6.53)

$$+ \sum_{k=1}^{n} \underset{\sim}{L}_k \left(\sum_{\ell=0}^{n-k} (-1)^{n-k-\ell} \binom{n-k}{\ell} \frac{\partial^{2n-2k} U}{\partial Z^{\ell} \partial Z^{*\ell} \partial X^{2n-2k-2\ell}} \right) = 0 ,$$

with

$$\underset{\sim}{L}_k \equiv \sum_{m,p,q=0}^{m+p+q \leq k} A^{(k)}_{m,p,q}(X,Z,Z^*) \frac{\partial^{m+p+q}}{\partial X^m \partial Z^p \partial Z^{*q}} .$$

Let us consider the principal part of this differential equation, and let A be the set of multi-indices $(j,j,2n-2j)$, in the principal part. We notice that the convex hull of the set of points $\bigcup_{j=0}^{n-1} \{(j,j,2n-2j)\}$ does not contain the point $(n,n,0)$. Hence, by a theorem due to Hörmander [Ho 1], pg. 118, 119, there exists a unique analytic solution to the Goursat problem for the differential equation (6.53) with data (6.51-6.52). We have already shown, however, that the data (6.57) is dependent on the data (6.51), and hence, we have the desired conclusion.

As in the case of Tjong's operator, Colton also seeks a solution of (6.12) in the form (6.16). Following the same type of approach outlined above, Colton obtains the following differential equation for the E-function in terms of his ξ_i variables,

$$\mu t\left(\frac{\partial^2\hat{E}}{\partial\xi_2^2} + \frac{\partial^2\hat{E}}{\partial\xi_3^2} - 4\frac{\partial^2\hat{E}}{\partial\xi_1\partial\xi_3} - 2\frac{\partial^2\hat{E}}{\partial\xi_2\partial\xi_3} + \hat{Q}\hat{E}\right)$$

$$(6.54) \qquad + (1-t^2)\frac{\partial^2\hat{E}}{\partial\xi_1\partial t} - \frac{1}{t}\frac{\partial\hat{E}}{\partial\xi_1} = 0 .$$

He then seeks a solution to this equation in the form (6.18), which leads to a recursive scheme for the $p^{(n)}(\underline{\xi};\zeta)$, namely

$$p_1^{(n+1)} = \frac{-1}{2n+1}\{p_{22}^{(n)} + p_{33}^{(n)} - 4p_{13}^{(n)} - 2p_{23}^{(n)} + \hat{Q}p^{(n)}\} ,$$

(6.55)

$$\text{with } p^{(0)} \equiv 1, \text{ and } p^{(n+1)}(0,\xi_2,\xi_3,\zeta) = 0 .$$

That Colton's solution of (6.55) exists in the form (6.18) is proved using the method of majorants and the recursive equations (6.55) [Co 3,5].

Colton [Co 5] is able to show that his operator maps onto to the class of solutions of (6.12) (without putting any restrictions on Q other than analyticity and realness) in a rather direct and elegant manner. Since for the case of real solutions it is sufficient to give Goursat data on $Z = 0$, one is led to consider those solutions of (6.15), which satisfy

.56) $$U(X,0,Z^*) = \varphi(X,Z^*) \ .$$

Since we are dealing with real solutions here, Colton considers the representations

$$u(\underset{\sim}{x}) = \text{Re } \underset{\sim}{C}_3\{f\} = \frac{1}{4\pi i} \int_{|\zeta|=1} \int_{\gamma} E(X,Z,Z^*,\zeta,t) f(\nu,\zeta) \frac{dt}{\sqrt{1-t^2}} \cdot \frac{d\zeta}{\zeta}$$

.57) $$+ \frac{1}{4\pi i} \int_{|\zeta|=1} \int_{\gamma} \bar{E}(X,-Z^*,-Z,\zeta,t) \bar{f}(\bar{\nu},\zeta) \frac{dt}{\sqrt{1-t^2}} \frac{d\zeta}{\zeta} \ .$$

Now, as in the case of Tjong's operator, the E-function becomes identically 1 when $\xi_1 = 0$; hence, we consider the soluability of

$$U(X,0,Z^*) = \varphi(X,Z^*) = \sum_{n,m=0}^{\infty} c_{n,m} X^n Z^{*m}$$

$$= \frac{1}{4\pi i} \int_{|\zeta|=1} \int_{\gamma} 1 \cdot f(\nu_1,\zeta) \frac{dt}{\sqrt{1-t^2}} \frac{d\zeta}{\zeta}$$

$$(6.58) \qquad + \frac{1}{4\pi i} \int_{|\zeta|=1} \int_{\gamma} \bar{E}(X,-Z^*,\, 0,\zeta,t)\bar{f}(\nu_2,\zeta) \frac{dt}{\sqrt{1-t^2}} \frac{d\zeta}{\zeta} .$$

Here $\mu_1 = X + Z^*\zeta^{-1}$, and $\mu_2 = X - \zeta Z^*$, with $\nu_k = \mu_k[1-t^2]$. Colton shows that the terms $p^{(n)}(\xi;\zeta)$ have zeros of at least order two at $\zeta = 0$; this plus the fact that if $f(\mu,\zeta)$ is taken in the form

$$f(\mu,\zeta) = \sum_{n=0}^{\infty} \sum_{m=0}^{n} a_{n,m}\mu^n\zeta^m$$

$\bar{f}(\nu_2,\zeta)$ will only contain non-negative powers of ζ leads him to conclude that

$$\frac{1}{4\pi i} \int_{|\zeta|=1} \int_{\gamma} \bar{E}(X,-Z^*,0,\zeta,t)\bar{f}(\nu_2,\zeta) \frac{dt}{\sqrt{1-t^2}} \frac{d\zeta}{\zeta}$$

$$= \frac{1}{4\pi i} \int_{|\zeta|=1} \int_{\gamma} \bar{f}(\nu_2,\zeta) \frac{dt}{\sqrt{1-t^2}} \frac{d\zeta}{\zeta} .$$

Consequently (6.58) becomes

$$\varphi(X,Z^*) = \frac{1}{4\pi i}\int_{|\zeta|=1}\int_{\gamma}[f(\nu_1,\zeta)+\bar{f}(\nu_2,\zeta)]\cdot\frac{dt}{\sqrt{1-t^2}}\frac{d\zeta}{\zeta}\,.$$

.59)
$$= \frac{1}{4\pi i}\int_{|\zeta|=1}\int_{\gamma}[g(\mu_1,\zeta)+\bar{g}(\mu_2,\zeta)]\frac{d\zeta}{\zeta}\,.$$

Since $U(X,Z,Z^*)$ is real for real $x_k\,(k=1,2,3)$, $\varphi(X,0) = u(X,0,0) = \sum_{n=0}^{\infty} c_{n,0}X^n = \sum_{n=0}^{\infty}\overline{c_{n,0}}X^n$, i.e. the $c_{n,0}$ are real.

Putting the expansions for $f(\mu,\zeta)$ and $\varphi(X,Z^*)$ into (5.59) yields upon integrating (recall that $\mu_2 = X-\zeta Z^*$),

$$\sum_{n=0}^{\infty}\sum_{m=0}^{\infty} c_{n,m}X^nZ^{*m} = \tfrac{1}{2}\sum_{n=0}^{\infty}\sum_{m=0}^{n} a_{n,m}\binom{n}{m}X^{n-m}Z^{*m} + \tfrac{1}{2}\sum_{n=0}^{\infty}\overline{a_{n,0}}\,X^n\,.$$

Upon equating coefficients of like powers we get the conditions

.60)
$$c_{n,0} = \tfrac{1}{2}[a_{n,0}+\overline{a_{n,0}}]\,,\quad (n=0,1,\ldots)$$

$$c_{n,m} = \frac{(n+m)!}{n!\,m!}\,a_{n+m,m}\,,\quad (n=0,1,\ldots),\,(m\ge 1)\,,$$

which yields

$$\varphi(X,Z^*) = \tfrac{1}{2} \sum_{n=0}^{\infty} (a_{n,0} + \overline{a_{n,0}}) X^n + \tfrac{1}{2} \sum_{n=0}^{\infty} \sum_{m=1}^{n} \frac{(n+m)!}{n!\, m!} a_{n+m,\, m} X^n Z^m .$$

Recalling the integral definition of the beta-function one has

$$\int_0^1 \varphi(tX, (1-t)Z^*)\,dt = \sum_{n=0}^{\infty} \frac{a_{n,0}}{n+1} X^n + \tfrac{1}{2} \sum_{n=0}^{\infty} \sum_{m=1}^{n} \frac{a_{n,m}}{n+1} X^{n-m} Z^{*m} ;$$

this in turn leads to

$$\frac{\partial}{\partial\mu}\Big[\mu \int_0^1 \varphi(t\mu, (1-t)\mu\zeta)\,dt\Big]$$

$$= \tfrac{1}{2} \sum_{n=0}^{\infty} \sum_{m=0}^{n} a_{n,m} \mu^n \zeta^m + \tfrac{1}{2} \sum_{n=0}^{\infty} a_{n,0} \mu^n$$

$$= \tfrac{1}{2}\, g(\mu,\zeta) + \tfrac{1}{2}\, \varphi(\mu,0) .$$

Consequently, one has the inverse operator representation

[Co 5] pg. 756.

(5.61) $$g(\mu,\zeta) = (\underset{\sim}{C}_3^{-1}U)(\mu,\zeta) \equiv 2\frac{\partial}{2\mu}[\mu\int_0^1 U(t\mu,0,(1-t)\mu\zeta)dt] - \tfrac{1}{2}U(\mu,0,0)$$

Remark: When $Q \equiv 0$, Colton's formula (5.61) gives an alternate representation for $\underset{\sim}{B}_3^{-1}$.

We now show how it is possible to develop an analogue to the Whittaker-Bergman operator for the case of self-adjoint strongly-elliptic, second order equations with Hölder continuous coefficients

(5.62) $$A(x,D)u \equiv D_i(a_{ij}(\underset{\sim}{x})D_j u) \equiv \sum_{i,j=1}^{3} \frac{\partial}{\partial x_i}(a_{ij}(\underset{\sim}{x})\frac{\partial u}{\partial x_j}) = 0 .$$

Here the summation convention for repeated indices is used in the middle term; this notation will be continued in what follows. Let $A \equiv (a_{ij})$ be a three-by-three matrix of coefficients, P^* = transpose of P, and $P \equiv (p_{ij})'$ be a matrix such that $PAP^* = I$. Let $\underset{\sim}{\xi}, \underset{\sim}{x}$ be arbitrary vectors in $\mathbb{R}^3$ and $\underset{\sim}{Y} = P(\underset{\sim}{x} - \underset{\sim}{\xi})$.

It may be shown that a natural choice of an auxiliary variable, for the constant coefficient equation

(5.63) $$A(\underset{\sim}{\xi},D)u \equiv a_{ij}(\underset{\sim}{\xi})D_i D_j u = 0 ,$$

is given by

$$\mu = Y_i N_i(\zeta) = [P(\underset{\sim}{x} - \underset{\sim}{\xi})]_i N_i(\zeta)$$

$$= p_{ij}(x_j - \xi_j) H_i(\zeta) \ , \quad \text{with} \quad N_i N_i \equiv 0 \ .$$

To see this we observe that if $f(\mu,\zeta)$ is a holomorphic function of two complex variables defined in the Reinhardt circular domain

$$\mathcal{D} \equiv \{(\mu,\zeta) \mid |\mu| \le a \ , 1-\varepsilon \le |\zeta| \le 1+\varepsilon \ , \quad 0<\varepsilon<1\} \ ,$$

then for $\underset{\sim}{x}, \zeta$ chosen such that $(\mu,\zeta) \in \mathcal{D}$

$$\text{(6.64)} \qquad A(\xi,D) f(\mu,\zeta) = [a_{\ell m}(\xi) p_{i\ell} p_{jm}] N_j(\zeta) N_i(\zeta) \frac{\partial^2 f}{\partial \mu^2} \ .$$

However,

$$a_{\ell m}(\xi) p_{i\ell}(\xi) p_{jm}(\xi) = p_{i\ell} a_{\ell m} p^*_{mj} = (PAP^*)_{ij} = \delta_{ij} \ ,$$

so

$$\text{(6.65)} \qquad A(\xi,D) f(\mu,\zeta) = \delta_{ij} N_i N_j \frac{\partial^2 f}{\partial \mu^2} = N_i N_i \frac{\partial^2 f}{\partial \mu^2} \equiv 0 \ ,$$

Since the $N_i(\zeta)$ have been chosen as components of an isotropic vector. The Whittaker-Bergman operator for (6.63) is then seen to have the form

(6.66) $$H(\underset{\sim}{x}) = (\underset{\sim}{B}_3 f)(\underset{\sim}{x}, \underset{\sim}{\xi}) \equiv \frac{1}{2\pi i} \int_{\mathfrak{L}} f(\mu, \zeta) \frac{d\zeta}{\zeta} ,$$

with $\mu = p_{ij}(x_j - \xi_j) N_i$, and $\mathfrak{L}$ an arbitrary, rectifiable closed curve.

In order to generate solutions to (6.62), it is helpful first to find the $\underset{\sim}{B}_3$ - associate for the reciprocal distance function. Recalling that the components of $\underset{\sim}{N}$ are $N_1 = \frac{1}{2}(\zeta - 1/\zeta)$, $N_2 = i/2(\zeta + 1/\zeta)$, $N_3 = 1$, one has

$$p_{kj}(x_j - \xi_j) N_k = (x_j - \xi_j)\{\tfrac{1}{2} p_{ij}(\zeta - 1/\zeta) + i/2\ p_{2j}(\zeta + 1/\zeta) + p_{3j}\} .$$

Setting $\alpha_1 = \frac{1}{2}(x_j - \xi_j)(p_{1j} + ip_{2j})$,

(6.67) $$\alpha_2 = \tfrac{1}{2}(x_j - \xi_j)(-p_{1j} + ip_{2j}) ,$$

$$\alpha_3 = (x_j - \xi_j) p_{3j} ,$$

the roots of $\zeta p_{ij}(x_j - \xi_j)N_i = 0$ may be written as

$$\zeta = \frac{1}{2\alpha_i}\,[-\alpha_3 \pm \sqrt{\alpha_3^2 - 4\alpha_1\alpha_2}]\ .$$

The product of the two roots is α_2/α_1 ; so if the x_j, p_{kj} $(k,j = 1,2,3)$ are real, then

$$\frac{|\alpha_2|}{|\alpha_1|} = \left|\frac{(x_j - \xi_j)(p_{1j} - ip_{2j})}{(x_j - \xi_j)(p_{1j} + p_{2j})}\right| = 1\ .$$

Consequently, one root lies inside $|\zeta| = 1$ and the other outside, unless, of course, they are both equal and lie on $|\zeta| = 1$. We conclude that

$$\frac{1}{2\pi i}\int_{|\zeta|=1} \frac{d\zeta}{\zeta p_{kj}(x_j - \xi_j)N_k(\zeta)} = \underset{|\zeta|<1}{\text{Residue}}\ \frac{1}{\zeta p_{kj}(x_j - \xi_j)N_k(\zeta)}$$

$$= \frac{\pm 1}{\sqrt{\alpha_3^2 - 4\alpha_1\alpha_2}} = \frac{\pm 1}{\sqrt{p_{kj}p_{k\ell}(x_j - \xi_j)(x_\ell - \xi_\ell)}}$$

$$= \frac{\pm 1}{\sqrt{a^{j\ell}(x_j - \xi_j)(x_\ell - \xi_\ell)}} \equiv 4\pi P(x,\xi) = \frac{1}{|\underset{\sim}{Y}|}\ , \tag{6.68}$$

Where $(a^{j\ell}) = A^{-1} = P^*P = (p^*_{jk}p_{k\ell})$. Hence, we have found the $\underset{\sim}{B}_3$-associate for the parametrix, $P(\underset{\sim}{x},\underset{\sim}{\xi})$.

To obtain a Bergman-Whittaker operator for the variable coefficient case one now seeks a solution of (6.62) in a domain $\Omega \ni \underset{\sim}{0}$ (without loss of generality we seek local solutions about $(\underset{\sim}{0})$

(6.69) $$u(\underset{\sim}{x}) = \int_\Omega P(\underset{\sim}{x},\underset{\sim}{\xi})Q(\underset{\sim}{\xi})d\underset{\sim}{\xi} + (\underset{\sim}{B}_3 f)(\underset{\sim}{x};\underset{\sim}{0}) .$$

Replacing the parametrix by (6.68) and changing the orders of integration by Fubini's theorem yields an alternate representation

(6.70) $$u(\underset{\sim}{x}) = \frac{1}{2\pi i}\int_{\mathcal{L}} \frac{d\zeta}{\zeta}\left[\frac{1}{4\pi}\int_\Omega \frac{Q(\underset{\sim}{\xi})d\underset{\sim}{\xi}}{p_{kj}(\underset{\sim}{\xi})(x_j-\xi_j)N_k(\zeta)}\right] + (\underset{\sim}{B}_3 f)(\underset{\sim}{x};\underset{\sim}{0})$$

$$= \underset{\sim}{B}_3[F(\underset{\sim}{x};\zeta) + f_0(\mu,\zeta)] .$$

Even though (6.70) has the <u>form</u> of a Bergman-Whittaker operator, for this to be <u>really</u> meaningful we should replace $Q(\underset{\sim}{\xi})$ by a known function. To this end, we rewrite (6.62) in the form

$$A(\underset{\sim}{x}, D_x)u \equiv A(\underset{\sim}{0}, D_x)u + [a_{kj}(\underset{\sim}{x}) - a_{kj}(\underset{\sim}{0})\ \frac{\partial^2 u}{\partial x_k \partial x_j} + \frac{\partial a_{kj}}{\partial x_k}\frac{\partial u}{\partial x_j}$$

(6.71)

$$\equiv A(\underset{\sim}{0}, D_x)u + C(\underset{\sim}{x}, D_x)u = 0 \ ,$$

and substituting into this (6.69) for u leads to a singular integral equation for $Q(\underset{\sim}{x})$, namely [See Garabedian [Ga 1] pg. 170]

(6.72) $$Q(\underset{\sim}{x}) + \int_{\Omega} Q(\underset{\sim}{\xi}) A(\underset{\sim}{x}, D_x) P(\underset{\sim}{x}, \underset{\sim}{\xi}) d\underset{\sim}{\xi} + C(\underset{\sim}{x}, D_x) h_0(\underset{\sim}{x}) = 0 \ ,$$

with $h_0(\underset{\sim}{x}) = \underset{\sim}{B}_3 f_0(\mu, \zeta)$.

The integral equation (6.72) may be solved by an iteration procedure; namely set

(6.73) $$Q_{\ell+1}(\underset{\sim}{x}) = -\phi(\underset{\sim}{x}) + \int_{\Omega} K(\underset{\sim}{X}, \underset{\sim}{\xi}) Q_{\ell}(\underset{\sim}{\xi}) d\underset{\sim}{\xi} \ , \ \ell \geq 0 \ ,$$

where $Q_0(\underset{\sim}{x}) = -\phi(\underset{\sim}{x}) \equiv C(\underset{\sim}{x}, D_x) h_0(\underset{\sim}{x})$,

and $K(\underset{\sim}{x}, \underset{\sim}{\xi}) \equiv A(\underset{\sim}{x}, D_x) P(\underset{\sim}{x}, \underset{\sim}{\xi})$.

Then one has

$$Q_{\ell+1}(\underset{\sim}{x}) = -\phi(\underset{\sim}{x}) + \sum_{m=1}^{\ell+1} (-1)^{m+1} \int_{\Omega} K_m(\underset{\sim}{x}, \underset{\sim}{\xi})\phi(\underset{\sim}{\xi})\, d\underset{\sim}{\xi} \ ,$$

with $K_{m+1}(\underset{\sim}{x}, \underset{\sim}{\xi}) = \int_{\Omega} K(\underset{\sim}{x}, \underset{\sim}{\eta}) K_m(\underset{\sim}{\eta}, \underset{\sim}{\xi})\, d\underset{\sim}{\eta}$,

$$K_1(\underset{\sim}{x}, \underset{\sim}{\xi}) \equiv K(\underset{\sim}{x}, \underset{\sim}{\xi}) \ .$$

To show that as $\ell \to \infty$ the $Q_\ell(\underset{\sim}{x})$ tend uniformly to a continuous limit function in Ω . We consider the iterated integrals

$$I_m(\underset{\sim}{x}) \equiv \int_{\Omega} K_m(\underset{\sim}{x}, \underset{\sim}{\xi})\phi(\underset{\sim}{\xi})\, d\underset{\sim}{\xi} \ ,$$

and the formal series

$$\mathcal{K}(\underset{\sim}{x}, \underset{\sim}{\xi}) \equiv \sum_{m=1}^{\infty} (-1)^{m+1} K_m(\underset{\sim}{x}, \underset{\sim}{\xi}) \ .$$

For $m = 1$ a theorem in Kellog [Ke] (pg. 153-156) states that $I_1(\underset{\sim}{x}) \in C^0(\Omega)$, since $\phi(\underset{\sim}{x})$ is in the Hölder class $C^\alpha(\Omega)$

(it is actually analytic), and since by hypothesis the coefficients of $C(\underset{\sim}{x}, D_x)$ are also in $C^{\alpha}(\Omega)$. For $m \geq 2$ our argument is based on the following Claim:

$$\int_{\Omega} (a^{\ell m}(\underset{\sim}{\xi})(x_\ell - \xi_\ell)(x_m - \xi_m))^{-\alpha'} (a^{ij}(\underset{\sim}{\xi})(y_i - \xi_i)(y_j - \xi_j))^{-\beta'} d\xi \tag{6.74}$$

$$\leq C(a^{\ell m}(\underset{\sim}{y})(x_\ell - y_\ell)(x_m - y_m))^{3/2 - \alpha' - \beta'}$$

when $\alpha' + \beta' > 3$.

Now if the $a_{\ell m}(x) \in C^{1,\alpha}(\Omega)$ then

$$|K(\underset{\sim}{x}, \underset{\sim}{\xi})| \leq C_1 \cdot (a^{\ell m}(\underset{\sim}{\xi})(x_\ell - \xi_\ell)(x_m - \xi_m))^{(\alpha - 3)/2}, \tag{6.75}$$

and by (6.74) above one has

$$|K_2(\underset{\sim}{x}, \underset{\sim}{\xi})| \leq C_2 \cdot (a^{\ell m}(\underset{\sim}{\xi})(x_\ell - \xi_\ell)(x_m - \xi_m))^{\alpha - 3/2}$$

and indeed more generally

$$|K_n(\underset{\sim}{x}, \underset{\sim}{\xi})| \leq C_2 (a^{\ell m}(\underset{\sim}{\xi})(x_\ell - \xi_\ell)(x_m - \xi_m))^{(n\alpha - 3)/2}. \tag{6.76}$$

Eventually an iterate occurs where $n\alpha > 1$, and in this case $\int K_m(\underset{\sim}{x},\underset{\sim}{\xi})\phi(\underset{\sim}{\xi})d\underset{\sim}{\xi}$ is Hölder continuous. To see this we consider

$$(6.77)\qquad \left|\int_\Omega [K_m(\underset{\sim}{x},\underset{\sim}{\xi}) - K_m(\underset{\sim}{y},\underset{\sim}{\xi})]\phi(\underset{\sim}{\xi})d\underset{\sim}{\xi}\right| \le C_1\int_\Omega \frac{|r(\underset{\sim}{x},\underset{\sim}{\xi}) - r(\underset{\sim}{y},\underset{\sim}{\xi})|d\underset{\sim}{\xi}}{[r(\underset{\sim}{x},\underset{\sim}{\xi})r(\underset{\sim}{y},\underset{\sim}{\xi})]^{3-n\alpha}}$$

where $r(\underset{\sim}{x},\underset{\sim}{\xi}) = (a^{\ell m}(\underset{\sim}{\xi})(x_\ell - \xi_\ell)(x_m - \xi_m))^{\frac{1}{2}}$.

But $|r(\underset{\sim}{x},\underset{\sim}{\xi}) - r(\underset{\sim}{y},\underset{\sim}{\xi})| \le r(\underset{\sim}{x},\underset{\sim}{y})$ by the triangle inequality, hence one obtains by (6.74) the following upper bound for the integral in (6.77)

$$C_1 r(\underset{\sim}{x},\underset{\sim}{y})\int_\Omega \frac{d\underset{\sim}{\xi}}{[r(\underset{\sim}{x},\underset{\sim}{\xi})r(\underset{\sim}{y},\underset{\sim}{\xi})]^{3-n\alpha}}$$

$$6.78)\qquad \le C_1[r(\underset{\sim}{x},\underset{\sim}{y})]^{2(n\alpha-1)},$$

(We may choose $n\alpha < \frac{3}{2}$ by taking α sufficiently small, i.e. one may always use a smaller Hölder coefficient.) The same reasoning shows us that the iterated kernels $K_n(\underset{\sim}{x},\underset{\sim}{\xi})$ are also eventually Hölder continuous, say for $n \ge N$. Therefore, when $m \ge N$ one then has the estimate

$$\sup_{\underset{\sim}{x} \in \Omega} |K_{m+1}(\underset{\sim}{x},\underset{\sim}{\xi})| \le \sup_{\underset{\sim}{x} \in \Omega} |K_m(\underset{\sim}{x},\underset{\sim}{\xi})| \int_{\Omega} |K_1(\underset{\sim}{x},\underset{\sim}{\xi})|\, d\xi$$

$$\le C_0 \text{ meas } (\Omega) \sup_{\underset{\sim}{x} \in \Omega} |K_m(\underset{\sim}{x},\underset{\sim}{\xi})| \ .$$

The integral of $|K_1|$ converges, since the coefficients of the principal part of $C(\underset{\sim}{x},D_x)$ not only are Hölder continuous, they also vanish at $\underset{\sim}{x} = \underset{\sim}{\xi}$. Consequently,

$$\sup_{\underset{\sim}{x},\underset{\sim}{\xi} \in \Omega} |K_m(\underset{\sim}{x},\underset{\sim}{\xi})| \le \lambda^{m-N} C_1(\Omega), \quad m \ge N$$

and the series

$$\mathcal{K}(\underset{\sim}{x},\underset{\sim}{\xi}) = \sum_{n=0}^{\infty} K_n(\underset{\sim}{x},\underset{\sim}{\xi}) \tag{6.79}$$

is seen to converge uniformly to a $C^0(\Omega)$ function for $\underset{\sim}{x} \ne \underset{\sim}{\xi}$. If one defines the function $Q(\underset{\sim}{x})$ as the limit

$$Q(\underset{\sim}{x}) \equiv \lim_{n \to \infty} Q_n(\underset{\sim}{x}) = \int_{\Omega} \mathcal{K}(\underset{\sim}{x},\underset{\sim}{\xi})\, \phi(\underset{\sim}{\xi})\, d\underset{\sim}{\xi}$$

then this is a $C^0(\Omega)$ function, and furthermore, it is also a solution of (6.72). Consequently, one has

$$u(\underset{\sim}{x}) = h_0(\underset{\sim}{x}) + \int_\Omega d\underset{\sim}{\xi}\ P(\underset{\sim}{x},\underset{\sim}{\xi}) \int_\Omega d\underset{\sim}{\eta}\ \mathcal{K}(\underset{\sim}{\xi},\underset{\sim}{\eta})\phi(\underset{\sim}{\eta})$$

(6.80) $$= h_0(\underset{\sim}{x}) + \int_\Omega R(\underset{\sim}{x},\underset{\sim}{\xi})\phi(\underset{\sim}{\xi})d\underset{\sim}{\xi}\ ,$$

with $$R(\underset{\sim}{x},\underset{\sim}{\xi}) \equiv \int_\Omega P(\underset{\sim}{x},\underset{\sim}{\eta})\ \mathcal{K}(\underset{\sim}{\eta},\underset{\sim}{\xi})d\underset{\sim}{\eta}\ ,$$

and $$h_0(\underset{\sim}{x}) = (\underset{\sim}{B}_3 f_0)(\underset{\sim}{x}) \equiv (\underset{\sim}{B}_3 f)(\underset{\sim}{x};\underset{\sim}{0})\ .$$

Again, by the Claim (4.16) one has that

(6.81) $$|R(\underset{\sim}{x},\underset{\sim}{\xi})| \leq C[r(\underset{\sim}{x},\underset{\sim}{\xi})]^{\alpha - 1}\ ,$$

and since $\phi(\underset{\sim}{\xi}) \in C^\alpha(\Omega)$, it follows from Kellog (pp. 153-156) that (6.80) defines a $u(\underset{\sim}{x}) \in C^2(\Omega)$. The representation (6.80) then permits a Bergman-Whittaker type representation for the variable coefficient case, namely

$$u(\underset{\sim}{x}) = (\underset{\sim}{R}_3 f_0)(\underset{\sim}{x}) \equiv (\underset{\sim}{B}_3 f_0)(\underset{\sim}{x}) + \int_\Omega R(\underset{\sim}{x},\underset{\sim}{\xi}) C(\underset{\sim}{\xi}, D_\xi) \underset{\sim}{B}_3 f_0 d\xi$$

$$\equiv \underset{\sim}{B}_3 \{f_0(\mu,\zeta) + \int_\Omega R(\underset{\sim}{x},\underset{\sim}{\xi}) C(\underset{\sim}{\xi}, D_\xi) f_0(\mu,\zeta) d\xi\}$$

with $\mu = \frac{\xi_1}{2}(\zeta - \frac{1}{\zeta}) + \frac{i\xi_2}{2}(\zeta + \frac{1}{\zeta}) + \xi_3$, and $R(\underset{\sim}{x},\underset{\sim}{\xi})$ known.

We next turn to verifying the <u>Claim</u> (6.74). To this end we note that this inequality is the three-dimensional analogue in a Riemannian space of Hadamard's inequality [He 1] pg. 198. Since (4.1) is assumed to be strongly elliptic in Ω, there exist positive constants $\bar{\lambda}_0$, and $\bar{\lambda}_1$ such that

$$\bar{\lambda}_0 |\underset{\sim}{\eta}|^2 \le a_{\ell m}(\underset{\sim}{\xi}) \eta_\ell \eta_m \le \bar{\lambda}_1 |\underset{\sim}{\eta}|^2 , \tag{6.83}$$

for all points $\underset{\sim}{\xi} \in \Omega$ and any real vector $\underset{\sim}{\eta}$. The condition (6.83), likewise implies the existence of positive constants λ_0, λ_1, such that

$$\lambda_0 |\underset{\sim}{\eta}|^2 \le a^{\ell m}(\underset{\sim}{\xi})\eta_\ell \eta_m \le \lambda_1 |\underset{\sim}{\eta}|^2 ,$$

for all $\underset{\sim}{\xi} \in \Omega$ and any real vector $\underset{\sim}{\eta}$.

Therefore the left-hand side of (6.74) may be bounded by

$$\int_\Omega (a^{\ell m}(\underset{\sim}{\xi})(x_\ell - \xi_\ell (x_m - \xi_m))^{-\alpha'} \quad (a^{ij}(\underset{\sim}{\xi})(y_i - \xi_i)(y_j - \xi_j))^{-\beta'} \, d\xi$$

$$\le \lambda_0^{-\alpha' - \beta'} \int_\Omega |\underset{\sim}{x} - \underset{\sim}{\xi}|^{-\alpha'} |y - \xi|^{-\beta'} d\underset{\sim}{\xi} \tag{6.84}$$

The integral on the right-hand side of (6.84) may now be treated in essentially the same manner as the two dimensional case. (The reader is referred to the above two references for details.) In fact, for arbitrary dimension $n \ge 2$ one may easily establish the following estimate for strongly elliptic differential operators

$$\int_\Omega [r(\underset{\sim}{x}, \underset{\sim}{\xi})]^{-\alpha'} [r(\underset{\sim}{y}, \underset{\sim}{\xi})]^{-\beta'} d\underset{\sim}{\xi}$$

$$(6.85)\qquad \le \begin{cases} c_0[r(\underset{\sim}{x},\underset{\sim}{y})]^{n-\alpha'-\beta'}, & \alpha'+\beta'>n \\ c_0+c_1 \log |r(\underset{\sim}{x},\underset{\sim}{y})|, & \alpha'+\beta'=n \\ c_0, & \alpha'+\beta'<n . \end{cases}$$

We next turn to the problem of finding a locally complete system of solutions to (6.62).

Theorem (6.2): For Ω a simply-connected, regular domain, containing the origin, whose diameter is sufficiently small, the integral operator (6.82) generates a complete family of solutions.

Proof: In the Cartesian coordinates, the set of all homogeneous, harmonic polynomials $h_{nm}(x) = r^n P_n{}^m(\cos\theta)e^{im\varphi}$, $(n=0,1,\ldots)$, $(m=0,\pm1,\pm2,\ldots,\pm n)$ form a complete family of solutions for Laplace's equation. After a non-singular coordinate transformation these give us a complete system of solutions for $A(0,D)u=0$. Using these oblique axes, harmonic functions (6.82) provides us with a corresponding family.

Before concluding this chapter we wish to present Colton's treatment [Co 6] of the equation

6.86) $$\Delta_3 u + a(x,y,z)u_x + b(x,y,z)u_y + c(x,y,z)u_z + d(x,y,z)u = 0 ,$$

where the coefficients are analytic (in fact, for simplicity let us assume entire) functions of their independent variables. As before we introduce the new independent variables X,Z,Z^* and obtain an equation of the form,

6.87) $$U_{XX} - U_{ZZ^*} + A(X,Z,Z^*)U_X + B(X,Z,Z^*)U_Z$$
$$+ C(X,Z,Z^*)U_{Z^*} + D(X,Z,Z^*)U = 0 .$$

This equation can be treated somewhat more easily if we put it into "standard form" by making the substitution

(6.88) $$V(X,Z,Z^*) = U(X,Z,Z^*)\exp\{-\int_0^Z C(X,Z^1,Z^*)\,dZ^1\} .$$

It is easily seen that $V(X,Z,Z^*)$ must satisfy

6.89) $$V_{XX} - V_{ZZ^*} + \tilde{A}(X,Z,Z^*)V_X + \tilde{B}(X,Z,Z^*)V_Z + \tilde{D}(X,Z,Z^*)V = 0$$

where

$$\tilde{A}(X,Z,Z^*) = A(X,Z,Z^*) + 2\int_0^Z C_X(X,Z^1,Z^*)\,dZ^1 ,$$

$$\tilde{B}(X,Z,Z^*) = B(X,Z,Z^*) - \int_0^Z C_{Z^*}(X,Z^1,Z^*)\,dZ^1 ,$$

$$\tilde{D}(X,Z,Z^*) = D(X,Z,Z^*) + A(X,Z,Z^*)\int_0^Z C_X(X,Z^1,Z^*)\,dZ^1$$

$$+ B(X,Z,Z^*)C(X,Z,Z^*) - C_{Z^*}(X,Z,Z^*)$$

$$+ \int_0^Z C_{XX}(X,Z^1,Z^*)\,dZ^1 + \left(\int_0^Z C_X(X,Z^1,Z^*)\,dZ^1\right)^2 . \tag{6.90}$$

From Lemma (6.1) it is easy to see if $u(x,y,z)$ is a real valued C^2 solution of (6.86) in some neighborhood of the origin, then $U(X,Z,Z^*)$ is an analytic function in some neighborhood of the origin in $\mathbb{C}^3$, and is furthermore, uniquely determined by the data $U(X,0,Z^*)$.

By introducing the Colton-Tjong variables

$$\xi_1 = 2\zeta Z,\quad \xi_2 = X + 2\zeta Z,\quad \xi_3 = X + 2\zeta^{-1}Z^* ,$$

one can show by direct substitution that

$$V(X,Z,Z^*) = \underset{\sim}{C}_3^1\ \{f\}$$

(6.91) $$\equiv \frac{1}{2\pi i} \int_{|\zeta|=1} \int_{\gamma} E(X,Z,Z^*,\zeta,t) f(\mu[1-t^2],\zeta) \cdot \frac{dt}{\sqrt{1-t^2}} \frac{d\zeta}{\zeta}$$

where γ is a path in T joining -1 to 1, is a complex valued solution of (6.89) if E is a regular solution of the partial differential equation

$$\mu t(4E^*_{13} + 2E^*_{23} - E^*_{22} - E^*_{33} - \tilde{D}^* E^*)$$

$$+ (1-t^2)E^*_{1t} - \frac{1}{t} E^*_1 - \tilde{A}^*[(E^*_2 + E^*_3)\mu t + \tfrac{1}{2}(1-t^2)E^*_t$$

(6.92) $$- \frac{1}{2t} E^*] - \tilde{B}^* \zeta[(2E^*_1 + 2E^*_2)\mu t + \tfrac{1}{2}(1-t^2)E^*_t - \frac{1}{2t} E^*] = 0 ,$$

where $\tilde{A}^*(\xi_1,\xi_2,\xi_3,\zeta) \equiv \tilde{A}(X,Z,Z^*)$, etc. .

<u>Remark</u>: It is obvious that the function $U(X,Z,Z^*)$, defined by

(6.93) $$U(X,Z,Z^*) = \underset{\sim}{C}_3\{f\} \equiv \exp\left[\int_0^Z C(X,Z^1,Z^*)\,dZ^1\right] \underset{\sim}{C}{}^1_3\{f\} ,$$

is clearly a solution (6.87).

As in the earlier cases we seek a solution of the form

$$E^*(\xi_1,\xi_2,\xi_3,\zeta,t) = \sum_{n=1}^{\infty} t^{2n}\mu^n p^{(n)}(\xi_1,\xi_2,\xi_3,\zeta) \,. \tag{6.94}$$

However, because of the term $\frac{1}{2t}\, E^*(\tilde{A}^* + \zeta\tilde{B}^*)$ in (6.92) we can not, as before, take $p^{(0)} \equiv 1$. It turns out that one must take

$$E^*(0,\xi_2,\xi_3,\zeta,t) = \mu t^2 \,. \tag{6.95}$$

Substitution of (6.94) into (6.92) leads to the following recursion formulae for the coefficients $p^{(n)}$,

$$p_1^{(n+1)} - \tfrac{1}{2}(\tilde{A}^* + \tilde{B}^*\zeta)p^{(n+1)} = \frac{1}{2n+1}\{p_{22}^{(n)} + p_{33}^{(n)} - 4p_{13}^{(n)}$$

$$- 2p_{23}^{(n)} + (\tilde{A}^* + 2\tilde{B}^*\zeta)p_2^{(n)} + \tilde{A}^*p_3^{(n)} + 2\tilde{B}^*\zeta p_1^{(n)}$$

$$+ \tilde{D}^*p^{(n)}\} \,, \quad n \geq 1 \,,$$

$$p^{(1)}(\xi_1,\xi_2,\xi_3,\zeta) = \exp\,[\tfrac{1}{2}\int_0^{\xi_1}(\tilde{A}^* + \tilde{B}^*\zeta)\,d\tilde{\xi}_1$$

$$p^{(n+1)}(0,\xi_2,\xi_3,\zeta) = 0 \,, \quad (n \geq 1) \,. \tag{6.96}$$

In order to show that there exists an E-function in the form (6.94) we reduce the recursive system (6.96) to a case already treated. By setting

$$q^{(n)} \equiv p^{(n)} \exp\left[-\tfrac{1}{2} \int_0^{\xi_1} (\tilde{A}^* + \tilde{B}^* \zeta)\, d\xi_1^1 \right] ,$$

the recursive system reduces to the form [Co 6]

$$q_1^{(n+1)} = \frac{1}{2n+1} \{ q_{33}^{(n)} + q_{22}^{(n)} - 2q_{23}^{(n)} - 4q_{13}^{(n)} + R^* q_3^{(n)}$$

$$+ S^* q_2^{(n)} + T^* q_1^{(n)} + W^* q^{(n)} \}$$

(6.97)
$$q^{(1)}(\xi_1, \xi_2, \xi_3, \zeta) = 1 ,$$

$$q^{(n+1)}(0, \xi_2, \xi_3, \zeta) = 0 , \quad (n \geq 1) ,$$

where

$$R^* = \tilde{A}^* - 2(\tilde{A}^* + \tilde{B}^* \zeta) - \frac{\partial}{\partial \xi_2} \int_0^{\xi_1} (\tilde{A}^* + \tilde{B}^* \zeta)\, d\xi_1^1$$

$$+ \frac{\partial}{\partial \xi_3} \int_0^{\xi_1} (\tilde{A}^* + \tilde{B}^* \zeta)\, d\xi_1^1$$

$$S^* = \tilde{A}^* + 2\tilde{B}^* \zeta - \frac{\partial}{\partial \xi_3} \int_0^{\xi_1} (\tilde{A}^* + \tilde{B}^* \zeta)\, d\xi_1^1$$

$$+ \frac{\partial}{\partial \xi_2} \int_0^{\xi_1} (\tilde{A}^* + \tilde{B}^* \zeta)\, d\xi_1^1 ,$$

$$T^* = 2\tilde{B}^*\zeta - 2\frac{\partial}{\partial\xi_3}\int_0^{\xi_1}(\tilde{A}^* + \tilde{B}^*\zeta)\,d\xi_1^1 ,$$

$$W^* = \tilde{D}^* + \tilde{B}^*\zeta(\tilde{A}^* + \tilde{B}^*\zeta) + \tfrac{1}{2}(\tilde{A}^* + 2\tilde{B}^*\zeta)$$

$$\cdot \frac{\partial}{\partial\xi_2}\int_0^{\xi_1}(\tilde{A}^* + \tilde{B}^*\zeta)\,d\xi_1^1 + \tfrac{1}{2}\tilde{A}^*\frac{\partial}{\partial\xi_3}\int_0^{\xi_1}(\tilde{A}^* + \tilde{B}^*\zeta)\,d\xi_1^1$$

$$- 2\frac{\partial}{\partial\xi_3}(\tilde{A}^* + \tilde{B}^*\zeta) - (\tilde{A}^* + \tilde{B}^*\zeta)\frac{\partial}{\partial\xi_3}\int_0^{\xi_1}(\tilde{A}^* + \tilde{B}^*\zeta)\,d\xi_1^1$$

$$- \frac{\partial^2}{\partial\xi_2\partial\xi_3}\int_0^{\xi_1}(\tilde{A}^* + \tilde{B}^*\zeta)\,d\xi_1^1 - \tfrac{1}{2}\left(\frac{\partial}{\partial\xi_2}\int_0^{\xi_1}(\tilde{A}^* + \tilde{B}^*\zeta)\,d\xi_1^1\right)$$

$$\cdot \left(\frac{\partial}{\partial\xi_3}\int_0^{\xi_1}(\tilde{A}^* + \tilde{B}^*\zeta)\,d\xi_1^1\right) + \tfrac{1}{2}\frac{\partial^2}{\partial\xi_2^2}\int_0^{\xi_1}(\tilde{A}^* + \tilde{B}^*\zeta)\,d\xi_1^1$$

$$+ \tfrac{1}{4}\left(\frac{\partial}{\partial\xi_2}\int_0^{\xi_1}(\tilde{A}^* + \tilde{B}^*\zeta)\,d\xi_1^1\right)^2 + \tfrac{1}{2}\frac{\partial^2}{\partial\xi_3^2}\int_0^{\xi_1}(\tilde{A}^* + \tilde{B}^*\zeta)\,d\xi_1^1$$

$$+ \tfrac{1}{4}\left(\frac{\partial}{\partial\xi_3}\int_0^{\xi_1}(\tilde{A}^* + \tilde{B}^*\zeta d\xi_1^1\right) .$$

(6.98)

Since the systems of the form (6.97) have already been treated in this chapter we proceed to consider the problem of whether the mapping $\underset{\sim}{C}_3\{f\}$ can be used to construct an onto map from the class of holomorphic functions $f(\mu,\zeta)$ to the strong solutions of (6.86). To this end, Colton [Co6], decided to consider solutions written in the form

$$U(X,Z,Z^*) = U(0,0,0)\ U_0(X,Z,Z^*) + \operatorname{Re} \underset{\sim}{C}_3\{f\} , \tag{6.99}$$

where the function U_0 is the unique solution of (6.87) which satisfies the Goursat data $U_0(X,0,Z^*) = 1$, $U_0(X,Z,0) = 1$. It is easy to show [Co6] using Hörmander's generalized Cauchy-Kowelevski theorem [Ho1] pp. 116-119 that

$$\operatorname{Re} U_0(X,Z,Z^*) = U_0(X,Z,Z^*) .$$

In the special case where $D = 0$, $U_0(X,Z,Z^*) \equiv 1$; when $D \neq 0$, U_0 may be found by the iterative scheme

$$U_0(X,Z,Z^*) = 1 + \lim_{n\to\infty} W_n(X,Z,Z^*) ,$$

$$W_0(X,Z,Z^*) \equiv 0 ,$$

$$W_{n+1} = \int_0^Z dZ^1 \int_0^{Z^*} dZ^{*1} \left(\frac{\partial^2 W_n}{\partial X^2} + A \frac{\partial W_n}{\partial X} + B \frac{\partial W_n}{\partial Z} \right.$$

$$\left. + C \frac{\partial W_n}{\partial Z^*} + DW_n - D \right) , \quad (n \geq 0) . \tag{6.100}$$

Since we have made the assumption that the coefficients of the differential equation are entire, the sequence $W_n(X,Z,Z^*)$

converges uniformly to $U_0(X,Z,Z^*)$ in any compact polydisk of the X,Z,Z^*-space, i.e. $U_0(X,Z,Z^*)$ is an entire function of its arguments.

Colton [Co6] shows that if $U(X,Z,Z^*)$ is represented in the form (6.99) one can relate $f(\mu,\zeta)$ directly to the Goursat data $U(X,0,Z^*)$. Indeed,

$$f(\mu,\zeta) = \frac{3}{2\pi} \int_{\gamma^1} g(\mu[1-t^2],\zeta) \frac{(1-t^2)}{t^4} \, dt ,$$

with

$$g(\mu,\zeta) = \sum_{n=0}^{\infty} \sum_{m=0}^{n+1} a_{nm} \mu^n \zeta^m . \tag{6.101}$$

(Note that the summation does not include negative powers of ζ .) Here the coefficients a_{nm} are given by the formulae

$a_{n-1,0} = \gamma_{n0}$, $(n \geq 1)$

$$a_{n+m-1,m} = \frac{2\Gamma(n+1)\Gamma(m+1)}{\Gamma(n+m+1)} \gamma_{nm} - \sum_{k=0}^{n-1} \frac{\Gamma(n+1)\delta_{k,n}\gamma_{n-k,0}}{\Gamma(n+m-1)\Gamma(k+1)} \tag{6.102}$$

$(n \geq 0 , m > 0)$, where the $\delta_{k,m}$ are defined by

(6.103) $$\delta_{k,m} = \left(\frac{\partial^{k+m}}{\partial X^k \partial Z^{*m}} \exp\left[\int_0^{-Z^*} \bar{C}(X,Z^1,0)\,dZ^1 \right] \right)_{X = Z^* = 0} ,$$

and the γ_{nm} are the Taylor coefficients of $U(X,0,Z^*) - U(0,0,0)$.

We show this as follows: let $\bar{E}(X,Z,Z^*\zeta,t)$ be the generating function corresponding to the partial differential equation †

$$V_{XX} - V_{ZZ^*} + \bar{A}(X,Z,Z^*)V_X + \bar{B}(X,Z,Z^*)V_Z + \bar{D}(X,Z,Z^*)V = 0 .$$

Then for real $x,y,z,$ one has that

$$\mathrm{Re}\,\tilde{\underset{\sim}{C}}_3\{f\} = \frac{1}{4\pi i} \int_{|\zeta|=1} \int_{\gamma} \exp\left[\int_0^{Z} C(X,Z^1,Z^*)\,dZ^1 \right]$$

$$\cdot\ E(X,Z,Z^*,\zeta,t)\, f(\mu[1-t^2],\zeta)\, \frac{dt}{\sqrt{1-t^2}} \frac{d\zeta}{\zeta}$$

$$+ \frac{1}{4\pi i} \int_{|\zeta|=1} \int_{\gamma} \exp\left[\int_0^{-Z^*} \bar{C}(X,Z^1,-Z)\,dZ^1 \right]$$

---

† If $B(X,Z,Z^*) \equiv \sum_{\ell,m,n=0}^{\infty} b_{\ell mn} X^\ell Z^m Z^{*n}$, then

$\bar{B}(X,Z,Z^*) \equiv \sum_{\ell,m,n=0}^{\infty} \overline{b_{\ell mn}}\, X^\ell Z^m Z^{*n}$, etc.

$$(6.104) \qquad \cdot\ \bar{E}(X,-Z^*,-Z,\zeta,t)\,\bar{f}(\bar{\mu}[1-t^2],\zeta)\,\frac{dt}{\sqrt{1-t^2}}\,\frac{d\zeta}{\zeta}\,.$$

As before, we attempt to determine $f(\mu,\zeta)$ by the Goursat data $U(X,0,Z^*) - U(0,0,0)U_0(X,0,Z^*)$; i.e. from the integral equation

$$U(X,0,Z^*) - U(0,0,0) = \frac{1}{4\pi i}\int_{|\zeta|=1}\int_{\gamma} \mu_1 f(\mu_1[1-t^2],\zeta)\,\frac{t^2 dt}{\sqrt{1-t^2}}\,\frac{d\zeta}{\zeta}$$

$$+\frac{1}{4\pi i}\int_{|\zeta|=1}\int_{\gamma}\exp\left[\int_0^{-Z^*}\bar{C}(X,Z^1,0)\,dZ^1\right]\bar{E}(X,-Z^*,0,\zeta,t)$$

$$(6.105) \qquad \bar{f}(\mu_2[1-t^2],\zeta)\,\frac{dt}{\sqrt{1-t^2}}\,\frac{d\zeta}{\zeta}\,,\quad \mu_1 = X+\zeta^{-1}Z^*\,,\quad \mu_2 = X-\zeta Z^*\,.$$

If one writes $\bar{E}^*(\xi_1,\xi_2,\xi_3,\zeta,t) \equiv \bar{E}(X,Z,Z^*,\zeta,t)$, in its series expansion

$$\bar{E}^*(\xi_1,\xi_2,\xi_3,\zeta,t) = \sum_{n=1}^{\infty} t^{2n}\mu^n\bar{p}^{(n)}(\xi_1,\xi_2,\xi_3,\zeta)\,,$$

then each $\bar{p}^{(n)}$, $(n \geq 2)$ may be shown [Co6] to be an entire function of X, Z, Z^*, and ζ, and moreover, vanishes when

$\zeta = 0$. We produce an analogous type argument, in detail, for the fourth order case in Chapter VII. From this we obtain

$$U(X,0,Z^*) - U(0,0,0) = \frac{1}{4\pi i} \int\limits_{|\zeta|=1} \int\limits_{\gamma} \mu_1 f(\mu_1[1-t^2],\zeta) \frac{t^2 dt}{\sqrt{1-t^2}} \frac{d\zeta}{\zeta}$$

$$+ \frac{1}{4\pi i} \int\limits_{|\zeta|=1} \int\limits_{\gamma} \exp\left[\int_0^{-Z^*} \bar{C}(X,Z^1,0)\,dZ^1\right] \tilde{p}^{(1)}(X,-Z^*,0,\zeta)$$

$$\cdot \mu_2 \bar{f}(\mu_2[1-t^2],\zeta) \frac{t^2 dt}{\sqrt{1-t^2}} \frac{d\zeta}{\zeta}$$

$$= \frac{1}{4\pi i} \int\limits_{|\zeta|=1} \mu_1\, g(\mu_1,\zeta) \frac{d\zeta}{\zeta}$$

$$+ \frac{1}{4\pi i} \int\limits_{|\zeta|=1} \mu_2 \exp\left[\cdot \int_0^{-Z^*} \bar{C}(X,Z^1,0)\,dZ^1\right] \tilde{p}^{(1)}(X,,-Z^*,0,\zeta)$$

$$\cdot \bar{g}(\mu_2,\zeta) \frac{d\zeta}{\zeta} ,$$

with $\tilde{p}^{(1)}(X,Z,Z^*,\zeta) = \exp\left[\zeta \int_0^Z \bar{A}(X+2\zeta Z\tau, \tau, \frac{\zeta}{2}(2\zeta^{-1}Z^* - 2\zeta Z + 2\zeta\tau))\,d\tau\right]$

(6.106) $$\cdot \exp\left[\zeta^2 \int_0^Z \bar{B}(X+2\zeta Z\tau, \tau, \tfrac{\zeta}{2}(2\zeta^{-1}Z^* - 2\zeta Z + 2\zeta\tau))\,d\tau\right],$$

where we have used the reciprocal identity

$$g(\mu,\zeta) = \int_\gamma t^2 f(\mu[1-t^2],\zeta)\,\frac{dt}{\sqrt{1-t^2}} .$$

We note that all terms involving derivatives w.r.t. X and Z* of $\tilde{p}^{(1)}$ and w.r.t. Z* of $\mu_2\bar{g}(\mu_2,\zeta)$ vanish because the integrands are analytic functions of ζ. One obtains

(6.107)
$$2n!\,m!\,\gamma_{nm} = (n+m)!\,a_{n+m-1,m} + \sum_{k=0}^{n-1} \frac{\Gamma(n+1)}{\Gamma(k+1)}\,\delta_{km}\overline{a_{n-k-1,0}},$$

$$\text{where}\quad \delta_{km} = \left(\frac{\partial^{k+m}}{\partial X^k \partial Z^{*m}} \exp\left[\int_0^{-Z^*} \bar{C}(X,Z^1,0)\,dZ^1\right]\right)_{X=Z^*=0} .$$

In (6. .24) the finite series is omitted when $n=0$. Also, note that for $k \geq 1$ we have $\delta_{k,0} = 0$, so

$$2\gamma_{n0} = a_{n-1,0} + \overline{a_{n-1,0}} .$$

Since U(X,0,0) is real one can assume without loss of generality that the $a_{n,0}$ are real, i.e.

$$\gamma_{n0} = a_{n-1,0} .$$

This establishes the equations (6.102), (6.103).

From the above discussion, it is clear that if $U(X,Z,Z^*)$ has polynomial Goursat data, then the associated analytic function $f(\mu,\zeta)$ is an entire function of μ and ζ. Since (6.86) has analytic coefficients it is well-known (See for example Browder [Br 2], Lax [La 1], Malgrange [Ma 1].) that (8.1.1) has the Runge approximation property, i.e. if $\mathfrak{S}_1 \subset \mathfrak{S}_2$ then any solution of (8.1.1) defined in $\mathfrak{S}_1$ may be uniformly approximated on compact subsets of $\mathfrak{S}_1$ by a sequence of solutions in $\mathfrak{S}_2$. Now it can be easily shown [Co] that if $\mathfrak{S}$ is a bounded, simply connected domain in $\mathbb{R}^3$, that the functions

(6.108)
$$\begin{aligned} &u_0(x,y,z) \equiv U_0(X,Z,Z^*) \\ &u_{2n,m}(x,y,z) = \operatorname{Re}\, \underset{\sim}{C}_3\{\mu^n\zeta^m\}, (0 \le m \le n+1), n \ge 0 , \\ &u_{2n+1,m}(x,y,z) = \operatorname{Im}\, \underset{\sim}{C}_3\{\mu^n\zeta^m\}, (0 \le m \le n+1), n \ge 0 , \end{aligned}$$

form a complete family of solutions for approximating strong

Solutions of equation (6.86) in compact subsets of $\mathfrak{S}$. Hence the operator $\underset{\sim}{C}_3$ permits us to obtain a constructive method for solving the differential equation (6.86) with its associated boundary value problems, and therefore is suggestive of a numerical approach to these problems.

VII. Singularity Properties of Solutions to Partial Differential Equations

Most of the results that will follow will be based in some manner on Hadamard's multiplication of singularity theorem, namely: *Let* $f(z)$ *and* $g(z)$ *be defined by the series*

$$f(z) = \sum_{n=0}^{\infty} a_n z^n , \quad |z| < R ,$$

$$g(z) = \sum_{n=0}^{\infty} b_n z^n , \quad |z| < R' ;$$

and moreover, suppose that $f(z)$ *has singularities at* $\alpha_1, \alpha_2, \ldots, \alpha_n$, *and* $g(z)$ *has singularities at* $\beta_1, \beta_2, \ldots, \beta_m$. *Then the singularities of*

$$F(z) = \sum_{n=0}^{\infty} a_n b_n z_n .$$

are to be found at the points $\alpha_p \beta_q$ $p = 1, \ldots, n; q = 1, \ldots, m)$. The proof of this theorem may be found in [Di 1], [Gi 1] pg. 23, or Titchmarsh [Ti 1] .

<u>Remark</u>: Let us define the following sets in C^{n+1}:

$G_K \equiv \{(z,\zeta): F_K(z,\zeta) = 0\}$, $(K = 1,2)$, $z = (z_1, \ldots, z_n)$. Then

$$\mathfrak{S} = (\mathfrak{S}_1 \cap \mathfrak{S}_2)^*$$

is meant to be the projection onto the z-space of the intersection $\mathfrak{S}_1 \cap \mathfrak{S}_2$.

The next theorem, which we refer to as the <u>envelope method</u>, is basic to our theory [Gi 0],[Gi 1], pg. 23.

<u>Theorem (7.1)</u>: <u>Let</u> $F(x)$ <u>be defined, in a neighborhood of</u> z^0, <u>by the integral representation</u>

$$F(z) = \int_{\mathfrak{L}} K(z;\zeta)\, d\zeta ,$$

<u>where</u> $K(z,\zeta)$ <u>is a holomorphic function of</u> $(n+1)$ <u>complex</u> variables in $C^{n+1}/\mathfrak{S}$, and where $\mathfrak{S}$, defined by $\mathfrak{S} \equiv \{(z,\zeta): S(z,\zeta) = 0\}$, <u>is an analytic set</u>. <u>Then</u> $F(z)$ <u>is</u>

<u>regular for all points which can be reached along a path from</u> z^0 <u>that does not pass through the set,</u>

$$\Sigma = (\mathfrak{S} \cap \mathfrak{S}_1)^* ,$$

<u>where</u> $\mathfrak{S}_1 \equiv \{(z,\zeta): \frac{\partial S}{\partial \zeta}(z;\zeta) = 0\}$.

<u>Proof</u>. Since $F(z)$ is holomorphic in a neighborhood of z^0, we consider the possibility of analytically continuing $F(z)$ to a larger region. This is done by attempting the continuation along contours γ initiating at z^0 . This may be done providing γ does not cross a singularity of the integrand when ζ is simultaneously restricted to the integration path. All such points z that may be reached in this way we refer to as the <u>initial domain of definition</u>. We note, however, that as $F(z)$ is continued along γ the singularities of $K(z;\zeta)$ move in the ζ-plane. It follows that the initial domain of definition consists of just those points which may be reached without a singularity in the ζ-plane crossing $\mathfrak{L}$. This definition, however, may be extended by deforming the path of integrating. This extended region we call the <u>domain of association</u> of $F(z)$.

Suppose a singularity is threatening to cross over the path of integration, and that it is no longer possible to avoid this singularity by deforming the contour. Then we have reached a boundary point of the domain of association. Whenever the singular point $\zeta = \alpha(z)$ is a simple zero of $S(z;\zeta) = 0$, we clearly may avoid such a point. This means the only possible boundary points must correspond to $S(z;\zeta) = 0$ and $\frac{\partial S}{\partial \zeta}(z;\zeta) = 0$ holding simultaneously.

<u>Remark</u>: The only situation for which a boundary point occurs is when a multiple zero occurs as z varies, and pinches $\mathfrak{L}$ between two of the zeros.

<u>Remark</u>: If $\mathfrak{L}$ is not a closed contour but has terminal points ζ_1 and ζ_2 then the additional set of points

$$\mathfrak{S}^+ \equiv \{z: \quad S(z;\zeta) = 0 \ , \ \zeta = \zeta_1 \text{ or } \zeta_2\}$$

may be singular points [Gi 1] pg. 24.

We next obtain a several variable analogue of the envelope method. Without loss of generality, we may take the dimensionality of the integration space to be two.

Theorem (7.2): Let $F(z)$ be defined, in a neighborhood of z^0, by the integral representation

$$F(z) = \int_{\mathfrak{L}_1} d\zeta_1 \int_{\mathfrak{L}_2} d\zeta_2 K(z;\zeta_1,\zeta_2) ,$$

where $K(z;\zeta)$ is a holomorphic function of $(n+2)$ variables in $C^{n+2}/\mathfrak{S}$. Then $F(z)$ is regular for all points which can be reached along a path from z^0 that does not pass through the set

$$\Sigma = (\mathfrak{S} \cap \mathfrak{S}_1 \cap \mathfrak{S}_2)^* , \quad \mathfrak{S}_k \equiv \{(z;\zeta) | \frac{\partial S}{\partial \zeta_k} = 0\} .$$

Proof: The proof of this theorem parallels the last one. Eventually one must consider the situation where an attempted continuation is no longer possible because one can not avoid a singularity of the integrand crossing the domain of integration. In this respect, we notice if $\nabla_\zeta S(z,\zeta) \neq 0$ at a point $\zeta = a$, then the first differential of S does not vanish (with respect to ζ), and there exist torii about $\zeta = a$ into which it is always possible to deform $\mathcal{L}_1 \times \mathcal{L}_2$ such that $S(z;\zeta) \neq 0$ for $\zeta \in \mathcal{L}_1 \times \mathcal{L}_2$. This follows because in a neighborhood of $\zeta = a$ S may be approximated by its first differential if $\nabla S \neq 0$.

We note one other result in this direction, namely:

Theorem (7.3): *If $F(z)$ is now defined, in a neighborhood of z^0, by*

$$F(z) = \int_{\mathcal{L}_1} d\zeta_1 \int_{\mathcal{L}_2} d\zeta_2 K_1(z;\zeta) K_2(z;\zeta)$$

and the functions $K_\nu(z;\zeta)$ are holomorphic in $C^{n+2}/\mathfrak{S}_{\nu,0}$ with $\mathfrak{S}_{\nu,0} \equiv \{(z,\zeta) \,|\, S_\nu(z;\zeta) = 0\}$. Then $F(z)$ is regular at all points which can be reached along a path from z^0 that does not pass through the set

$$\mathfrak{S} = \Big[\bigcup_{\nu=1,2} G^*_\nu\Big] \cup \mathfrak{H}^*,$$

where $\mathfrak{S}^*_\nu = (\mathfrak{S}_{\nu,0} \cap \mathfrak{S}_{\nu,1} \cap \mathfrak{S}_{\nu,2})^*$, *with*

$$\mathfrak{S}_{\nu,k} \equiv \{(z,\zeta) \mid \frac{\partial S}{\partial \zeta_k}(z;\zeta) = 0\} ,$$

and $$\mathfrak{H}^* \equiv [\mathfrak{S}_{1,0} \cap \mathfrak{S}_{2,0} \cap \mathfrak{H}]^* ,$$

with $$\mathfrak{H} \equiv \{(z,\zeta) \mid \frac{\partial S_1}{\partial \zeta_1}\frac{\partial S_2}{\partial \zeta_2} - \frac{\partial S_1}{\partial \zeta_2}\frac{\partial S_2}{\partial \zeta_1} = 0\} .$$

Proof: We notice first that the sets $\mathfrak{S}^*_\nu$ are the possible singular points which occur by applying the envelope method to the singularities of the $K_\nu(z;\zeta)$ separately. However, it is possible for a singularity of K_1 and a singularity of K_2 to pinch the integration domain between them, thus making it impossible to deform $\mathfrak{L}_1 \times \mathfrak{L}_2$ any further. Such singular points arise when one considers the set

$$\mathfrak{H}^* = [\mathfrak{S}_{1,0} \cap \mathfrak{S}_{2,0} \cap \quad]^*;$$

in other words, after one has obtained a confluence of singularities by K_1 and K_2, an envelope method is applied.

The above singularity theorems may be used to obtain necessary and sufficient criteria for solutions of differential equations to have singularities. The basic idea is to use an integral representation for the solution, which maps holomorphic functions (of one or several complex variables) onto the class of regular solutions. As an example we consider the case of generalized axially symmetric potential theory (GASPT)[†]. These are the solutions of the equation

$$\Delta_n u + \frac{\partial^2 u}{\partial y^2} + \frac{2\nu}{y}\frac{\partial u}{\partial y} = 0\,, \quad \Delta_n \equiv \frac{\partial^2}{\partial x_1^2} + \dots + \frac{\partial^2}{\partial x_n^2}\,. \tag{7.1}$$

We consider for simplicity, the special case where $n = 1$. It is well known [Gi 1] pg. 168, that the solutions regular about the origin may be represented in the form

$$u(\underset{\sim}{x}) = (\underset{\sim}{A}_\mu f)(\underset{\sim}{x}) = \alpha_\mu \int_{\mathfrak{L}} f(\sigma)\,(\zeta - \zeta^{-1})^{2\mu - 1}\,\frac{d\zeta}{\zeta}\,, \tag{7.2}$$

[†] For a detailed account of GASPT the reader is directed to the works of Weinstein [We 3-6], who developed the theory. Also see Huber [Hu 1,2], Weinacht [We 1,2], Henrici [He 2-4], Gilbert [Gi 1], Colton [Co 1,2], Colton-Gilbert [CG 3], and Parter [Pa 1].

where $\mathfrak{L}$ is the upper arc on the unit circle from 1 to -1, $\sigma = x_1 + \frac{i}{2} x_2(\zeta + \zeta^{-1})$, $\underset{\sim}{x} = (x_1, x_2)$, and α_μ is a constant which we choose for purposes of normalization to be $4\Gamma(2\mu)(4i)^{-2\mu}/\Gamma(\mu)^2$. If we introduce polar coordinates then it is easy to see how the operator $\underset{\sim}{A}_\mu$ maps holomorphic functions onto solutions. For example, if we take $f(\sigma)$ to be defined by the power series

$$f(\sigma) = \sum_{n=0}^{\infty} a_n \sigma^n \, , \tag{7.3}$$

then the corresponding solution $u(r,\theta) = \underset{\sim}{A}_\mu f$ has the representation [Gi 1] Chapter V,

$$u(r,\theta) = (2\mu - 1) \sum_{n=0}^{\infty} B(2\mu - 1, n+1) a_n r^n C_n^\mu (\cos\theta) \, , \tag{7.4}$$

where $C_n^\mu(\xi)$ is an ultraspherical harmonic. There are many possibilities for representing the inverse operator $(\underset{\sim}{A}_\mu^{-1} u)(\sigma) = f(\sigma)$. One such possibility makes use of the

Poisson kernel; we have from this [Gi 1]pg. 174, the formula

$$(7.5) \qquad (\underset{\sim}{A}_{\mu}^{-1} u)(\sigma) \equiv \int_{C} w(r,\xi)K(\sigma/r,\xi)\,d\xi ,$$

where C is a smooth curve joining -1 to 1, $w(r\cos\theta) \equiv u(r,\theta)$, and the kernel K is defined as

$$K(\eta,\xi) = \beta_{\mu} \frac{(1-\xi^2)^{\mu-\frac{1}{2}}(1-\eta^2)}{[1-2\xi\eta+\eta^2]^{\mu+1}} , \quad \beta_{\mu} = \frac{\mu}{\pi}\,\Gamma(\mu)^2\, 2^{2\mu-1} .$$

Theorem (7.4): If the only finite singularity of $f(\sigma)$ is at $\sigma=\alpha$, then the only possible singularities of $u(\underset{\sim}{x})$ on its first Riemann sheet lie at α and $\bar{\alpha}$.

Proof: We use the envelope method. Since $\mathfrak{S} \equiv \{(\underset{\sim}{x};\zeta): (x_1-\alpha)\zeta + \frac{1}{2}ix_2(\zeta^2+1)$, one may easily compute that

$$G^* \equiv (\mathfrak{S} \cap \mathfrak{S}_1)^* \equiv \{(x-\alpha)^2 + y^2 \equiv (z-\alpha)(z-\bar{\alpha}) = 0\},$$

By using our representation for $\underset{\sim}{A}_{\mu}^{-1}$ we may also obtain <u>necessary</u> conditions for $u(\underset{\sim}{x})$ to be singular.

<u>Lemma</u>: <u>Let</u> $u(\underset{\sim}{x}) = (\underset{\sim}{A}_{\mu} f)(\underset{\sim}{x})$ <u>be defined in a neighborhood of the origin</u>. <u>Let</u> $w(r,\xi)$ <u>be regular in</u> $\mathbb{C}^2/\mathfrak{S}$, <u>where</u>

$$\mathfrak{S} \equiv \{(r,\xi):\quad \xi = \psi(r)\} \,.$$

<u>Then</u> $f(\sigma)$ <u>is regular for all points which may be connected to the origin by an arc not passing through</u>

$$\mathfrak{S}^* = (\mathfrak{S}_0 \cap \mathfrak{S}_1)^* \,, \text{ with } \mathfrak{S}_0 \equiv \{r,\xi,\sigma):\ r^2 - 2\sigma r\psi(r) + \sigma^2 = 0\} \,,$$

$$\mathfrak{S}_1 \equiv \{(r,\xi,\sigma):\ r - \sigma(r\psi)' = 0\} \,.$$

<u>Proof</u>: This follows as a consequence of the reasoning in the proof of Theorem (7.3).

<u>Theorem (7.5)</u>: <u>Under the hypotheses of the above lemma, if in addition</u>, $\xi = \frac{1}{2}(r/\alpha + \alpha/r)$, i.e. $z = \alpha, \bar{\alpha}$ <u>are the only singularities of</u> $u(\underset{\sim}{x})$, <u>then</u> $f(\sigma)$ <u>is singular only at</u> $\sigma = \alpha$.

Proof:

$$\mathfrak{S}^* = (\mathfrak{S}_0 \cap \mathfrak{S}_1)^* = (\{r^2 - \sigma(r^2/\alpha + \alpha) + \sigma^2 = 0\} \cap \{2r(1 - \sigma/\alpha) = 0\})^*$$

$= \{\sigma\colon \sigma = 0, \alpha\}$. However, $f(\sigma)$ is analytic at $\sigma = 0$; hence $\sigma = \alpha$ is the only possible singularity.

Remark: Since $(z - \alpha)(\bar{z} - \alpha) = z\bar{z} - \alpha(z + \bar{z}) + \alpha^2 = x_1^2 + x_2^2 - 2\alpha x_1 + \alpha^2$

$= r^2 - 2\alpha r\xi + \alpha^2$, we note that if $u(\underset{\sim}{x})$ is singular $z = \alpha, \bar{\alpha}$, $f(\sigma)$ may have a singularity at $\sigma = \alpha$. However, since the singularity manifold may also be written as $(z - \bar{\alpha})(\bar{z} - \bar{\alpha}) = r^2 - 2\bar{\alpha}r\xi + \bar{\alpha}^2 = 0$, which implies $f(\sigma)$ may have a singularity at $\sigma = \bar{\alpha}$. Putting these ideas together we have the

Theorem (7.6): The necessary and sufficient conditions for $u(\underset{\sim}{x}) = (\underset{\sim}{A}_\mu f)(\underset{\sim}{x})$ to be singular at $z = \alpha$, on its first Riemann sheet, is for $f(\sigma)$ to be singular at $\sigma = \alpha$ or $\bar{\alpha}$.

Problem: There are numerous other types of singularity theorems for singular differential operators. The reader is referred to Chapter IV of [Gi 1] for a selection of these.

We turn next to considering singularity theorems for harmonic functions of three variables. To this end we recall the Bergmann-Whittaker operator, which maps holomorphic functions of two complex variables onto harmonic functions, i.e.

(7.6) $$H(\underset{\sim}{x}) = (\underset{\sim}{B}_3 f)(\underset{\sim}{x}) \quad ; \quad \underset{\sim}{B}_3 f = \frac{1}{2\pi i} \int_{\mathcal{L}} f(\mu, \zeta) \frac{d\zeta}{\zeta} ,$$

where $\mu = X + Z\zeta + Z^* \zeta^{-1}$, $\mathcal{L}$ is a rectifiable closed curve with index 1 with respect to the origin, and $X = x_3$, $Z = x_1 + ix_2$, and $Z^* = -x_1 + ix_2$.

If $f(\mu, \eta)$ is defined formally by the power series

(7.7) $$f(\mu,\zeta) = \sum_{n=0}^{\infty} \sum_{m=-n}^{+n} a_{nm}\mu^n \zeta^m ,$$

then the operator $\underset{\sim}{B}_3$ associates with this series the formal series of harmonic polynomials

(7.8) $$H(\underset{\sim}{x}) = \sum_{n=0}^{\infty} \sum_{m=-n}^{n} a_{nm} h_{nm}(\underset{\sim}{x}) ,$$

where the $h_{nm}(\underset{\sim}{x})$ may be written in terms of the spherical polar coordinates as [Gi 1] pg. 49.

(7.9) $$h_{nm}(\underset{\sim}{x}) = \frac{n!}{(n+m)!} i^m r^n P_n^m (\cos\theta) e^{im\varphi} .$$

It is well-known that the family of functions $\underset{\sim}{B}_3(\mu^n\zeta^m)$ are complete with respect to uniform approximation in compact, simply-connected, regions in $\mathbb{E}^3$.

The following sufficient conditions for $H(\underset{\sim}{x})$ to be singular are a consequence of the envelope method [Gi1] pg 55, and [Gi0].

Theorem (7.7): Let the defining function for the set of singularities of $f(\mu,\zeta)\zeta^{-1}$ be a global defining function in $\mathbb{C}^2$, and let this defining function be given by $h(\mu,\zeta) \equiv S(\underset{\sim}{x};\zeta) = 0$. Then $H(\underset{\sim}{x}) = (\underset{\sim}{B}_3 f)(\underset{\sim}{x})$ is regular for all points $\underset{\sim}{x}$ which may be reached along a contour γ initiating at $\underset{\sim}{0}$ providing γ does not meet $\mathfrak{S} \equiv (\{\underset{\sim}{x}: s(\underset{\sim}{x};\zeta) = 0\} \cap \{\underset{\sim}{x}: \frac{\partial s}{\partial \zeta} = 0\})^*$.

As a non-trivial example we consider the case where $f(\mu,\zeta) = \zeta F(\mu^{-1}(\zeta - 1/\zeta))$, and $F(z)$ is singular only at $z = \beta$.

However, in this case the initial point must be contained in a neighborhood of infinity, i.e. this f generates a harmonic function regular at infinity. The singularity manifold of the integrand can be defined as

$$S(\underset{\sim}{x};\zeta) \equiv \zeta[\beta(x_1 + ix_2) - 2] + 2\beta x_3 - \zeta^{-1}[\beta(x_1 - ix_2) - 2] = 0 \ .$$

Eliminating ζ between $S = 0$, and

$$\frac{\partial S}{\partial \zeta}(\underset{\sim}{x};\zeta) \equiv [\beta(x_1 + ix_2) - 2] + \zeta^{-2}[\beta(x_1 - ix_2) - 2] = 0 \ ,$$

yields as the set of possible singular points of $H(\underset{\sim}{x})$

$$\mathfrak{S} \equiv \{\underset{\sim}{x}: \ (x_1 - \frac{2}{\beta})^2 + x_2^2 + x_3^2 = 0\} \subset \mathbb{C}^3 \ .$$

It is possible to show that these are actual singular points if we introduce a representation for the inverse operator

B_3^{-1} . In [Gi 1] pg. 50 the following representation is given for harmonic functions regular at infinity,

10) $$(B_3^{-1} U)(\hat{\mu},\zeta) \equiv \int_{-1}^{1} d\xi \int_{|\eta|=1} \frac{d\eta}{\eta} \frac{r(\hat{\mu}+\mu)}{(\hat{\mu}-\mu)^2} U(r,\xi,\eta) ,$$

where $U(r,\cos\theta,e^{i\varphi})$ is $H(x)$ expressed in terms of the polar spherical coordinates, r,θ, and φ .

---

† A representation for B_3^{-1} may be obtained for harmonic functions regular about the origin using reflection. One obtains [GL 1], pp. 24-25,

$$f(\nu,\zeta) = B_3^{-1} H)(\nu,\zeta)$$

$$\equiv \frac{-a}{2\pi^2} \int_0^1 d\alpha \int_0^1 \frac{d\beta\sqrt{\beta}}{\sqrt{1-\beta}} \left\{ \int_{0, |x'|=R}^{2\pi} d\varphi' \int_0^{2\pi} \sin\theta' \, H(x' \frac{a^2}{R^2}) \right.$$

$$\left. \cdot \left(\frac{12 s\alpha\beta(1-\alpha)\hat{\mu}+a}{[r s\alpha\beta(1-\alpha)\hat{\mu}-a]^3} \right) d\theta' \right\}$$

where $\hat{\mu} = X' + \zeta(1-\frac{1}{\alpha})Z' + \zeta^{-1}(1-\frac{1}{\alpha})^{-1} Z^{*\prime}$; R is chosen so that $\{x : |x| \le \frac{1}{R}\} \subset\subset D$.

Theorem (7.8): Let $U(r, \cos\theta, e^{i\varphi})$ be a harmonic function regular at infinity and let $\rho = \chi(\xi, \eta)$ be a global defining function for the singularities of $U(\rho, \xi, \eta)$ in C^3. Then the function of two complex variables defined by (7.10) as

$$f(\nu, \zeta) \equiv (\underset{\sim}{B}_3^{-1} U)(\nu, \zeta) ,$$

is regular for all points (ν, ζ) which may be reached from $(\infty, 1)$ by an arc γ which does not meet the set

(7.11) $$\mathfrak{S} \equiv (\mathfrak{S}_0 \cap \mathfrak{S}_1 \cap \mathfrak{S}_2)^* ,$$

where

$$\mathfrak{S} \equiv \{(\nu, \zeta):\ s(\nu, \zeta; \xi, \eta) \equiv \chi(\xi, \eta)[\xi + \tfrac{i}{2}(1 - \xi^2)^{\frac{1}{2}}(\tfrac{\zeta}{\eta} + \tfrac{\eta}{\zeta})] - \nu = 0$$

$$\mathfrak{S}_1 \equiv \{(\nu, \zeta):\ \frac{\partial S}{\partial \xi} = 0\} , \quad \mathfrak{S}_2 \equiv \{(\nu, \zeta):\ \frac{\partial S}{\partial \eta} = 0\} ,$$

or the set of coordinate planes $\nu = \nu_n$ $(n = 1, 2, \ldots)$ with $\nu_{k_j} = \pm\chi(\pm 1, \eta_k)$ with η_k a root of $\frac{\partial \Phi}{\partial \eta}(\pm 1, \eta) = 0$.

Proof: In spherical polar coordinates we may write μ as

7.12) $$\mu = r[\cos\theta + \frac{i}{2}\sin\theta\,(\frac{e^{i\varphi}}{\zeta} + \zeta e^{-i\varphi})] ;$$

consequently only when

$$\nu - \mu \equiv \nu - \chi(\xi,\eta)[\xi + \frac{i}{2}(1-\xi^2)^{\frac{1}{2}}(\frac{\zeta}{\eta} + \frac{\eta}{\zeta})] = 0$$

can singularities occur(by collescence of the singularities of the kernel and those of U). These are the singularities which give rise to $\mathfrak{S}$ by means of the envelope method. The singular coordinate planes occur because of the end points ±1 in the ξ-integration. Since these points cannot be varied, they may correspond to singularities, i.e. we must consider as possible singularities those points ν, ζ which satisfy simultaneously

$$S(\nu,\zeta;\pm 1,\eta) \equiv \pm\chi(\pm 1,\eta) - \nu = 0 \text{ , and } \frac{\partial S}{\partial\eta} \equiv \pm\chi_\eta(\pm 1,\eta) = 0 .$$

It was shown in [Gi 1] pg. 61-62, that the particular elimination used above is not necessary. For instance, if the singularities of $H(\underset{\sim}{x})$ are given instead in Cartesian coordinates, i.e. as say, $x_3 = P(x_1, x_2)$ or as $F(\underset{\sim}{x}) = 0$, then one may write

$$S(\nu, \zeta; x_1, x_2) \equiv \frac{\zeta}{2}(x_1 + ix_2) + P(x_1, x_2) - \frac{1}{2\zeta}(x_1 - ix_2) - \nu =$$

In which case one eliminates x_1 and x_2 between S, and

$$\frac{\partial S}{\partial x_1} \equiv \frac{1}{2}\left(\zeta - \frac{1}{\zeta}\right) + \frac{\partial P}{\partial x_1} = 0 ,$$

$$\frac{\partial X}{\partial x_2} \equiv \frac{i}{2}\left(\zeta + \frac{1}{\zeta}\right) + \frac{\partial P}{\partial x_2} = 0 .$$

As an example of this idea let us consider as the singularities of $H(\underset{\sim}{x})$ the set

(7.13) $$\mathfrak{S} \equiv \{x\colon (x_1 - \frac{2}{\beta})^2 + x_2^2 + x_3^2 = 0\} \subset \mathbb{C}^3 .$$

The restriction of this set to $\mathbb{E}^3$ falls into two possible cases (1) If $\beta = \text{real}$ then $\mathfrak{S}$ degenerates into the point

$(2/\beta,0,0) \in \mathbb{E}^3$; (2) if β is complex then $\mathfrak{S} \cap E^3$ becomes a circle

$$x_1 = \frac{2}{|\beta|} \operatorname{Re} \beta, \quad x_2^2 + x_3^2 = \frac{4}{|\beta|^4} (\operatorname{Im} \beta)^2 . \tag{7.14}$$

The circle (7.14) (or degenerate circle as the case may be) is just a subset of (7.13). It is interesting nevertheless, that this set generates all the singularities of that $f(\mu,\zeta)$ which was seen to provide the set (7.13) as possible singularities for $H(\underset{\sim}{x})$. We note that in this case we must have

$$S(\nu,\zeta;x_3) \equiv \frac{2\operatorname{Re}\beta}{|\beta|^2}[\zeta + \frac{1}{\zeta}] \pm \frac{i}{2}\left(\frac{4}{|\beta|^4}(\operatorname{Im}\beta)^2 - x_3^2\right)^{\frac{1}{2}}(\zeta + \frac{1}{\zeta})$$

$$+ x_3 - \nu = 0 .$$

Eliminating x_3 between this and

$$\frac{\partial S}{\partial x_3} \equiv \pm \frac{ix_3}{2}[\zeta + \frac{1}{\zeta}]\left(\frac{4}{|\beta|^4}(\operatorname{Im}\beta)^2 - x_3^2\right)^{-\frac{1}{2}} + 1 = 0 ,$$

yields

$$\pm \{ \mp i(\mathrm{Im}\ \beta)[\zeta + \frac{1}{\zeta}]^2 + (\mathrm{Re}\ \beta)[\zeta - \frac{1}{\zeta}]^2 + 4i\ \mathrm{Im}\ \beta\} = \nu|\beta|^2[\zeta - \frac{1}{\zeta}] .$$

A proper choice of signs provides $(\zeta - \frac{1}{\zeta}) = \beta\nu$ which was the given singularity set of $g(\nu,\zeta)\zeta^{-1}$.

The above example suggests the conjecture, that the possible singularities of $H(\underset{\sim}{x})$, as given by Theorem (7.7) are actual singularities.

We now turn to discussing necessary and sufficient conditions for a harmonic function (in three variables) to be singular. Our results may be extended to the case of solutions to (6.12), since the envelope method is applied in essentially the same way to determine the singularities. Hence, without loss of generaility, we restrict our discussion to the case of harmonic functions. If $H(\underset{\sim}{x})$ is a harmonic function with the Cauchy data,

(7.15) $$H(x_1,x_2,0) = \hat{\varphi}_1(x_1,x_2) \equiv \varphi_1(Z,Z^*) ,$$

$$H_{x_3}(x_1,x_2,0) = \hat{\varphi}_2(x_1,x_2) \equiv \varphi_2(Z,Z^*) ,$$

then by the preceeding discussion in Chapter 6,

(7.16) $$H(x) = h_1(x) + \int_0^{x_3} h_2(x_1,x_2,s)\,ds,$$

where $h_k(\underset{\sim}{x})$ are harmonic functions satisfying the data

(7.17) $$h_k(x_1,x_2,0) = \Phi_k(x_1,x_2)\ ,$$
$$\frac{\partial h_k}{\partial x_3}(x_1,x_2,0) = 0\ , \quad (k=1,2)\ .$$

(7.18) $$h_k(x) = (\underset{\sim}{B}_3 g)(x) \equiv \frac{1}{2\pi i}\int_{\mathfrak{L}} g_k(\mu,\zeta)\frac{d\zeta}{\zeta}\ , \quad (k=1,2)\ ,$$

where $\mathfrak{L}$ is a rectifiable curve homotopic to $|\zeta| = 1$.
Consequently, the prescribed harmonic function is given by

(7.19) $$H(x) = (\underset{\sim}{B}_3 g)(\underset{\sim}{x})\ , \text{ with } g(\mu,\zeta) = g_1(\mu,\zeta) + \int_0^{\mu} g_2(\mu,\zeta)\,d\mu\ .$$

We recall it was shown, that if the singularities of $\zeta^{-1}g(\mu,\zeta)$ are given by a global defining function,

$$S(\underset{\sim}{x}, \zeta) \equiv \sigma(\mu, \zeta) = 0 ,$$

then $H(x)$ is regular at each point $\underset{\sim}{x}$ which may be reached by a path γ which does not meet the set

$$\mathfrak{S}^* \equiv \{\underset{\sim}{x} \mid S(\underset{\sim}{x}, \zeta) = 0 , \text{ and } \frac{\partial S}{\partial \zeta}(\underset{\sim}{x}, \zeta) = 0 \quad \text{for} \quad \zeta \in \mathbb{C}^1\} .$$

In what follows, we shall show that the points of $\mathfrak{S}^*$ are not only the "possible" singularities of $H(x)$, but are indeed the actual singular points. We consider the special case where $\sigma = 0$ may be solved for μ ,i.e. we are not considering ramifications in the μ-variable. Then

$$S(\underset{\sim}{x}, \zeta) \equiv \mu - \Phi(\zeta) \equiv Z\zeta + X + Z^*\zeta^{-1} - \Phi(\zeta) = 0 , \tag{7.20}$$

and

$$S_{\zeta}(\underset{\sim}{x}, \zeta) \equiv Z - Z^*\zeta^{-2} - \Phi'(\zeta) = 0 ,$$

are the equations from which ζ must be eliminated. Recombining (7.20) one obtains for $X \equiv x_3 = 0$,

$$\begin{aligned} 2Z^* &= \zeta\Phi(\zeta) - \zeta^2\Phi'(\zeta) \equiv F(\zeta) , \\ F(\zeta) &= \sum_{n=1}^{\infty} p_n \zeta^n . \end{aligned} \tag{7.21}$$

The representation (7.21) may now be inverted locally, about $\zeta = 0$ thereby leading to

$$(7.22) \qquad \zeta = \Psi(Z^*) \equiv \sum_{n=1}^{\infty} \frac{2^n}{n!} Z^{*n} p_n \ , \ p_n = \left\{ \frac{d^{n-1}}{d\zeta^{n-1}} \left(\frac{\zeta}{F(\zeta)} \right)^n \right\}_{\zeta=0} .$$

We treat the global elimination of ζ as follows. Let D be the domain of holomorphy of the function $F(\zeta)$ defined above. Then $D \subset \mathbb{C}$, and we may cover D by $\bigcup_{\zeta_0 \in D} N(\zeta_0; \varepsilon_{\zeta_0})$, where the $N(\zeta_0; \varepsilon_{\zeta_0})$ are neighborhoods of ζ_0 in which $F(\zeta)$ is invertible. Using the Lindelöf theorem we may extract a countable subcover of D, namely $\bigcup_{i=1}^{\infty} \theta_i$, $\theta_i \equiv \{\zeta: |\zeta - \zeta_i| < \delta_i\}$. In each θ_i we have,

$$(7.23) \qquad 2(Z^* - Z_i^*) = F(\zeta) - 2Z_i^* \equiv Q_i(\zeta) \equiv \sum_{n=1}^{\infty} q_n^{(i)} (\zeta - \zeta_i)^n ,$$

where $2Z_i^* = F(\zeta_i)$. Therefore we may invert (7.23) locally and obtain

(7.24) $$\zeta = \psi_i(Z^*) \equiv \zeta_i + \sum_{n=1}^{\infty} \frac{2^n}{n!} p_n^{(i)} (Z^* - Z^*_i)^n \; ;$$

where the $p_n^{(i)}$ are now given by

(7.25) $$p_n^{(i)} = \left\{ \frac{d^{n-1}}{d\zeta^{n-1}} \left(\frac{\zeta - \zeta_i}{Q(\zeta)} \right)^n \right\}_{\zeta = \zeta_i}$$

For each θ_i we perform the elimination of ζ as indicated by our inversion formulae (7.24-7.25) and thereby obtain the following "component" of the restricted (to $X = x_3 = 0$) set of possible singularities,

(7.26) $$Z\,\psi_i(Z^*) + Z^*/\psi_i(Z^*) - \Phi \circ \psi_i(Z^*) = 0 \; .$$

We notice that in $\theta_i \cap \theta_j$, $\psi_i(Z^*) \equiv \psi_j(Z^*)$; hence, ψ_i and ψ_j may be thought of as direct analytic continuations of each other. If one considers <u>all</u> possible analytic continuations of ψ_i one obtains a global analytic function ψ , which is single-valued on a suitable Riemann surface over the Z^*-plane. The correct interpretation of the restriction (to $X = 0$) of the set of possible singularities is obtained by replacing ψ_i in (7.26) by the globally defined function ψ, i.e. it is the multi-layered set over (Z,Z^*) which is given by

7) $$Z\psi(Z^*) + Z^*/\psi_i(Z^*) - \Phi\circ\psi(Z^*) = 0 \,.$$

We next recall the inversion formula given by Gilbert and Kukral for the operator $\underset{\sim}{B}_3$

8) $$g(\mu,\zeta) = (\underset{\sim}{B}_3^{-1}H)(\mu,\zeta) \equiv \int_0^1 \left\{\frac{\partial}{\partial\mu}\left[\mu\varphi_1\left(\frac{\mu t}{\zeta}, \mu\zeta[1-t]\right)\right] + \mu\varphi_2\left(\frac{\mu t}{\zeta}, \mu\zeta[1-t]\right)\right\}dt.$$

To simplify our discussion of the singularities which arise from using this operator, we first consider the special case with $g(\mu,\zeta)$ given by

9) $$g(\mu,\zeta) = \mu\int_0^1 \varphi\left(\frac{\mu t}{\zeta}, \mu\zeta[1-t]\right)dt \,;$$

here $\varphi(Z,Z^*) = H^*(0,Z,Z^*) \equiv H(Z-Z^*, -i(Z+Z^*),0)$, and $H^*(X,Z,Z^*) = (\underset{\sim}{B}_3 g)(X,Z,Z^*)$.

Now if $g(\mu,\zeta)$ <u>has</u> singularities of the form $\sigma = 0$, then H^* <u>may</u> have singularities on $\mathfrak{S}^*$. To ensure that these points are actual singularities we must see if they map back under the transformation (7.29) onto the set given by $\sigma = 0$.

According to (7.27) the singularities of $\varphi(Z,Z^*)$ must lie on the set of points which satisfies the relation (7.27); hence, the singularities of the integrand of (7.29) are, for fixed μ, ζ, those t which satisfy

$$\chi(\mu,\zeta,t) \equiv \frac{\mu t}{\zeta} p(t) + \frac{\mu\zeta[1-t]}{p(t)} - \Phi\circ p(t) = 0 , \tag{7.30}$$

and

$$\chi_t(\mu,\zeta,t) \equiv \frac{\mu}{\zeta} p(t) + \frac{\mu t}{\zeta} p'(t) - \frac{\mu\zeta}{p(t)} - \frac{\mu\zeta[1-t]}{[p(t)]^2} p'(t) - \Phi'\circ p(t) p'(t) = 0 , \tag{7.31}$$

where $p(t) \equiv \Psi(\mu\zeta[1-t])$, and $\Phi\circ p(t) \equiv \Phi(p(t))$.

Equation (7.31) can be rewritten as

$$\frac{\mu}{\zeta}[p(t)]^3 + \frac{\mu t}{\zeta}[p(t)]^2 p'(t) - \mu\zeta p(t) - \mu\zeta p'(t) + \mu t\zeta p'(t) + F\circ p(t) p'(t) - p(t) p'(t) \Phi\circ p(t) = 0 ;$$

however, since $F\circ p(t) \doteq 2\mu\zeta[1-t]$, this becomes

$$(7.32)\qquad \frac{\mu}{\zeta}[p(t)]^3 + \frac{\mu t}{\zeta}[p(t)]^2 p'(t) - \mu\zeta p(t) + \mu\zeta[1-t]\phi'(t)$$

$$- p(t)p'(t)\Phi\circ p(t) = 0 \;.$$

If one multiplies (7.30) by pp' and subtracts this from (7.32) we have

$$\mu\left(\frac{[p(t)]^3}{\zeta} - \zeta p(t)\right) = 0 \;,$$

as possible singularities†, or $p(t) = \pm\zeta$. These have as their inverse image (recall that p^{-1} is a function),

$$t_1 = p^{-1}(\zeta) \quad \text{and} \quad t_2 = p^{-1}(-\zeta) \;.$$

From the definition of $p(t)$, and $F(\zeta)$ one has

$$t_1 = 1 - \frac{1}{2\mu}[\Phi(\zeta) - \zeta\Phi'(\zeta)] \;,$$

(7.33) and

$$t_2 = 1 - \frac{1}{2\mu}[\Phi(-\zeta) + \zeta\Phi'(-\zeta)] \;.$$

If we put $t = t_k$ into $\chi(\mu,\zeta,t) = 0$ one obtains a set of possible singularities for $g(\mu,\zeta)$. These are the

---

† The set $\mu = 0$ cannot be an actual singularity by hypothesis; hence, there is no need of further discussion here.

singularities which lie on the first sheet of $g(\mu,\zeta)$ and cannot be avoided as we travel along an arbitrary path [starting at say $(\mu,\zeta) = (0,1)$], while at the same time we vary the path of integration. Putting $t = t_1$ into (7.30) yields

$$\chi(\mu,\zeta,t_1) = \frac{\mu}{\zeta}\left(1 - \frac{1}{2\mu}[\Phi(\zeta) - \zeta\Phi'(\zeta)]\right)p(t_1)$$

$$+ \frac{\zeta}{2p(t_1)}(\Phi(\zeta) - \zeta\Phi'(\zeta)) - \Phi_0 p(t_1) = 0 \ ;$$

noting that $p(t_1) = \zeta$, this reduces to $\mu = \Phi(\zeta)$, which are the known singularities of $g(\mu,t)$. On the other hand if one puts $t = t_2$ into (7.30), a similar computation yields the spurious set of singularities $\mu = -f(-\zeta)$. The end point singularities at $t = 0,1$ do not add any interesting cases here. We summarize the above discussion by

Theorem (7.9): *In (7.19) let $g(\mu,\zeta) \equiv g_1(\mu,\zeta)$. Let the defining function for the set of singularities of $g(\mu,\zeta)\zeta^{-1}$ be a global defining function in C^2, and furthermore of the form $\mu = \Phi(\zeta)$. Then the harmonic function $H(\underset{\sim}{x}) = (\underset{\sim}{B}_3 g)(\underset{\sim}{x})$*

is not only regular for all points $\underset{\sim}{x}$, which may be reached by continuation along a curve γ (starting at some initial point of definition $\underset{\sim}{x}^0$) provided $\gamma \cap \mathfrak{S}^* = \emptyset$, it is actually singular at all those points of the first Riemann sheet lying over $\mathfrak{S}^*$.

Example 1: If $g(\mu, \zeta) = (\mu - \alpha)^{-1}$, then by computing the residue of $\zeta^{-1} g(\mu, \zeta)$ for the singularity inside $|\zeta| < 1$, one has

$$(7.34) \qquad H(\underset{\sim}{x}) = \frac{\pm 1}{\sqrt{(X-\alpha)^2 - 4ZZ^*}} = \frac{\pm 1}{\sqrt{x_1^2 + x_2^2 + (x_3 - \alpha)^2}} ,$$

where the appropriate sign is chosen depending upon which branch is indicated by the residue computation. More precisely, the singularity set of the integrand has two roots; for real x_1, x_2, x_3, one root lies in $|\zeta| < 1$ and the other outside, unless they are both equal and lie on $|\zeta| = 1$.

The envelope method applied to the singularity manifold for the integrand

$$S(\underset{\sim}{x}, \zeta) \equiv Z\zeta^2 + (X - \alpha)\zeta + Z^* = 0 ,$$

yields exactly the singularities of (7.34) (as it should); hence,

(7.35) $$\chi(\mu,\zeta,t) \equiv \alpha^2 - 4\mu^2 t(1-t) = 0 \ .$$

The end point singularities (at $t=0,1$) contribute the singularity $\mu=\infty$. Eliminating t from (7.35) by using

$$\chi_t(\mu,\zeta,t) \equiv -4\mu^2(1-2t) = 0$$

yields $\mu = \pm\alpha$, which returns to us the actual singularity of $\zeta^{-1}g(\mu,\zeta)$ at $\mu=\alpha$. It is interesting to note that if instead,

$$H(\underset{\sim}{x}) = \frac{1}{\sqrt{(x-\alpha)^2 - 4ZZ^*}} + \frac{1}{\sqrt{(x+\alpha)^2 - 4ZZ^*}} \ ,$$

then on $X = x_3 = 0$

$$H = \frac{2}{\sqrt{\alpha^2 - 4ZZ^*}} = \frac{2}{\sqrt{\alpha^2 + x_1^2 + x_2^2}} \ ,$$

and $H_x \equiv 0$.

This corresponds to the case of (7.17) with $\varphi_2 \equiv 0$, which _is_ the case our theorem applies to.

For the present problem we must also consider singularities which arise from H_x on $x=0$, and see if these cancel the ones generated by $\varphi_1(Z,Z^*)$.

In the present case they cannot do this.

<u>Example 2</u>: If $g(\mu,\zeta) = (\mu - \alpha\zeta^{-1})^{-1}$, then the integrand of $\underset{\sim}{B}_3 g$ has singularities at

$$\zeta = \frac{1}{2z}\,[-X \pm \sqrt{X^2 - 4Z(Z^* - \alpha)}\,] .$$

The residue of $\zeta^{-1} g(\mu,\zeta)$ inside $\mathfrak{L}$ is $\pm\,(X - 4Z(Z^* - \alpha))^{-\frac{1}{2}}$; hence, we may represent χ by

$$\chi(\mu,\zeta,t) \equiv \frac{\mu t}{\zeta}\,(\mu\zeta[1-t] - \alpha) = 0 .$$

Eliminating t, between $\chi = 0$ and $\chi_t = 0$ yields

$$\mu\zeta^{-1}(1 - \frac{\alpha}{\mu\zeta})(\mu\zeta[1 + \frac{\alpha}{\mu\zeta}] - 2\alpha) = 0 \text{ , or } \mu\zeta = \alpha \text{ , } \mu = 0 \text{ ,}$$

$\zeta = \infty$ as possible singularities. Hence, we have reproduced again our original singularity set.

VIII. HIGHER ORDER AND HIGHER DIMENSIONAL ELLIPTIC PROBLEMS

In this chapter we extend the methodology of Chapter VI to the case of higher order, and also higher dimensional elliptic problems. In particular we will discuss the function theoretic methods derived by the author with Kukral [Ku1][GK2-6] for constructing solutions to the equations,

(8.7) $\Delta_3^2 u + Q(x_1, x_2, x_3)u = 0,$

(8.2) $\Delta_4^2 u + Q(x_1, x_2, x_3, x_4)u = 0,$

(8.3) $\Delta_n u + Q(x_1, \ldots, x_n)u = 0, \; n \geq 5,$

(8.4) $\Delta_n^2 u + Q(x_1, \ldots, x_n)u = 0, \; n \geq 5.$

The subdivision of the problems into the cases outlined above is not an artificial one. The fourth order equations in three and four dimensional space can be treated using methods corresponding to those employed for the second order case by Colton [Co 3 - 6]. One can show also in these instances, that an operator can be found which generates a complete family of solutions. The problem for dimension

$n \geq 5$ is much more difficult due to the unhappy fact that <u>there is no Whittaker-Bergman operator for</u> $n \geq 5$. This result was obtained by Kukral [Ku 1,2] in his thesis, and will be reported on subsequently. This does not preclude the possibility of other types of operators which would generate a complete family of solutions, or operators which might generate a <u>linearly dependent set</u> of solutions, that however, span the space. The Whittaker-Bergman operators have the property of generating exactly the correct number of harmonic polynomials of degree m. On the other hand, it may be possible to construct Whittaker-Bergman <u>type</u> operators for $n \geq 5$ which only generate harmonic polynomials of $K < n$ independent variables. This latter situation is worth investigating.

We now discuss the equation $\Delta_3^2 u + Q(x_1, x_2, x_3)u = 0$. As in [GK 2] we introduce the Bergman variables X, Z, Z^* and thereby change the equation

$$\Delta_3^2 u + Q(x_1, x_2, x_3)u = 0$$

into an equation of the form

(8.5) $U_{XXXX} - 2U_{XXZZ^*} + U_{ZZ^*Z^*} + Q(X,Z,Z^*)U = 0.$

As before, we assume Q is entire, and we seek a solution of (8.1.2) in the form

$$U(X,Z,Z^*) = \underset{\sim}{K}_3\{f\} \equiv \frac{1}{2\pi i}\int_{|\zeta|=1}\int_{\gamma} E(X,Z,Z^*,\zeta,t)\, f(\omega,\zeta)\frac{dt}{\sqrt{1-t^2}}\,\frac{d\zeta}{\zeta}$$

where $\omega = \mu[1-t^2]$, and γ is a path joining $t=-1$ to 1.

In order to find the differential equality E must satisfy, we substitute this expression into (8.1.2) and integrate by parts. In this case we will have to integrate by parts twice [G K 1]; the reader should see [Ku 1] for explicit details. By introducing the Colton-Tjong coordinates

(8.5) $\xi_1 = 2\zeta Z$, $\xi_2 = X + 2\zeta Z$, $\xi_3 = X + 2\zeta^{-1}Z^*$, [MT 1], [Co 3,5], and

defining $E^*(\xi_1, \xi_2, \xi_3, \zeta, t) \equiv E(X,Z,Z^*,\zeta,t)$ one obtains, after tedious but direct computation, the following differential equation for E^*:

$$E^*_{2222} + E^*_{3333} + 6E^*_{2233} - 8E^*_{1223} - 4E^*_{2223}$$

$$-8E^*_{1333} - 4E^*_{2333} + 16E^*_{1133} + 16E^*_{1233}$$

$$-\frac{2(1-t^2)}{\mu t}\left(E^*_{122t} + E^*_{133t} - 4E^*_{113t} - 2E^*_{123t}\right)$$

$$+\frac{2}{\mu t^2}\left(E^*_{122} + E^*_{133} - 4E^*_{113} - 2E^*_{123}\right)$$

$$+\frac{1}{\mu^2 t^2}\left((1-t^2)^2 E^*_{11tt} + \frac{E^*_{11t}}{t}(-3+3t^4) + \frac{3E^*_{11}}{t^2}\right) \tag{8.7}$$

$$+ Q^*E^* = 0 \quad .$$

For the fourth order equations it will turn out that we will need two independent solutions of (8.7). (Please note the existence of two E-functions in the radial and two variable cases). To show the existence of these solutions we attempt to find them in the form

$$E^* = 1 + \sum_{n=1}^{\infty} t^{2n}\mu^n p^{(n)}(\xi_1, \xi_2, \xi_3, \zeta), \tag{8.8}$$

and

$$\hat{E}^* = \frac{\xi_1}{2\zeta} + \sum_{n=1}^{\infty} t^{2n}\mu^n q^{(n)}(\xi_1, \xi_2, \xi_3, \zeta). \tag{8.9}$$

It occurs that a sufficient condition for E^* and $\hat{E}^*$ to formally satisfy (8.7) is, that

$$p_{11}^{(1)} = 0 \ , \ q_{11}^{(1)} = 0 \ ,$$

$$p_{11}^{(2)} = \frac{2}{3}\left(p_{122}^{(1)} + p_{133}^{(1)} - 4p_{113}^{(1)} - 2p_{123}^{(1)}\right) - \frac{1}{3} Q^* \ ,$$

$$q_{11}^{(2)} = \frac{2}{3}\left(q_{122}^{(1)} + q_{133}^{(1)} - 4\, q_{113}^{(1)} - 2\, q_{123}^{(1)}\right) - \frac{1}{3}\frac{\xi_1}{2\zeta} Q^* \ , \tag{8.10}$$

where both $p^{(n)}$ and $q^{(n)}$ satisfy the following equation for $n \geq 1$:

$$p_{11}^{(n+2)} = \frac{1}{(2n+1)(2n+3)}\Big\{2(2n+1)\left(p_{122}^{(n+1)} + p_{133}^{(n+1)} - 4p_{113}^{(n+1)} - 2p_{123}^{(n+1)}\right)$$

$$- \left(p_{2222}^{(n)} + p_{3333}^{(n)} + 6\, p_{2233}^{(n)} + 16\, p_{1233}^{(n)}\right)$$

$$-4\, p_{2223}^{(n)} - 4\, p_{2333}^{(n)} - 8\, p_{1223}^{(n)} - 8\, p_{1333}^{(n)} + 16\, p_{1133}^{(n)}\Big) - Q^* p^{(n)}\Big\}, \tag{8.11}$$

with $p^{(n)}(0, \xi_2, \xi_3, \zeta) = 0 = q^{(n)}(0, \xi_2, \xi_3, \zeta)$,

$$\text{and } \ p_1^{(n)}(0, \xi_2, \xi_3, \zeta) = 0 = q_1^{(n)}(0, \xi_2, \xi_3, \zeta). \tag{8.12}$$

This follows from direct computation; the interested reader is directed to [Ku 1] for complete details. One is now able to use majorization with the recursive system (8.1.7) to prove that the E-functions E^* and $\hat{E}^*$ are solutions of the differential equation (8.7), and, furthermore, they are regular in the product domain $G_R \times B \times T$ where R is an arbitrary positive number, and

$$G_R = \{(\xi_1, \xi_2, \xi_3) : |\xi_i| < R, \ i = 1, 2, 3\},$$

$$B_\epsilon = \{\zeta : 1-\epsilon < |\zeta| < 1+\epsilon, \ 0 < \epsilon < \tfrac{1}{2}\},$$

$T = \{t : |t| \le 1\}$. We can also normalize one E-functions such that

(8.13) $$E^*(0,\xi_2, \xi_3, \zeta, t) = 1, \quad E_1^*(0,\xi_2, \xi_3, \zeta, t) = 0 ,$$

$$\hat{E}^*(0,\xi_2, \xi_3, \zeta, t) = 0, \quad \hat{E}_1^*(0,\xi_2, \xi_3, \zeta, t) = \frac{1}{2\zeta} .$$

We provide some arguments that the above is indeed the case. First, note that for $n = 1$, $p^{(1)} \equiv q^{(1)} \equiv 0$ satisfy the requirements. For $n = 2$, the equations (8.10) can then be solved as

$$p^{(2)}(\xi_1, \xi_2, \xi_3, \zeta) = -1/3 \int_0^{\xi_1}\int_0^{\xi_1'} Q^*(\xi_1'', \xi_2, \xi_3, \zeta) d\xi_1'' \, d\xi_1' ,$$

and

$$(8.14)\quad q^{(2)}(\xi_1, \xi_2, \xi_3, \zeta) = -1/3 \int_0^{\xi_1} \int_0^{\xi_1'} \frac{\xi_1''}{2\zeta} Q^*(\xi_1'', \xi_2, \xi_3, \zeta)\, d\xi_1''\, d\xi_1' .$$

By induction it is easy to show that the $p^{(n)}$ and $q^{(n)}$ are uniquely determined and regular in $\overline{G_R} \times \overline{B_{2\epsilon}}$. We can now show using majorization and standard arguments that the formal series for E^* and $\hat{E}^*$ converge uniformly in $G_R \times B_\epsilon \times T$ to a holomorphic function. We omit the details [Ku 1] of this argument which are direct but very tedious. That the E-functions satisfy the initial conditions (8.13) follows from the initial conditions for the coefficients $p^{(n)}$ and $q^{(n)}$ (8.12), and our particular choice for the leading terms of E and $\hat{E}$. If one wishes real solutions for (8.1) then we note that the operator

$$u(x,y,z) = U(X,Z,Z^*) \equiv \operatorname{Re} \underset{\sim}{K}_3^{(2)}\{f,\hat{f}\}$$

$$\equiv \operatorname{Re}\left\{\frac{1}{2\pi i}\int_{|\zeta|=1}\int_{\gamma} E(X,Z,Z^*,\zeta,t) f(\omega,\zeta)\frac{dt}{\sqrt{1-t^2}}\frac{d\zeta}{\zeta}\right.$$

$$\left. + \frac{1}{2\pi i}\int_{|\zeta|=1}\int_{\gamma} \hat{E}(X,Z,Z^*,\zeta,t)\hat{f}(\omega,\zeta)\frac{dt}{\sqrt{1-t^2}}\frac{d\zeta}{\zeta}\right\}$$

carries pairs of analytic functions of the two complex variables μ (generating variable) and ζ into real valued solutions. We want to show this map is onto, i.e. that every strong solution of (8.1.1), in (10), can be written in this form.

From Lemma (6.1) we know that the real solutions of (8.1) are uniquely determined (in a complex neighborhood of the origin) by the Goursat data

$$U(X,0,Z^*) = F(X,Z),\ U_Z(X,0,Z^*). \tag{8.16}$$

We wish to show that we can uniquely determine the pair of analytic functions $\{f,\hat{f}\}$ in terms of this data. Indeed, we shall show that

$$f(\mu,\zeta) = -\frac{1}{2\pi}\int_{\gamma'} g(\mu[1-t^2],\zeta)\frac{dt}{t^2},$$

$$\hat{f}(\mu,\zeta) = -\frac{1}{2\pi}\int_{\gamma'} \hat{g}(\mu[1-t^2],\zeta)\frac{dt}{t^2}, \tag{8.17}$$

where γ' is a rectifiable arc joining the points $t=-1$ and 1, and not passing through the origin. Here g and $\hat{g}$ are defined by

$$g(\mu,\zeta) = 2\frac{\partial}{\partial\mu}\left\{\mu\int_0^1 F(t\mu,\mu\zeta[1-t])dt\right\} - F(\mu,0)$$

and

$$\hat{g}(\mu,\zeta) = 2\,\frac{\partial}{\partial\mu}\Big\{\mu\int_0^1 G(t\mu,\mu\zeta[1-t])\,dt\Big\}$$

$$-\ \zeta\,\frac{\partial}{\partial\mu}\Big\{g(\mu,\zeta) - \frac{1}{2\pi i}\int_{|\alpha|=1} g\Big(\frac{\mu}{\alpha},\ \zeta\alpha\Big)\frac{d\alpha}{\alpha}\ +$$

$$-\ \frac{1}{2\pi i}\int_{|\alpha|=1} g\Big(\frac{\mu}{2},\ \zeta\alpha\Big)d\alpha\Big\} + g_{\zeta\mu}(\mu,0) - \zeta\int_0^{\mu} G_{\zeta}(\mu,0)\,d\mu\ .$$

We establish this result by formal operations with power series, our approach being patterned after Colton's [Co 3,5] for the second order case. To this end suppose that locally (about the origin) we have the following expansions

$$g(\mu,\zeta) = \sum_{n=0}^{\infty}\sum_{m=0}^{n} a_{nm}\,\mu^n\,\zeta^m\ ,$$

$$\hat{g}(\mu,\zeta) = \sum_{n=0}^{\infty}\sum_{m=0}^{n} \hat{a}_{nm}\,\mu^n\,\zeta^m\ ,$$

(8.18)

$$F(X,Z^*) = \sum_{n=0}^{\infty}\sum_{m=0}^{\infty} c_{nm}\,X^n\,Z^{*m},$$

$$G(X,Z^*) = \sum_{n=0}^{\infty}\sum_{m=0}^{\infty} d_{n,m}\,X^n\,Z^{*m}\ ,$$

and make the following definitions

$$\bar{g}(\mu,\zeta) \equiv \overline{g(\bar{\mu},\bar{\zeta})}\ ,\ \bar{\hat{g}}(\mu,\zeta) = \overline{\hat{g}(\bar{\mu},\bar{\zeta})}\ ,\ \text{etc.}\ ;$$

furthermore, let $\bar{E}(X,Z,Z^*,\zeta,t)$ and $\bar{\hat{E}}(X,Z,Z^*,\zeta,t)$ be the generating functions corresponding to the differential equation

$$U_{XXXX} - 2U_{XXZZ^*} + U_{ZZZ^*Z^*} + \bar{Q}(X,Z,Z^*)U = 0\ .$$

Then one may write

$$\begin{aligned}\operatorname{Re} \underset{\sim}{K}_3^{(2)}\{f,\hat{f}\} &= \frac{1}{4\pi i}\int_{|\zeta|=1}\int_{\gamma} E(X,Z,Z^*,\zeta,t)f(\omega,\zeta)\frac{dt}{\sqrt{1-t^2}}\frac{d\zeta}{\zeta}\\
&+ \frac{1}{4\pi i}\int_{|\zeta|=1}\int_{\gamma} \bar{E}(X,-Z^*,-Z,\zeta,t)\bar{f}(\bar{\omega},\zeta)\frac{dt}{\sqrt{1-t^2}}\frac{d\zeta}{\zeta}\\
&+ \frac{1}{4\pi i}\int_{|\zeta|=1}\int_{\gamma} \hat{E}(X,Z,Z^*,\zeta,t)f(\omega,\zeta)\frac{dt}{\sqrt{1-t^2}}\frac{d\zeta}{\zeta}\\
&+ \frac{1}{4\pi i}\int_{|\zeta|=1}\int_{\gamma} \bar{\hat{E}}(X,-Z^*,-Z,\zeta,t)\bar{\hat{f}}(\bar{\omega},\zeta)\frac{dt}{\sqrt{1-t^2}}\frac{d\zeta}{\zeta}\ ,\end{aligned}$$

where $\bar{\omega} = \bar{\mu}[1-t^2]$, $\bar{\mu} = (x - \zeta Z^* - \zeta^{-1}Z)$. We notice next that the E-functions simplify on the characteristic plane $Z = 0$, i.e.

$$E(x,0,Z^*,\zeta,t) = E^*(0,\xi_2,\xi_3,\zeta,t) = 1\,,$$

(8.20)

$$\hat{E}(X,0,Z^*,\zeta,t) = \hat{E}^*(0,\xi_2,\xi_3,\zeta,t) = 0.$$

Next we must consider the first conjugate E-function,

$$\bar{E}(X,Z,Z^*,\zeta,t) = \bar{E}^*(\xi_1,\xi_2,\xi_3,\zeta,t) = 1 + \sum_{n=2}^{\infty} t^{2n}\mu^{n}\bar{p}^{(n)}\,,$$

The first non-vanishing component of $\bar{E}$ is

$$\bar{p}^{(2)} = -\frac{1}{3}\int_0^{\xi_1}\int_0^{\xi_1'} \overline{Q^*(\xi_1'',\xi_2,\xi_3,\zeta)}\,d\xi_1''\,d\xi_1'$$

$$= -\frac{1}{3}\int_0^{\xi_1}\int_0^{\xi_1'} \bar{Q}\left(\xi_2-\xi_1'',\frac{\xi_1''}{2\zeta},\frac{\zeta}{2}(\xi_3-\xi_2+\xi_1'')\right)d\xi_1''\,d\xi_1'$$

$$= -\frac{2\zeta}{3}\int_0^{\xi_1}\int_0^{\xi_1/2\zeta} \bar{Q}\left(\xi_2-2\zeta\tau,\tau,\frac{\zeta}{2}(\xi_3-\xi_2+2\zeta\tau)\right)d\tau\,d\xi_1'$$

$$= -\frac{4\zeta^2}{3}\int_0^{Z}\int_0^{\sigma} \bar{Q}\left(X+2\zeta Z-2\zeta\tau,\tau,\frac{\zeta}{2}\left(\frac{2Z^*}{\zeta}-2\zeta Z+2\zeta\tau\right)\right)d\tau\,d\sigma\,.$$

From the above computation we may conclude that $\bar{p}^{(2)}$ is an entire function of X,Z,Z^*,ζ and has a zero of order at least two at $\zeta = 0$. It can be shown by induction, that for all $n \geq 2$, $\bar{p}^{(n)}$ is entire in X, Z, Z^*, ζ and has a zero of order at least two at $\zeta = 0$.

We also note that when $Z = 0$ we have $\bar{\mu} = X - \zeta Z^*$ so that $\bar{f}(\bar{\mu}[1-t^2], \zeta)$ contains only non negative powers of ζ in its series expansion. Consequently, along $Z = 0$

$$\frac{1}{4\pi i} \int_{|\zeta|=1} \int_{\gamma} \bar{E}(X, -Z^*, 0, \zeta, t)\, \bar{f}(\bar{\mu}[1-t^2], \zeta) \frac{dt}{\sqrt{1-t^2}} \frac{d\zeta}{\zeta} =$$

$$\frac{1}{4\pi i} \int_{|\zeta|=1} \int_{\gamma} \bar{f}(\bar{\mu}[1-t^2], \zeta) \frac{dt}{\sqrt{1-t^2}} \frac{d\zeta}{\zeta} .$$

We turn next to $\hat{\bar{E}}(X, Z, Z^*, \zeta, t) = \hat{\bar{E}}^*(\xi_1, \xi_2, \xi_3, \zeta, t)$

$$= \frac{\xi_1}{2\zeta} + \sum_{n=2}^{\infty} t^{2n} \mu^n \bar{q}^{(n)} .$$

By a similar argument to the one given above, we may show that each $\bar{q}^{(n)}$ has a zero of order at least two at $\zeta = 0$. This yields

$$\frac{1}{4\pi i} \int_{|\zeta|=1} \int_{\gamma} \hat{\bar{E}}(X, -Z^*, 0, \zeta, t)\, \hat{\bar{f}}(\bar{\mu}[1-t^2], \zeta) \frac{dt}{\sqrt{1-t^2}} \frac{d\zeta}{\zeta}$$

$$= -\frac{Z^*}{4\pi i}\int_{|\zeta|=1}\bar{\hat{f}}(\bar{\mu}[1-t^2],\zeta)\frac{dt}{\sqrt{1-t^2}}\frac{d\zeta}{\zeta} \quad .$$

Starting with (8.15) and evaluating this on $Z=0$ leads then to the following equation for the data

$$U(X,0,Z^*) = F(X,Z^*) = \frac{1}{4\pi i}\int_{|\zeta|=1} f(\mu_1[1-t^2],\zeta)\frac{dt}{\sqrt{1-t^2}}\frac{d\zeta}{\zeta}$$

$$+ \frac{1}{4\pi i}\int_{|\zeta|=1}\int_{\gamma}\bar{f}(\mu_2[1-t^2],\zeta)\frac{dt}{\sqrt{1-t^2}}\frac{d\zeta}{\zeta}$$

$$- \frac{Z^*}{4\pi i}\int_{|\zeta|=1}\int_{\gamma}\bar{\hat{f}}(\mu_2[1-t^2],\zeta)\frac{dt}{\sqrt{1-t^2}}\frac{d\zeta}{\zeta} ,$$

with $\mu_1 = X+\zeta^{-1}Z^*$, $\mu_2 = X-\zeta Z^*$. Using the reciprocal integral relation for the $g(\mu,\zeta)$, $\hat{g}(\mu,\zeta)$ one has

$$F(X,Z^*) = \frac{1}{4\pi i}\int_{|\zeta|=1} g(\mu_1,\zeta)\frac{d\zeta}{\zeta} + \frac{1}{4\pi i}\int_{|\zeta|=1}\bar{g}(\mu_2,\zeta)\frac{d\zeta}{\zeta}$$

(8.21)

$$- \frac{Z^*}{4\pi i}\int_{|\zeta|=1}\bar{\hat{g}}(\mu_2,\zeta)\frac{d\zeta}{\zeta} \quad .$$

Since $U(X,Z,Z^*)$ is real for real x,y,z, one has $F(X,0) = U(X,0,0) = \sum_{n=0}^{\infty} c_{n,o}X^n = \sum_{n=0}^{\infty} \overline{c_{n,o}}X^n$, so that the $c_{n,o}$ are all real. Furthermore, since

$$\sum_{n=0}^{\infty} d_{n,1} X^n = G_{Z^*}(X,0) = U_{ZZ^*}(X,0,0)$$

$$= -\left\{u_{x_2x_2}(x_1\,,0,0) + u_{x_3x_3}(x_1\,,0,0)\right\}$$

is real for real $x_1 = X$ we can conclude $d_{n,1}$ is real for $n \geq 0$. If we let

$$U(X,Z,Z^*) = \sum_{i,j,k=0}^{\infty} \alpha_{ijk} X^i Z^j Z^{*k}$$

then

$$c_{n,1} = \alpha_{n,0,1} = (-1)^{0+1}\overline{\alpha_{n,1,0}} = -\overline{\alpha_{n,1,0}} = -\overline{d_{n,0}}\ ,$$

so that quantity $(c_{n,1} + d_{n,0})$ is pure imaginary $(n \geq 0)$. Putting the series expansions into (8.21) and performing the integrations yields

$$\sum_{n=0}^{\infty}\sum_{m=0}^{\infty} c_{nm} X^n Z^{*m} = \tfrac{1}{2}\sum_{n=0}^{\infty}\sum_{m=0}^{n} a_{nm}\binom{n}{m} X^{n-m} Z^{*m}$$

$$+ \tfrac{1}{2}\sum_{n=0}^{\infty} \overline{a_{n,0}}\, X^n - \frac{Z^*}{2}\sum_{n=0}^{\infty} \overline{\hat{a}_{n,0}}\, X^n$$

$$= \tfrac{1}{2}\sum_{n=0}^{\infty}\sum_{m=0}^{\infty} \binom{n+m}{m} a_{n+m,m}\, X^n Z^{*m}$$

$$(8.22)\qquad + \tfrac{1}{2}\sum_{n=0}^{\infty} \overline{a_{n,0}}\, X^n - Z^*/2 \sum_{n=0}^{\infty} \overline{\hat{a}_{n,0}}\, X^n \ .$$

Equating coefficients of like powers gives the conditions

$$c_{n,0} = \frac{a_{n,0} + \overline{a_{n,0}}}{2}\ , \quad n = 0,1,2,\ldots$$

$$c_{n,1} = (n+1)\, a_{n+1,1} - \overline{\hat{a}_{n,0}}\ , \quad n = 0,1,2,\ldots$$

$$c_{n,m} = \tfrac{1}{2}\, \frac{(n+m)!}{n!\,m!}\, a_{n+m,m}\ , \quad n = 0,1,\ldots,\ m \geq 2$$

Since the $c_{n\,0}$ are real, we may without loss of generality assume that $a_{n,0}$ is real, i.e. $a_{n0} = c_{n0}$.

We consider next, the second piece of Goursat data. To this end we have

$$\frac{\partial}{\partial Z} \operatorname{Re} \underset{\sim}{K}_3^{(2)}\{f,\hat{f}\} = \frac{1}{4\pi i} \int_{|\zeta|=1} \int_{\gamma} E_2(X,Z,Z^*,\zeta,t)\, f(\mu[1-t^2],\zeta) \frac{dt}{\sqrt{1-t^2}}\, \frac{d\zeta}{\zeta}$$

$$+ \frac{1}{4\pi i} \int_{|\zeta|=1} \int_{\gamma} E(X,Z,Z^*,\zeta,t)\, f_{\omega}(\mu[1-t^2],\zeta) \sqrt{1-t^2}\ dt\ d\zeta$$

$$- \frac{1}{4\pi i} \int_{|\zeta|=1} \int_{\gamma} \bar{E}_3(X,-Z^*,-Z,\zeta,t)\, \bar{f}(\bar{\mu}[1-t^2],\zeta) \frac{dt}{\sqrt{1-t^2}}\, \frac{d\zeta}{\zeta}$$

$$- \frac{1}{4\pi i} \int_{|\zeta|=1} \int_{\gamma} \bar{E}(X,-Z^*,-Z,\zeta,t) \bar{f}_{\bar{\omega}}(\bar{\mu}[1-t^2],\zeta) \frac{(1-t^2)}{\zeta} \frac{dt}{\sqrt{1-t^2}} \frac{d\zeta}{\zeta}$$

$$+ \frac{1}{4\pi i} \int_{|\zeta|=1} \int_{\gamma} \hat{E}_2(X,Z,Z^*,\zeta,t) \hat{f}(\mu[1-t^2],\zeta) \frac{dt}{\sqrt{1-t^2}} \frac{d\zeta}{\zeta}$$

$$+ \frac{1}{4\pi i} \int_{|\zeta|=1} \int_{\gamma} \hat{E}(X,Z,Z^*,\zeta,t) \hat{f}_{\omega}(\mu[1-t^2],\zeta) \zeta (1-t^2) \frac{dt}{\sqrt{1-t^2}} \frac{d\zeta}{\zeta}$$

$$- \frac{1}{4\pi i} \int_{|\zeta|=1} \int_{\gamma} \bar{\hat{E}}_3(X,-Z^*,-Z,\zeta,t) \bar{\hat{f}}(\bar{\mu}[1-t^2],\zeta) \frac{dt}{\sqrt{1-t^2}} \frac{d\zeta}{\zeta}$$

$$(8.23) \quad - \frac{1}{4\pi i} \int_{|\zeta|=1} \int_{\gamma} \hat{E}(X,-Z^*,-Z,\zeta,t) \bar{f}_{\bar{\omega}}(\bar{\mu}[1-t^2],\zeta) (1-t^2) \frac{1}{\zeta} \frac{dt}{\sqrt{1-t^2}} \frac{d\zeta}{\zeta} .$$

We note that $E_2 = \frac{\partial}{\partial Z} E^* = (E_1^* + E_2^*) 2\zeta$, and that for $Z = 0$ we have $E_1^* = E_2^* = 0$ from $E^*(0, \xi_2, \xi_3, \zeta, t) = 0$. So $E_2(X,0,Z^*,\zeta,t) = 0$. Also we have that $E(X,0,Z^*,\zeta,t) = E^*(0,\xi_2,\xi_3,\zeta,t) = 1$. Since the $\bar{p}^{(n)}$ have zeros of order at least <u>two</u> at $\zeta = 0$ we may also show that $\bar{E}_3$ vanishes at $\zeta = 0$ also. The analytic function $\bar{f}(\bar{\mu}[1-t^2],\zeta)$ contains only non-negative powers of ζ for $Z = 0$; consequently,

$$\frac{1}{4\pi i} \int_{|\zeta|=1} \int_{\gamma} \bar{E}_3(X,-Z^*,0,\zeta,t) \bar{f}(\mu_2[1-t^2],\zeta) \frac{dt}{\sqrt{1-t^2}} \frac{d\zeta}{\zeta} = 0 .$$

Again, because the $\bar{p}^{(n)}$ all have zeros of order at least two at $\zeta = 0$ we have

$$\frac{1}{4\pi i}\int_{|\zeta|=1}\int_{\gamma}\bar{E}(X,-Z^*,0,\zeta,t)\bar{f}_{\bar{\omega}}(\mu_2[1-t^2],\zeta)(t^2-1)^{\frac{1}{2}}dt\,\frac{d\zeta}{\zeta^2}$$

$$= \frac{1}{4\pi i}\int_{|\zeta|=1}\int_{\gamma}\bar{f}_{\bar{\omega}}(\mu_2[1-t^2],\zeta)(t^2-1)^{\frac{1}{2}}dt\,\frac{d\zeta}{\zeta^2}\quad .$$

We now analyze the contribution from the $\hat{E}$-terms. Along $Z = 0$ we have $\hat{E}(X,0,Z^*,\zeta,t) = \hat{E}(0,\xi_2,\xi_3,\zeta,t) = 0$. By the same arguments used previously regarding the zeros of $p^{(n)}$, one can conclude

$$\frac{1}{4\pi i}\int_{|\zeta|=1}\int_{\gamma}\bar{\hat{E}}(X,-Z^*,0,\zeta,t)\bar{f}_{\bar{\omega}}(\mu_2[1-t^2],\zeta)\sqrt{1-t^2}\,dt\,\frac{d\zeta}{\zeta^2}$$

$$= \frac{-Z^*}{4\pi i}\int_{|\zeta|=1}\int_{\gamma}\bar{\hat{f}}_{\bar{\omega}}(\mu_2[1-t^2],\zeta)\sqrt{1-t^2}\,dt\,\frac{d\zeta}{\zeta^2}\quad .$$

Also we have along $Z = 0$,

$$\hat{E}_2 = 2\zeta(\hat{E}_1^* + \hat{E}_2^*) = 2\zeta\left(\frac{1}{2\zeta} + 0\right) = 1.$$

Again, using our arguments regarding the $p^{(n)}$ we get

$$\frac{1}{4\pi i}\int_{|\zeta|=1}\int_{\gamma}\bar{\hat{E}}_3(X,-Z^*,0,\zeta,t)\bar{f}(\bar{\mu}[1-t^2],\zeta)\frac{dt}{\sqrt{1-t^2}}\frac{d\zeta}{\zeta} = 0 .$$

Putting all of this information into the equation (8.23), we obtain a condition for the data to satisfy, namely

$$G(X,Z^*) = \frac{1}{4\pi i}\int_{|\zeta|=1}\int_{\gamma} f_{\omega}(\mu_1[1-t^2],\zeta)\sqrt{1-t^2}\,dt\,d\zeta$$

$$-\frac{1}{4\pi i}\int_{|\zeta|=1}\int_{\gamma}\bar{f}_{\bar{\omega}}(\mu_2[1-t^2],\zeta)\sqrt{1-t^2}\,dt\,\frac{d\zeta}{\zeta^2}$$

$$+\frac{Z^*}{4\pi i}\int_{|\zeta|=1}\int_{\gamma}\bar{\hat{f}}_{\bar{\omega}}(\mu_2[1-t^2],\zeta)\sqrt{1-t^2}\,dt\,\frac{d\zeta}{\zeta^2}$$

$$(8.24)\qquad +\frac{1}{4\pi i}\int_{|\zeta|=1}\int_{\gamma}\hat{f}(\mu_1[1-t^2],\zeta)\frac{dt}{\sqrt{1-t^2}}\frac{d\zeta}{\zeta} ,$$

which becomes, upon performing the t-integration,

$$G(X,Z^*) = \frac{1}{4\pi i}\int_{|\zeta|=1} g_{\mu}(\mu_1,\zeta)\,d\zeta - \frac{1}{4\pi i}\int_{|\zeta|=1}\bar{g}_{\mu}(\mu_2,\zeta)\frac{d\zeta}{\zeta^2}$$

$$(8.25) \qquad + \frac{Z^*}{4\pi i} \int_{|\zeta|=1} \bar{\hat{g}}(\mu_2,\zeta)\frac{d\zeta}{\zeta^2} + \frac{1}{4\pi i} \int_{|\zeta|=1} \hat{g}(\mu_1,\zeta)\frac{d\zeta}{\zeta} .$$

From the series expansions for g and $\hat{g}$ we obtain the following

$$\frac{1}{4\pi i} \int_{|\zeta|=1} g_\mu(\mu_1,\zeta)d\zeta = \frac{1}{4\pi i} \int_{|\zeta|=1} \sum_{n=1}^{\infty} \sum_{m=0}^{n} n\, a_{n,m}\mu_1^{n-1}\zeta^m d\zeta$$

$$= \frac{1}{4\pi i} \int_{|\zeta|=1} \sum_{n=1}^{\infty} \sum_{m=0}^{n} n\, a_{n,m} \sum_{j=0}^{n-1} \binom{n-1}{j} X^{n-1-j} Z^{*j} \zeta^{m-j} d\zeta$$

$$= \tfrac{1}{2} \sum_{n=2}^{\infty} \sum_{m=0}^{n-2} n\, a_{n,m} \binom{n-1}{m+1} X^{n-m-2} Z^{*m+1} ,$$

and

$$\frac{1}{4\pi i} \int_{|\zeta|=1} \bar{g}_\mu(\mu_2,\zeta)\frac{d\zeta}{\zeta^2} = \int_{|\zeta|=1} \sum_{n=1}^{\infty} \sum_{m=0}^{n} n\, \overline{a_{n,m}}\, \mu_2^{n-1} \zeta^{m-2} d\zeta$$

$$= \frac{1}{4\pi i} \int_{|\zeta|=1} \sum_{n=1}^{\infty} \sum_{m=0}^{n} n\, \overline{a_{nm}} \sum_{j=0}^{n-1} \binom{n-1}{j} X^{n-1-j} (-Z^*)^j \zeta^{m+j-1}$$

$$= \tfrac{1}{2}\Big[\sum_{n=1}^{\infty} (n\, \overline{a_{n,0}}\, (n-1)X^{n-2}(-Z^*) + n\, \overline{a_{n,1}}\, X^{n-1}\Big] ,$$

and

$$\frac{Z^*}{4\pi i}\int_{|\zeta|=1}\bar{\hat{g}}_\mu(\mu_2,\zeta)\frac{d\zeta}{\zeta^2} = \frac{Z^*}{4\pi i}\int_{|\zeta|=1}\sum_{n=1}^{\infty}\sum_{m=0}^{n} n\,\overline{\hat{a}_{n,m}}\,\mu_2^{n-1}\zeta^m\frac{d\zeta}{\zeta^2}$$

$$= \frac{Z^*}{4\pi i}\int_{|\zeta|=1}\sum_{n=1}^{\infty}\sum_{m=0}^{n} n\,\bar{\hat{a}}_{n,m}\sum_{j=0}^{n-1} X^{n-1-j}\binom{n-1}{j}(-Z^*)^j\zeta^{m+j-1}\frac{d\zeta}{\zeta}$$

$$= \frac{Z^*}{2}\sum_{n=1}^{\infty}\left(n\,\overline{\hat{a}_{n0}}\,X^{n-2}(-Z^*)(n-1) + n\,\overline{\hat{a}_{n,1}}\,X^{n-1}\right),$$

and finally

$$\frac{1}{4\pi i}\int_{|\zeta|=1}\hat{g}(\mu_1,\zeta)\frac{d\zeta}{\zeta}$$

$$= \frac{1}{4\pi i}\int_{|\zeta|=1}\sum_{n=0}^{\infty}\sum_{m=0}^{n}\hat{a}_{n,m}\sum_{j=0}^{n}\binom{n}{j}X^{n-j}\,Z^{*j}\,\zeta^{m-j}\frac{d\zeta}{\zeta}$$

$$(8.26)\qquad = \tfrac{1}{2}\sum_{n=0}^{\infty}\sum_{m=0}^{\infty}\binom{n}{m}\hat{a}_{n,m}\,X^{n-m}\,Z^{*m}.$$

Substituting the expressions (8.26) into the data equation (8.25) yields

$$G(X,Z^*) \equiv \sum_{n=0}^{\infty}\sum_{m=0}^{\infty} d_{n,m}\,X^n\,Z^{*m} = \tfrac{1}{2}\sum_{n=2}^{\infty}\sum_{m=0}^{n-2} n\,a_{n,m}\binom{n-1}{m+1}X^{n-m-2}Z^{*m+1}$$

$$+ \tfrac{1}{2} \sum_{n=1}^{\infty} \left(n(n-1)\overline{a_{n,0}}\, X^{n-2}\, Z^* - n\, \overline{a_{n,1}}\, X^{n-1} \right)$$

$$+ \frac{Z^*}{2} \sum_{n=1}^{\infty} \left(n(n-1)\overline{\hat{a}_{n,0}}\, X^{n-2} (-Z^*) + n\, \overline{\hat{a}_{n,1}}\, X^{n-1} \right)$$

$$+ \tfrac{1}{2} \sum_{n=0}^{\infty} \sum_{m=0}^{n} \binom{n}{m} \hat{a}_{n,m}\, X^{n-m}\, Z^{*m} ,$$

or

$$\sum_{n=0}^{\infty} \sum_{m=0}^{\infty} d_{n,m} X^n\, Z^{*m} = \tfrac{1}{2} \sum_{n=0}^{\infty} \sum_{m=0}^{\infty} \frac{(n+m+2)!}{n!(m+1)!}\, a_{n+m+2,m}\, X^n\, Z^{*m+1}$$

$$+ \tfrac{1}{2} \sum_{n=0}^{\infty} \sum_{m=0}^{\infty} \frac{(n+m)!}{n!m!}\, \hat{a}_{n+m,m}\, X^n\, Z^{*m}$$

$$+ \tfrac{1}{2} \sum_{n=0}^{\infty} (n+2)(n+1)\, \overline{a_{n+2,0}}\, X^n\, Z^*$$

$$- \tfrac{1}{2} \sum_{n=0}^{\infty} (n+1)\overline{a_{n+1,1}}\, X^n$$

$$- \tfrac{1}{2} \sum_{n=0}^{\infty} (n+2)(n+1)\, \overline{\hat{a}_{n+2,0}}\, X^n\, Z^{*2}$$

(8.27) $$+ \frac{1}{2} \sum_{n=0}^{\infty} (n+1) \overline{\hat{a}_{n+1,1}} X^n Z^* .$$

From (8.27) we get the following conditions on the coefficients in g and $\hat{g}$

$$d_{n,0} = \frac{1}{2}\left[\hat{a}_{n,0} - (n+1) \overline{a_{n+1,1}}\right] , \quad n \geq 0 .$$

$$d_{n,1} = \frac{1}{2}\Big[(n+2)(n+1)a_{n+2,0} + (n+2)(n+1)\overline{a_{n+2,0}}$$

$$+ (n+1) \overline{\hat{a}_{n+1,1}} + (n+1) \hat{a}_{n+1,1}\Big] , \quad n \geq 0 .$$

Since we showed earlier that $a_{n,0}$ can be taken to be real, and $d_{n,1}$ is real, w.o.l.o.g. we may assume $\hat{a}_{n+1,1}$ is real for $n \geq 0$; hence,

(8.28) $$d_{n,1} = (n+2)(n+1)a_{n+2,0} + (n+1)\hat{a}_{n+1,1} , \quad n \geq 0.$$

Furthermore, we have

$$d_{n,2} = \frac{1}{2}\Big[\frac{(n+3)!}{n!2} a_{n+3,1} - (n+2)(n+1)\overline{\hat{a}_{n+2,0}}$$

(8.29) $$+ \frac{(n+2)!}{n!2} \hat{a}_{n+2,2}\Big] , \quad n \geq 0. \quad ,$$

$$d_{n,m} = \frac{1}{2}\left[\frac{(n+m+1)!}{n!m!} a_{n+m+1,m-1} + \frac{(n+m)!}{n!m!} \hat{a}_{n+m,m}\right]$$

(8.30) $\quad m \geq 3 \ , \ n \geq 0 \ .$

Since, as we have shown earlier, $c_{n,1} = \overline{-d_{n,0}}$, we may w.o.l.o.g. take $\hat{a}_{n,0} = 0 \ , \ n \geq 0$.

Returning to equation (8.22) we see that this reduces to the form

$$F(X,Z^*) = \sum_{n=0}^{\infty} \sum_{m=0}^{\infty} c_{n\,m} X^n Z^{*m}$$

$$= \sum_{n=0}^{\infty} \left[a_{n,0} + \frac{(n+1)}{2} a_{n+1,1} Z^* + \tfrac{1}{2} \sum_{m=2}^{\infty} \binom{n+m}{m} a_{n+m,m} Z^{*m} \right] X^n$$

(8.31) $$= \tfrac{1}{2} \sum_{n=0}^{\infty} \sum_{m=0}^{\infty} \binom{n+m}{m} a_{n+m,m} X^n Z^{*m} + \tfrac{1}{2} \sum_{n=0}^{\infty} a_{n,0} X^n \ .$$

Now by making use of the integral definition of the beta function we can rewrite (8.31) as

$$\int_0^1 F(tX, (1-t)Z^*)dt = \tfrac{1}{2} \sum_{n=0}^{\infty} \sum_{m=0}^{n} \frac{a_{n,m}}{n+1} X^{n-m} Z^{*m}$$

$$+ \tfrac{1}{2} \sum_{n=0}^{\infty} \frac{a_{n,0}}{n+1} X^n \ ,$$

from which one obtains

$$\frac{\partial}{\partial\mu}\left[\mu \int_0^1 F(t\mu,[1-t]\mu\,\zeta)\,dt\right]$$

$$= \tfrac{1}{2}\sum_{n=0}^{\infty}\sum_{m=0}^{n} a_{n,m}\,\mu^n\,\zeta^m + \tfrac{1}{2}\sum_{n=0}^{\infty} a_{n,0}\,\mu^m$$

$$= \tfrac{1}{2}\,g(\mu,\zeta) + \tfrac{1}{2}\,F(\mu,0)\ ,$$

or

$$g(\mu,\zeta) = 2\,\frac{\partial}{\partial\mu}\left[\mu\int_0^1 F(t\mu,(1-t)\,\mu\,\zeta)\,dt\right] - F(\mu,0)\ . \tag{8.32}$$

We next turn to seeking an expression for $\hat{g}(\mu,\zeta)$ using (8.27). One has, remembering that $\hat{a}_{n,0} = 0$,

$$G(X,Z^*) = \tfrac{1}{2}\sum_{n=0}^{\infty}\Big[-(n+1)\overline{a_{n+1,1}} + 2\,Z^*\Big((n+2)(n+1)a_{n+2,0} + (n+1)\hat{a}_{n+1,1}\Big)$$

$$+ Z^{*2}\left(\frac{(n+3)!}{n!2}\,a_{n+3,1} + \frac{(n+2)(n+1)}{2}\,\hat{a}_{n+2,2}\right)$$

$$+ \sum_{m=3}^{\infty} Z^{*m}\left(\frac{(n+m+1)!}{n!m!}\,a_{n+m+1,m-1} + \frac{(n+m)!}{n!m!}\,\hat{a}_{n+m,m}\right)\Big]\,X^n\ .$$

As before, using the beta-integrals, one obtains

$$\int_0^1 G(tx,\ (1-t)Z^*)dt$$

$$= \tfrac{1}{2}\sum_{n=0}^{\infty} \Big[- \overline{a_{n+1,1}} + Z^*\Big(2\ a_{n+2,0} + \frac{2\ \overline{\hat{a}_{n+1,1}}}{n+2}\Big)$$

$$+ Z^{*2}\Big(a_{n+3,1} + \frac{\hat{a}_{n+2,2}}{n+3}\Big)$$

$$+ \sum_{m=3}^{\infty} Z^{*m}\Big(a_{n+m+1,m-1} + \frac{\hat{a}_{n+m,m}}{n+m+1}\Big)\Big]\ x^n\ ,$$

and hence

$$2\ \frac{\partial}{\partial\mu}\Big[\mu\int_0^1 G(t\mu,\ (1-t)\mu\ \zeta)\ dt\Big]$$

$$= \sum_{n=0}^{\infty} \Big[-(n+1)\overline{a_{n+1,1}} + 2\ \mu\ \zeta\Big(a_{n+2,0}(n+2) + \hat{a}_{n+1,1}\Big) +$$

$$\mu^2\zeta^2\Big((n+3)a_{n+3,1} + \hat{a}_{n+2,2}\Big) + \sum_{m=3}^{\infty} \mu^m\zeta^m\Big((n+m+1)a_{n+m+1,m-1}$$

$$+ \hat{a}_{n+m,m}\Big)\Big]\ \mu^n$$

$$= \sum_{n=0}^{\infty} \sum_{m=1}^{\infty} (n+m+1) a_{n+m+1,m-1} \mu^{n+m} \zeta^{m} + \sum_{n=1}^{\infty} \sum_{m=1}^{\infty} \hat{a}_{n+m,m} \mu^{n+m} \zeta^{m}$$

$$- \sum_{n=0}^{\infty} (n+1) \overline{a_{n+1,1}} \mu^{n} + \sum_{n=0}^{\infty} (n+2) a_{n+2,0} \mu^{n+1} \zeta$$

$$(8.33) \quad + \sum_{n=0}^{\infty} \hat{a}_{n+1,1} \mu^{n+1} \zeta .$$

Noticing that $\sum_{n=1}^{\infty} \sum_{m=1}^{\infty} \hat{a}_{n+m,m} \mu^{n+m} \zeta^{m} = \sum_{n=0}^{\infty} \sum_{m=0}^{n} \hat{a}_{n,m} \mu^{n} \zeta^{m}$,

$(\hat{a}_{n,0} = 0)$, etc.

and the identities

$$\sum_{n=0}^{\infty} \hat{a}_{n+1,1} \mu^{n+1} \zeta = \sum_{n=0}^{\infty} \frac{(d_{n,1} - (n+2)(n+1) a_{n+2,0})}{n+1} \mu^{n+1} \zeta$$

$$= \sum_{n=0}^{\infty} \frac{d_{n,1}}{n+1} \mu^{n+1} \zeta - \sum_{n=0}^{\infty} (n+2) a_{n+2,0} \mu^{n+1} \zeta$$

$$= \int_{0}^{\mu} \sum_{n=0}^{\infty} d_{n,1} \mu^{n} \zeta d\mu - \frac{\partial}{\partial \mu} \sum_{n=0}^{\infty} a_{n+2,0} \mu^{n+2} \zeta .$$

$$= \zeta \int_{0}^{\mu} G_{\zeta}(\mu,0) d\mu - \frac{\partial}{\partial \mu} \left(\zeta g(\mu,0) - \mu \zeta g_{\mu}(0,0) \right) ,$$

and

$$\sum_{n=0}^{\infty}\sum_{m=1}^{\infty}(n+m+2)a_{n+m+2,m}\,\mu^{n+m+1}\,\zeta^{m+1} = \sum_{m=0}^{\infty}\sum_{n=m+2}^{\infty} n\,a_{n,m}\,\mu^{n-1}\,\zeta^{m+1}$$

$$= \sum_{n=2}^{\infty}\sum_{m=0}^{n-2} n\,a_{n,m}\,\mu^{n-1}\,\zeta^{m+1} = \zeta\,\frac{\partial}{\partial\mu}\Big(\sum_{n=2}^{\infty}\sum_{m=0}^{n-2} a_{n,m}\,\mu^{n}\,\zeta^{m}\Big)$$

$$= \zeta\,\frac{\partial}{\partial\mu}\Big(g(\mu,\zeta) - \frac{1}{2\pi i}\int_{|\alpha|=1} g(\mu\alpha^{-1},\zeta\alpha)\frac{d\alpha}{\alpha} - \frac{1}{2\pi i}\int_{|\alpha|=1} g(\mu\alpha^{-1},\zeta\alpha)\,d\alpha\Big)\ ,$$

(Note that the terms which contain μ^1 and μ^0 in $g(\mu,\zeta)$ are cancelled by subtraction.).

Putting the above into our expression for $\hat{g}(\mu,\zeta)$ we obtain

$$\hat{g}(\mu,\zeta) = 2\,\frac{\partial}{\partial\mu}\Big[\mu\int_0^1 G(t\mu,\,(1-t)\mu\,\zeta)\,dt\Big]$$

$$-\ \zeta\,\frac{\partial}{\partial\mu}\Big[g(\mu,\zeta) - \frac{1}{2\pi i}\int_{|\alpha|=1} g(\mu\alpha^{-1},\zeta\alpha)\frac{d\alpha}{\alpha}$$

$$-\ \frac{1}{2\pi i}\int_{|\alpha|=1} g(\mu\alpha^{-1},\zeta\alpha)\,d\alpha\Big] + g_{\zeta\mu}(\mu,0)$$

$$+ \zeta\left(g_\mu(0,0) - g_\mu(\mu,0)\right) - \zeta \int_0^\mu G_\zeta(\mu,0)\,d\mu$$

$$+ \frac{\partial}{\partial\mu}\left(\zeta g(\mu,0) - \mu\zeta g_\mu(0,0)\right)$$

from which we obtain the stated result.

Definition: Solutions of an equation $Lu = 0$ are said to have the Runge approximation property if, whenever $\mathfrak{S}_1$ and $\mathfrak{S}_2$ are two simply-connected domains, $\mathfrak{S}_1 \subset \mathfrak{S}_2$, and any solution in $\mathfrak{S}_1$ can be approximated uniformly in compact subsets of $\mathfrak{S}_1$ by sequences of solutions in $\mathfrak{S}_2$.

Definition: Solutions of the equation $Lu = 0$ are said to have the unique continuation property if whenever a solution vanishes in an open set it vanishes identically in its domain of definition.

It has been shown by Malgrange [Ma 1], Lax [La 1] and Browder [Br 3] that solutions of $Lu = 0$ have the Runge approximation property iff solutions of the formal adjoint equation have the unique continuation property. Since elliptic equations with analytic coefficients have the unique continuation property as a consequence of the analyticity of their solutions the equations (8.1 - 8.4) have the Runge approximation property. Indeed, we may easily show that the following sequence of functions is a complete family for (8.1) for a bounded, simply-connected domain $\mathfrak{S}$ in $\mathbb{R}^3$:

$$U_{4n,m} = \operatorname{Re} \underset{\sim}{K}_3^{(2)}\{\mu^n\zeta^m ; 0\} , \quad n \geq 0 , \quad 0 \leq m \leq n ;$$

$$U_{4n+1,m} = \operatorname{Re} \underset{\sim}{K}_3^{(2)}\{0 ; \mu^n\zeta^m\} , \quad n \geq 0 , \quad 0 \leq m \leq n ;$$

$$U_{4n+2,m} = \operatorname{Im} \underset{\sim}{K}_3^{(2)}\{\mu^n\zeta^m ; 0\} , \quad n \geq 0 , \quad 0 \leq m \leq n ;$$

$$U_{4n+3,m} = \operatorname{Im} \underset{\sim}{K}_3^{(2)}\{0 ; \mu^n\zeta^m\} , \quad n \geq 0 , \quad 0 \leq m \leq n .$$

See the thesis by Kukral [Ku 1] for further details.

We next turn to the discussion of equation

$$(8.2) \qquad \Delta_4^2\, u + \hat{Q}(x_1 , x_2 , x_3 , x_4)u = 0 .$$

We shall omit much of the details here; the interested reader is directed to [Ku 1] or [G K 1] for a more complete discussion.

Using methods previously employed we can establish without too much difficulty the following

Theorem (8.1): *Let* $Y = \frac{1}{2}(x_1 + i x_2)$, $Y^* = \frac{1}{2}(x_1 - i x_2)$, $Z = \frac{1}{2}(x_3 + i x_4)$, $Z^* = \frac{1}{2}(-x_3 + i x_4)$, *and let* $u(x_1, x_2, x_3, x_4)$ *be a real valued,* C^4 *solution of* (8.2) *in a neighborhood of the origin. Then* $U(Y,Y^*,Z,Z^*) = u(x_1, x_2, x_3, x_4)$ *is an analytic function of* Y,Y^*,Z,Z^* *in some neighborhood of the origin in* $\mathbb{C}^4$ *and is uniquely determined by the Goursat data*

$$F(Y,Z,Z^*) = U(Y,0,Z,Z^*) \quad \text{and} \quad G(Y,Z,Z^*) = U_{Y^*}(Y,0,Z,Z^*).$$

In terms of the Y,Y^*,Z,Z^* coordinates (8.2) becomes

$$U_{YYY^*Y^*} = 2\,U_{YY^*ZZ^*} - U_{ZZZ^*Z^*} + Q(Y,Y^*,Z,Z^*)U. \tag{8.34}$$

Introducing the Colton variation of the Colton-Gilbert coordinates,

$$\xi_1 = Y^*/\eta\zeta \;,\; \xi_2 = Y^*/\eta\zeta + Z^*/\eta \;,$$

$$\xi_3 = Z^*/\eta + Y \;,\; \xi_4 = Z/\zeta + Y \;, \tag{8.35}$$

one has $\mu = \xi_2 + \xi_4 = Y + Z\zeta^{-1} + Z^*\eta^{-1} + Y^*\zeta^{-1}\eta^{-1}$, and $\omega = (1-t^2)\mu$.

Then using our previous approach we can establish [Ku 1], [G K 4]

Theorem (8.2): *Let* D *be a neighborhood of the origin in the* μ *plane,* $B \equiv \{(\zeta,\eta) : 1-\varepsilon < |\zeta| < 1+\varepsilon \ , \ 1-\varepsilon < |\eta| < 1+\varepsilon\}$, $\mathfrak{S}$ *a neighborhood of the origin in the* $\xi_1, \xi_2, \xi_3, \xi_4$ *space, and* $T \equiv \{t : |t| \leq 1\}$. *Let* $f(\mu,\zeta,\eta)$ *be a function of three complex variables, analytic in the product domain* $D \times B$, *and*

$$E^*(\xi_1, \xi_2, \xi_3, \xi_4, \zeta, \eta, t) = E(Y, Y^*, Z, Z^*, \zeta, \eta, t) \tag{8.36}$$

be a regular solution of the partial differential equation

$$\begin{aligned}
&E^*_{1133} + E^*_{1144} + E^*_{2233} + E^*_{3344} + 2\,E^*_{1134} + 2\,E^*_{1233} \\
&\quad - 2\,E^*_{1334} + 2\,E^*_{1234} - 2\,E^*_{1344} - 2\,E^*_{2334} \\
&\quad - \frac{1}{t^2\mu}\left\{E^*_{113} + E^*_{114} + E^*_{123} - E^*_{134}\right\} \\
&\quad + \frac{(1-t^2)}{t\mu}\left\{E^*_{113t} + E^*_{114t} + E^*_{123t} - E^*_{134t}\right\} \\
&\quad + \frac{3\,E^*_{11}}{4t^4\mu^2} - \frac{3(1-t^4)}{4t^3\mu^2}\,E^*_{11t} + \frac{(1-t^2)^2}{4t^2\mu^2}\,E^*_{11tt} \\
&\quad + \eta^2\,\zeta^2\,Q^*\,E^* = 0 \ ,
\end{aligned} \tag{8.37}$$

<u>in</u> $\mathfrak{S} \times B \times T$ <u>where</u> $Q^*(\boldsymbol{\xi},\zeta,\eta) \equiv Q(Y,Y^*,Z,Z^*)$, <u>and subscripts denote differentiation with respect to</u> ξ_i <u>or</u> t. <u>Then</u>

$U(Y,Y^*,Z,Z^*) =$

$$= -\frac{1}{4\pi^2} \int_{|\zeta|=1} \int_{|\eta|=1} E(Y,Y^*,Z,Z^*,\zeta,\eta,t) f(\omega,\zeta,\eta) \frac{dt}{\sqrt{1-t^2}} \frac{d\eta}{\eta} \frac{d\zeta}{\zeta} \tag{8.38}$$

<u>where</u> γ <u>is a path in</u> T <u>joining</u> $t=-1$ <u>and</u> $t=1$ <u>is a complex valued, solution of</u> (8.2) <u>which is regular in a neighborhood of the origin in</u> C^4.

The proof involves a detailed, but direct analytical computation.

As in the fourth order, three variable case, we seek two E-functions of the form

$$E^* = 1 + \sum_{n=1}^{\infty} t^{2n} \mu^n p^{(n)}(\boldsymbol{\xi};\zeta,\eta), \quad \boldsymbol{\xi} \equiv (\xi_1,\xi_2,\xi_3,\xi_4),$$

$$\hat{E}^* = \zeta\eta\xi_1 + \sum_{n=1}^{\infty} t^{2n} \mu^n g^{(n)}(\boldsymbol{\xi};\zeta,\eta).$$

It turns out that E^* and $\hat{E}^*$ must satisfy the following recursion relations,

$$p_{11}^{(1)} = 0 = q_{11}^{(1)} ,$$

$$p_{11}^{(2)} = -\frac{4}{3}\eta^2\zeta^2 Q^* ,$$

$$q_{11}^{(2)} = -\frac{4}{3}\eta^3\zeta^3\xi_1 Q^* ,$$

and that both $p^{(n)}$ and $q^{(n)}$ satisfy the following equations for $n \geq 1$

$$p_{11}^{(n+2)} = \frac{-4}{(2n+3)(2n+1)}\Big\{(2n+1)\Big(p_{113}^{(n+1)} + p_{114}^{(n+1)} + p_{123}^{(n+1)} - p_{134}^{(n+1)}\Big)$$

$$+ \zeta^2\eta^2 Q^* p^{(n)} + p_{1133}^{(n)} + p_{1144}^{(n)} + p_{2233}^{(n)} + p_{3344}^{(n)} + 2\,p_{1134}^{(n)}$$

$$(8.41)\qquad + 2\,p_{1233}^{(n)} + 2\,p_{1234}^{(n)} - 2\,p_{1334}^{(n)} - 2\,p_{1344}^{(n)} - 2\,p_{2334}^{(n)}\Big\}$$

Again the details are omitted and the reader is directed to [Ku 1]. We assume, moreover, that the $p^{(n)}$, $q^{(n)}$, satisfy the initial conditions

$$p^{(n)}(0,\xi_2,\xi_3,\xi_4,\zeta,\eta) = q^{(n)}(0,\xi_2,\xi_3,\xi_4,\zeta,\eta) = 0 ,$$

(8.42)

$$p_1^{(n)}(0,\xi_2,\xi_3,\xi_4,\zeta,\eta) = q_1^{(n)}(0,\xi_2,\xi_3,\xi_4,\zeta,\eta) = 0.$$

Using the method of majorants it can be shown that such E-functions exist [Ku 1], and that they satisfy the initial conditions

$$E^*(0,\xi_2,\xi_3,\xi_4,\zeta,\eta,t) = 1\,,\ E_1^*(0,\xi_2,\xi_3,\xi_4,\zeta,\eta,t) = 0\,,$$

$$\hat{E}^*(0,\xi_2,\xi_3,\xi_4,\zeta,\eta,t) = 0\,,\ \hat{E}_1^*(0,\xi_2,\xi_3,\xi_4,\zeta,\eta,t) = \eta\zeta. \tag{8.43}$$

Since $\hat{Q}(x_1,x_2,x_3,x_4)$ is real valued for real x_i, the operator

$$u(x_1,x_2,x_3,x_4) \equiv U(Y,Y^*,Z,Z^*) = \mathrm{Re}\ \underset{\sim}{K}_4^{(2)}\{f,\hat{f}\}$$

$$\equiv \mathrm{Re}\Big\{\frac{-1}{4\pi^2}\int_{|\zeta|=1}\int_{|\eta|=1}\int_{\gamma} E(Y,Y^*,Z,Z^*,\zeta,\eta,t)\,f(\omega,\zeta,\eta)\ \cdot$$

$$\cdot\ \frac{dt}{\sqrt{1-t^2}}\frac{d\eta}{\eta}\frac{d\zeta}{\zeta} - \frac{1}{4\pi^2}\int_{|\zeta|=1}\int_{|\eta|=1}\int_{\gamma}\hat{E}(Y,Y^*,Z,Z^*,\zeta,\eta,t)\,\hat{f}(\omega,\zeta,\eta)$$

$$\cdot\ \frac{dt}{\sqrt{1-t^2}}\frac{d\eta}{\eta}\frac{d\zeta}{\zeta}$$

carries pairs of analytic functions of three variables into real valued solutions of (8.2). We wish to show that this mapping is onto the class of strong solutions of (8.2). In fact we can show that this is the case by recalling Theorem (8.1) and showing that the following relationships hold between the Goursat data $F(Y,Z,Z^*)$, $G(Y,Z,Z^*)$ and $g(\mu,\zeta)$, $\hat{g}(\mu,\zeta)$:

$$g(\mu,\zeta,\eta) \equiv \frac{\partial^2}{\partial\mu^2}\int_0^1\int_0^1 \Big\{2\, F(\mu t,\mu\zeta(1-t)s,\ \mu\eta(1-t)(1-s)) - F(0,\mu\zeta(1-t)s,\ \mu\eta(1-t)(1-s))\Big\}\,\mu^2(1-t)\,dt\,ds\ ,$$

and

$$\hat{g}(\mu,\zeta,\eta) \equiv \frac{\partial^2}{\partial\mu^2}\int_0^1\int_0^1 \mu^2(1-t)\Big\{2\, G(\mu t,\mu\zeta(1-t)s,\mu\eta(1-t)(1-s)) - 2\, G(0,(1-t)s\mu\zeta,(1-t)(1-s)\mu\eta) - \mu t\, G_1(0,(1-t)s\mu\zeta,(1-t)(1-s)\mu\eta)\Big\}\,dt\,ds$$

$$- \frac{1}{\zeta\eta}\frac{\partial}{\partial\mu}\Big\{g(\mu,\zeta,\eta) - g(\mu,0,\eta) - g(\mu,\zeta,0) + g(\mu,0,0)\Big\}\ , \tag{8.45}$$

where the $g,\hat{g}$ are related to the $f,\hat{f}$ by

$$f(\mu,\zeta,\eta) \equiv -\frac{1}{2\pi}\int_{\gamma'} g(\mu[1-t^2],\zeta,\eta)\frac{dt}{t^2} ,$$

$$\hat{f}(\mu,\zeta,\eta) \equiv -\frac{1}{2\pi}\int_{\gamma'} \hat{g}(\mu[1-t^2],\zeta,\eta)\frac{dt}{t^2} , \tag{8.46}$$

where γ' is a rectifiable arc joining the points $t=-1$ to $t=1$ but not passing through the origin. For details see [Ku 1] or [G K 4].

As in the three variable case we are able to construct, for a bounded simply connected domain, a complete family of solutions to (8.2), namely

$$U_{4n,m,\ell} = \text{Re}\ \underset{\sim}{K}_4^{(2)}\{\mu^n\zeta^m\eta^\ell ; 0\} ,$$

$$U_{4n+1,m,\ell} = \text{Re}\ \underset{\sim}{K}_4^{(2)}\{0 ; \mu^n\zeta^m\eta^\ell\} ,$$

$$U_{4n+2,m,\ell} = \text{I}_m\ \underset{\sim}{K}_4^{(2)}\{\mu^n\zeta^m\eta^\ell ; 0\} ,$$

$$U_{4n+3,m,\ell} = \text{I}_m\ \underset{\sim}{K}_4^{(2)}\{0 ; \mu^n\zeta^m\eta^\ell\} , \tag{8.47}$$

$$n \geq 0 , \quad 0 \leq m+\ell \leq n .$$

It would be very useful if the methods discussed in this and Chapter VI could be simply extended to the case of dimension $n > 4$. However, this is not possible because of a basic discovery due to Kukral [Ku 1,2], showing that there are no Bergman type operators for $n > 4$. We present below Kukral's important contribution to this subject.

Theorem (8.3): There is no auxillary variable μ, which in the sense of Bergman-Whittaker (for $n = 3$), and Gilbert (for $n = 4$), generates the homogeneous linearly independent harmonic polynomials in $n \geq 5$.

Proof: Suppose on the contrary there exists one, i.e.

(8.48) $\mu = A_1 N_1 + \ldots + A_n N_n$,

where the set $\{A_j\}$ is a set of nontrivial, linearly independent homogeneous polynomials of degree one in the variables $x_1, \ldots, x_n$, and the $\{N_i\}$ are independent products of $\zeta_1, \zeta_2, \ldots, \zeta_p$ and/or $\zeta_1^{-1}, \zeta_2^{-1}, \ldots, \zeta_p^{-1}$, $(n = p + 2)$. We must have μ^m harmonic for $m \geq 0$; in particular μ^2 is harmonic, and

(8.49) $$\mu^2 = A_1^2 N_1^2 + 2 A_1 A_2 N_1 N_2 = \dots + A_n^2 N_n^2 .$$

In (8.49) there can be at most

$$n + (n-1) + \dots + 3 + 2 + 1 = \frac{n(n+1)}{2}$$

distinct terms. However, there are

(8.50) $$h(2,p) = n(2,n-2) = \frac{(n-1)(n+2)}{2}$$

homogeneous linearly independent polynomials of degree two in n variables [Er 1, Vol II., pg. 237]. Since $\frac{1}{2}[n(n+1) - (n-1)(n+2)] = 1$, in order to have the proper number of polynomials of degree two as coefficients of the $N_i N_j$ in μ^2 exactly two terms must collapse into one term, no more and no less.

Problem: Show that for the Bergman operator and the Gilbert operator two terms in the expansion of μ^2 do collapse.

Since $n \geq 5$ and the A_j are distinct, there is at least one term (which without loss of generality we take to be A_1) such that the set

$$\{A_1 A_j : j = 1, 2, \dots, n\}$$

is a set of n homogeneous linearly independent harmonic polynomials of degree two, i.e. each member of the set will appear as a distinct coefficient in μ^2.

The terms A_j are of the form

$$A_j = a_1^{(j)} x_1 + a_2^{(j)} x_2 + \dots + a_n^{(j)} x_n .$$

Now since the set $\{A_j\}$ are non-trivial and linearly independent, the n vectors of the set

$$\{(a_1^{(j)}, \dots, a_n^{(j)}) : j = 1, \dots, n\}$$

are a basis for $\mathbb{C}^n$. However, the condition that $A_1 A_j$ is harmonic $(j = 1,\dots,n)$ is equivalent to the condition that the inner products

$$\overline{\left\langle \left(a_1^{(1)}, a_2^{(1)}, \dots, a_n^{(1)}\right), \left(a_1^{(j)}, \dots, a_n^{(j)}\right)\right\rangle} = 0$$

vanish for $j = 1,\dots,n$. This means that the vector $\left(a_1^{(1)}, a_2^{(1)}, \dots, a_n^{(1)}\right)$ is orthogonal to each member of a basis for $\mathbb{C}^n$, which is a contradiction.

<u>Problem</u>: Show explicitly why this argument does not apply to the cases $n = 2,3,4$, but must be true for $n = 5$.

Since it is still possible to find operators which generate complete families of harmonic functions, we should not abandon the project of finding operators for the differential equations (8.3) and (8.4). We consider first

(8.3) $$\Delta_n u + \hat{Q}(x_1, \ldots, x_n)u = 0 , \quad n \geq 5.$$

Our scheme [GK 5] is to introduce new Bergman type coordinates

(8.51) $$X_i = x_i , \quad i = 1,2,\ldots,n-2$$
$$Z = \tfrac{1}{2}\left(x_{n-1} + i\, x_n\right) , \quad Z^* = \tfrac{1}{2}\left(-x_{n-1} + i\, x_n\right) ,$$

so that (8.3) becomes

(8.52) $$\frac{\partial^2 U}{\partial X_1^2} + \ldots + \frac{\partial^2 U}{\partial X_{n-2}^2} - \frac{\partial^2 U}{\partial Z \partial Z^*} + Q(X_1,\ldots,X_{n-2},Z,Z^*)U = 0 ,$$

where $Q(X_1,\ldots,X_{n-2},Z,Z^*) \equiv \hat{Q}(x_1,\ldots,x_n)$,

and $U(X_1,\ldots,X_{n-2},Z,Z^*) \equiv u(x_1,\ldots,x_n)$.

The functions $N_p(\zeta_1,\ldots,\zeta_{n-3})$, $(p=1,\ldots,n-2)$, are to be chosen in such a way that they are regular and that

$$(8.53)\qquad \sum_{p=1}^{n-2} N_p^2 = 1 \ ,$$

$$(8.54)\qquad N_1 N_2 \cdots N_{n-2} \neq 0 \ ,$$

for $1-4\varepsilon < |\zeta_p| < 1+4\varepsilon$, $0<\varepsilon<1/8$, $(p=1,\ldots,n-3)$.

<u>Problem</u>: Show that such a collection of N_p are given, for example by

$$N_1 = \zeta_1 + \frac{1}{4\zeta_1} \ ,$$

$$N_2 = i\left(\zeta_1 - \frac{1}{4\zeta_1}\right)\left(\zeta_2 + \frac{1}{4\zeta_2}\right) \ ,$$

$$N_3 = i\left(\zeta_1 - \frac{1}{4\zeta_1}\right)\left(\zeta_2 - \frac{1}{4\zeta_2}\right)\left(\zeta_3 + \frac{1}{4\zeta_3}\right) \ ,$$

$$\vdots$$

$$N_{n-3} = i^{n-2}\left(\zeta_1 - \frac{1}{4\zeta_1}\right)\cdots\left(\zeta_{n-4} - \frac{1}{4\zeta_{n-4}}\right)\left(\zeta_{n-3} + \frac{1}{4\zeta_{n-3}}\right) \ ,$$

(8.55) $$N_{n-2} = i^{n-1}\left(\zeta_1 - \frac{1}{4\zeta_1}\right)\cdots\left(\zeta_{n-4} - \frac{1}{4\zeta_{n-4}}\right)\left(\zeta_{n-3} - \frac{1}{4\zeta_{n-3}}\right) .$$

Problem: Show that the set of $\{N_p\}$ given in [Gi 1] on page 82 do not satisfy the conditions (8.53).

Because of condition (8.54) the following transformation has non-zero Jacobian

(8.56)
$$\begin{aligned}
\xi_1 &= 2\zeta Z \\
\xi_p &= N_p X_p + 2\zeta N_p^2 Z \ , \quad 2 \le p \le n-2 \\
\xi_{n-1} &= N_{n-1} X_1 + 2\zeta N_{n-1}^2 Z \\
\xi_n &= \sum_{p=1}^{n-2} N_p X_p + 2\zeta^{-1} Z^* \ ,
\end{aligned}$$

where $1-\varepsilon < |\zeta| < 1+\varepsilon$, $\varepsilon < 1/8$,

and for convenience we have let $N_{n-1} \equiv N_1$, so that

$$\sum_{p=2}^{n-1} N_p^2 = \sum_{p=1}^{n-2} N_p^2 = 1 \quad .$$

Using the new coordinates (8.56) we can conveniently introduce the auxillary variable

$$\mu = \tfrac{1}{2}\left(\xi_2 + \cdots + \xi_n\right) = \sum_{p=1}^{n-2} N_p X_p + \zeta Z + \zeta^{-1} Z^* , \tag{8.57}$$

which has the property that $\Delta_n \mu^m \equiv 0$.

We are now in a position to consider whether solutions may be generated by an operator of the form

$$U(X_1,\ldots,X_n,Z,Z^*) \equiv \underset{\sim}{P}_n\{f\} \equiv$$

$$\left(\frac{1}{2\pi i}\right)^{n-2} \int\limits_{|\zeta_{n-3}|=1} \cdots \int\limits_{|\zeta_1|=1} \int\limits_{|\zeta|=1} E^*(\xi_1,\ldots,\xi_n,\zeta_1,\ldots,\zeta_{n-3},\zeta,t)$$

$$\cdot f\left(\mu[1-t^2];\zeta_1,\ldots,\zeta_{n-3},\zeta\right)\frac{dt}{\sqrt{1-t^2}}\frac{d\zeta}{\zeta}\frac{d\zeta_1}{\zeta_1}\cdots\frac{d\zeta_{n-3}}{\zeta_{n-3}} , \tag{8.58}$$

where γ is a path joining $t = -1$ to 1. It may be shown that a necessary condition on E^* is that it be a regular solution of the partial differential equation

$$-\frac{(1-t^2)}{t\mu} E^*_{1t} + \frac{E^*_1}{t^2\mu} - 2\sum_{p=2}^{n-1} E^*_{pn} N_p^2$$

(8.59)

$$+\sum_{p=2}^{n-1} E^*_{pp} N_p^2 - 4E^*_{1n} + E^*_{nn} + Q^* E^* = 0 .$$

Furthermore, it may be shown [GK 5] that the E-function has a formal expansion

(8.60) $$E^* = 1 + \sum_{\ell=1}^{\infty} t^{2\ell} \mu^{\ell} p^{(\ell)}(\underline{\xi}\,;\underline{\zeta}),$$

where the coefficients $p^{(\ell)}$ are defined recursively by

$$p_1^{(\ell+1)} = \frac{1}{2\ell+1}\Big\{Q^* p^{(\ell)} + \sum_{i=2}^{n-1} p_{ii}^{(\ell)} N_i^2 - 2\sum_{i=2}^{n-1} p_{in}^{(\ell)} N_i^2$$

(8.61) $$- 4p_{1n}^{(\ell)} + p_{nn}^{(\ell)}\Big\},\ (\ell = 0,1,2,\ldots),$$

(8.62) $$p^{(\ell+1)}(0,\xi_2,\ldots,\xi_n,\zeta_1,\ldots,\zeta) = 0,\quad (\ell = 0,1,2,\ldots).$$

Using (8.61) and the method of majorants the formal expansion (8.60) may be shown to converge uniformly to an actual solution of (8.59) when Q^* is holomorphic.

Moreover, the E-function is seen to satisfy the initial data

$$(8.63) \qquad E^{*}(0,\xi_2,\dots,\xi_n,\zeta_1,\dots,\zeta_{n-3},\zeta,t) = 1 .$$

If there was a Bergman-type operator for $n \geq 5$ one could extend our previous arguments for showing the mapping is onto. It is a conjecture that functions $N_p(\zeta)$ exist which generate a sufficient number of harmonic polynomials. We expect that the set (8.55) actually does this, but have been unable to establish this result to date.

We turn next to the fourth order equation

$$(8.64) \qquad \Delta_n^2 u + \hat{Q}(x_1,\dots,x_n)u = 0 .$$

Again we introduce the variables (8.57), which transforms this into

$$(8.65) \qquad \frac{\partial^2 U}{\partial X_1^2} + \dots + \frac{\partial^2 U}{\partial X_{n-2}^2} - \frac{\partial^2 U}{\partial Z \partial Z^*} + QU = 0 .$$

We seek a solution to (8.65) in the form (8.58). Introducing the coordinates (8.56), it results that the E-function must satisfy

$$\sum_{j=2}^{n-1}\sum_{i=2}^{n-1} N_j^2 N_i^2\left(E^*_{iijj} - 4E^*_{ijjn} + 4E^*_{ijnn}\right)$$

$$+ \sum_{j=2}^{n-1} N_j^2\Big\{\left(2E^*_{jjnn} - 8E^*_{1jjn} - 4E^*_{jnnn} + 16E^*_{1jnn}\right)$$

$$- \frac{(1-t^2)}{t\mu}\left(2E^*_{1jjt} - 4E^*_{1jnt}\right) + \frac{1}{t^2\mu}\left(2E^*_{1jj} - 4E^*_{1jn}\right)\Big\}$$

$$+ \left\{E^*_{nnnn} - 8E^*_{1nnn} + 16E^*_{11nn}\right\} - \frac{(1-t^2)}{t\mu}\left(2E^*_{1nnt} - 8E^*_{11nt}\right)$$

$$+ \frac{1}{t^2\mu}\left(2E^*_{1nn} - 8E^*_{11n}\right) + \frac{3(t^2-1)(t^2+1)}{t^3\mu^2}E^*_{11t}$$

$$+ \frac{3}{t^4\mu^2}E^*_{11} + \frac{(1-t^2)^2}{t^2\mu^2}E^*_{11tt} + Q^*E^* = 0\ . \tag{8.66}$$

The proof is by the usual methods, but is very detailed hence we omit it ([Kul] pp. 191 - 196). In order to show that two independent E-functions exist we seek solutions to (8.66) in the form (8.60). The functions $p^{(n)}$ are defined recursively by

$$p^{(0)} \equiv 1 \ , \ p^{(1)} \equiv 0 \ ,$$

$$p_{11}^{(\ell+2)} = \frac{-1}{(2\ell+1)(2\ell+3)} \Big\{ (2\ell+1) \Big[8p_{11n}^{(\ell+1)} - 2p_{1nn}^{(\ell+1)}$$

$$+ \sum_{j=2}^{n-1} N_j^2 \Big(4p_{1jn}^{(\ell+1)} - 2p_{1jj}^{(\ell+1)} \Big) \Big] + \sum_{j=2}^{n-1} N_j^2 \Big[2\, p_{jjnn}^{(\ell)}$$

$$- 8\, p_{1jjn}^{(\ell)} - 4\, p_{jnnn}^{(\ell)} + 16\, p_{1jnn}^{(\ell)} + \sum_{i=2}^{n-1} N_i^2 \Big(p_{iijj}^{(\ell)}$$

$$- 4\, p_{ijjn}^{(\ell)} + 4\, p_{ijnn}^{(\ell)} \Big) \Big] + p_{nnnn}^{(\ell)} - 8\, p_{1nnn}^{(\ell)}$$

$$+ 16\, p_{11nn}^{(\ell)} + Q^* p^{(\ell)} \Big\} \ , \ \ell \geq 0 \ , \tag{8.67}$$

with initial conditions

$$p^{(\ell+2)}(0, \xi_2, \ldots, \zeta) = 0 \ ,$$

$$p_1^{(\ell+2)}(0, \xi_2, \ldots, \zeta) = 0 \ , \ \ell \geq 0. \tag{8.68}$$

For the second E-fiјction we start the recursive scheme with $q^{(0)} \equiv \xi_1$, $q^{(1)} \equiv 0$, and replace Q^* in (8.67) by

$\xi_1 Q^*$. These formulae are obtained by, what is now for us, a standard method.

In the same manner, as previously demonstrated, one may show that the series solution converges uniformly to a holomorphic function [Ku 1] pg. 199 - 228.

IX. Initial Value Problems for Semilinear Elliptic Equations

Initial value problems for elliptic equations turn up quite naturally when one considers some of the inverse problems of mathematical physics. For instance, the study of shocks in compressible fluids [GL 1], [GK 1], and inclusions in elasticity give rise to such problems. The location of mineral deposits by graviometric means also suggest the investigation of initial value problems for an elliptic equation.

The difficulties which occur in constructing efficient numerical procedures for approximating solutions of initial value problems are due to the fact that these problems do not have the property that their solutions depend continuously on the boundary data, i.e. they are not well posed in the sense of Hadamard. Because of this, constructive schemes which provide stable, efficient numerical procedures for the computation of such solutions are extremely useful.

The author with A. K. Aziz and H. C. Howard [AG 1] [AGH 1], were the first to investigate holomorphic, semilinear, initial value problems using the function theoretic approach outlined below and also in the book [Gi 1]. Their first researches were concerned with iteration schemes for the semilinear elliptic equation.

Further general information concerning non-well-posed problems can be obtained from the interesting papers by L. E. Payne [Pa 1], D. Colton [Co 1,2] and the references therein.

$$\Delta u = f(x,y,u,u_x,u_y) \tag{9.1}$$

with generalized Goursat data given on the analytic arcs Γ_k, $(k=1,2)$,

$$u_x(x,y) = a_0^{(1)}(x)u(x,y) + a_1^{(1)}(x)u(x,y) + a_2^{(1)}(x), \tag{9.2}$$

on $\Gamma_1 = \{(x,y): y = f_1(x)\}$, and

$$u_y(x,y) = a_0^{(2)}(y)u(x,y) + a_1^{(2)}(y)u(x,y) + a_2^{(2)}(y), \tag{9.3}$$

on $\Gamma_2 = \{(x,y): x = f_2(y)\}$, with $f_1(0) = f_2(0) = 0$.

The approach taken was to analytically continue the differential equation into the complex domain, where it is formally hyperbolic, and then to solve the resulting operator equations by seeking a fixed point in a certain

Banach space of holomorphic functions. See the book [Gi 1] by the author for further details. This device has been subsequently applied by Colton [Co 1, 2] to investigate the Cauchy problem for the differential equations (9.1), and also the analogous equation of order $2n$.

The author with James Conlan [CG 4-6] has extended the above results to higher order, semi-linear elliptic equations with linear and also non-linear initial data. In this chapter we report solely on the, quite general cases, treated recently by the author with Conlan [CG 4-6]. For a discussion of earlier work see the references in the above cited works.

We consider first the fourth order equation

$$\Delta\Delta u = f(x,y,u,u_x,u_y,u_{xx},u_{xy},u_{yy},\Delta u_x,\Delta u_y) , \tag{9.4}$$

with initial data given on analytic carriers $y = f_k(x)$, $(k = 1,3)$, and $x = f_k(y)$, $(k = 2,4)$:

$$\begin{aligned} & a_1^{(k)} \Delta u_x + a_2^{(k)} \Delta u_y + a_3^{(k)} u_{xx} + a_4^{(k)} u_{yy} + a_5^{(k)} u_{xy} \\ & + a_6^{(k)} u_x + a_7^{(k)} u_y + a_8^{(k)} u + a_9^{(k)} = 0 \end{aligned} \tag{9.5}$$

where, for $k = 1,3, a_i^{(k)} = a_i^{(k)}(x)$, $(i = 1,\dots,9)$, on $y = f_k(x)$, and for $k = 2,4, a_i^{(k)} = a_i^{(k)}(y)$, $(i = 1,\dots,9)$, on $x = f_k(y)$. We assume, furthermore, that $f_k(0) = 0$, $k = 1,2,3,4$, and that the coefficients $a_i^{(k)}$ are analytic on their respective carriers.

The following assumptions are made concerning the data [CG 4],

H_1: No two carriers are tangent, and each carrier has an analytic extension into the bicylinder $\{|x| \le p\} \times \{|y| \le p\}$.

H_2: <u>The Goursat data is normalized* so that</u>

$$a_1^{(1)} + ia_2^{(1)} = a_1^{(2)} - ia_2^{(2)} = -\frac{1}{4} ,$$

$$a_3^{(4)} - a_4^{(4)} - ia_5^{(4)} = -1 = a_3^{(3)} - a_4^{(3)} + ia_5^{(3)} . \tag{9.6}$$

H_3: <u>The function</u> $f(x,y,p)$, $p = (p_1,\dots,p_8)$, <u>is holomorphic for</u> $(z,z^*,p) \in \mathfrak{G} \times \mathfrak{G}^* \times D$, $(z = x + iy,\ z^* = x - iy)$, <u>where</u> $\mathfrak{G}$ <u>is a simply connected domain in</u> $\mathbb{C}^1$, $\mathfrak{G}^*$ <u>is its conjugate domain, and</u> D <u>is a polydisk in</u> $\mathbb{C}^8$.

---

*This condition turns out to be the fourth order analogue of conditions (9.2), (9.3) for the second order case, and is primarily a convenience.

The basic idea is to introduce the complex variables $z = x + iy$, $\zeta = x - iy$ and to show that a solution exists to the corresponding hyperbolic equation by using fixed point arguments. The differential equation (9.1) becomes upon setting $U(z,\zeta) \equiv u\left(\frac{z+\zeta}{2}, \frac{z-\zeta}{2i}\right)$,

$$\frac{\partial^4 U}{\partial z^2 \partial \zeta^2} = F(z,\zeta,U,U_z,U_\zeta,U_{z\zeta},U_{zz},U_{\zeta\zeta},U_{zz\zeta},U_{z\zeta\zeta}) \tag{9.7}$$

and the data (9.6) becomes

$$\begin{aligned} &\alpha_1^{(k)} U_{zz\zeta} + \alpha_2^{(k)} U_{z\zeta\zeta} + \alpha_3^{(k)} U_{zz} + \alpha_4^{(k)} U_{\zeta\zeta} + \alpha_5^{(k)} U_{z\zeta} \\ &+ \alpha_6^{(k)} U_z + \alpha_7^{(k)} U_\zeta + \alpha_8^{(k)} U + \alpha_9^{(k)} = 0 , \end{aligned} \tag{9.8}$$

where $\alpha_i^{(k)} = \alpha_i^{(k)}(z)$ on $\zeta = g_k(z)$, $(k = 1,3)$, $(i = 1,\dots,9)$, and where $\alpha_i^{(k)} = \alpha_i^{(k)}(\zeta)$ on $z = g_k(\zeta)$, $(k = 2,4)$, $(i = 1,\dots,9)$.

The coefficients $\alpha_i^{(k)}$ are related to the $a_i^{(k)}$ by the following formulae when $k = 1,3$:

$$\begin{aligned} \alpha_1^{(k)}(z) &\equiv 4\left[a_1^{(k)} + ia_2^{(k)}\right]\left(\frac{z + g_k(z)}{2}\right) . \\ \alpha_2^{(k)}(z) &\equiv 4\left[a_1^{(k)} - ia_2^{(k)}\right]\left(\frac{z + g_k(z)}{2}\right). \end{aligned} \tag{9.9}$$

$$\alpha_3^{(k)}(z) \equiv \left[a_3^{(k)} - a_4^{(k)} + ia_5^{(k)}\right]\left(\frac{z + g_k(z)}{2}\right) ;$$

a similar relation holds in the cases $k = 2,4$.

The condition H_1 assures us there exists a conformal mapping that takes the curves into separate lines,

$$L_k \equiv \{(w,w^*) : w^* = we^{i\lambda_k}\} , \quad (k = 1,3) ,$$

(9.10)

$$L_k \equiv \{(w,w^*) : w = w^*e^{i\lambda_k}\} , \quad (k = 2,4) .$$

This mapping does not change the essential form of the differential equation (9.7) nor that of the data (9.8). Consequently, we will assume the equation has data prescribed on the rays (9.10).

Suppose a solution to our problem (9.7 , 9.8) exists, and let us define $s(z,\zeta) \equiv U_{zz\zeta\zeta}$. Then it follows that

$$U(z,\zeta) = \left(\int_0^z d\xi\right)^2 \left(\int_0^\zeta d\eta\right)^2 s(\xi,\eta)$$

$$+ \zeta\left(\int_0^z d\xi\right)^2 \Phi_1(\xi) + z\left(\int_0^\zeta d\eta\right)^2 \psi_1(\eta)$$

$$+ \left(\int_0^z d\xi \right)^2 \Phi_2(\xi) + \left(\int_0^\zeta d\eta \right)^2 \psi_2(\eta)$$

(9.11) $$+ c_1 z\zeta + c_2 \equiv \underset{\sim}{K}_{00}(s, \chi)$$

In a similar manner we introduce the integral operators $\underset{\sim}{K}_{pq}(s,\chi)$, $p+q \leq 3$, such that

(9.12) $$\frac{\partial^{p+q} U(z,\zeta)}{\partial z^p \partial \zeta^q} = \underset{\sim}{K}_{pq}(s,\chi),$$

and where χ is the fourtuple $\chi \equiv (\Phi_1, \psi_1, \Phi_2, \psi_2)$. Using these operators the initial conditions (9.8) may be written as an operator equation [CG 4]

(9.13) $$(\underset{\sim}{I} - \underset{\sim}{H} - \underset{\sim}{L})\chi = \Omega$$

where

$$(\underset{\sim}{I}\chi)(z,\zeta) \equiv \chi = \left(\Phi_1(z), \psi_1(\zeta), \Phi_2(z), \psi_2(\zeta)\right)$$

$$(\underset{\sim}{H}\chi)(Z,\zeta) = \left((\underset{\sim}{H}_1\chi)(z), (\underset{\sim}{H}_2\chi)(\zeta), (\underset{\sim}{H}_3\chi)(z), (\underset{\sim}{H}_4\chi)(\zeta) \right),$$

(9.14) $$(\underset{\sim}{L}\chi)(z,\zeta) = \left((\underset{\sim}{L}_1\chi)(z), (\underset{\sim}{L}_2\chi)(\zeta), (\underset{\sim}{L}_3\chi)(z), (\underset{\sim}{L}_4\chi)(\zeta) \right)$$

$$\Omega(z,\zeta) = \left(\Omega_1(z), \Omega_2(\zeta), \Omega_3(z), \Omega_4(\zeta)\right) .$$

Here the $\underset{\sim}{L}_k\chi$ $(k = 1,2,3,4)$ consists of all those terms that contain integrals of the Φ_i, ψ_i, $(i = 1,2)$. The Ω_k consist of terms not involving the Φ_i, ψ_i, and the $\underset{\sim}{H}_k$ involve linear combinations of the Φ_i, ψ_i and compositions.

<u>Problem</u>: Let X be the linear space of fourtuples χ. Show that X becomes a Banach space upon introducing the norm

$$\|\chi\|_\lambda \equiv \max\{\|\Phi_1\|_\lambda, \|\psi_1\|_\lambda \cdot \|\Phi_2\|_\lambda, \|\psi_2\|_\lambda\}, \quad \chi \in X,$$

$$(9.15) \qquad \|\Phi_i\|_\lambda \equiv \sup\{e^{-\lambda|z|}|\Phi_i(z)| : z \in \mathfrak{G}\}, \quad \lambda \geq 0,$$

$$\|\psi_i\|_\lambda \equiv \sup\{e^{-\lambda|\zeta|}|\psi_i(\zeta)| : \zeta \in \mathfrak{G}^*\}, \quad \lambda \geq 0.$$

We denote this Banach space by $A(\mathfrak{G}, \mathfrak{G}^*, \mathfrak{G}, \mathfrak{G}^*) = A$.

The problem of showing that (9.13) has a unique solution can be simplified by showing that $(\underset{\sim}{I} - \underset{\sim}{H})^{-1}$ exists, and reducing (9.13) to the equivalent operator equation,

$$(9.16) \qquad (\underset{\sim}{I} - (\underset{\sim}{I} - \underset{\sim}{H})^{-1}\underset{\sim}{L})\chi = (\underset{\sim}{I} - \underset{\sim}{H})^{-1}\Omega.$$

To show $(\underset{\sim}{I} - \underset{\sim}{H})^{-1}$ exists we show that $\underset{\sim}{H}^n$ is a contraction. This is a technical result; for further details see [CG 4]. It is also a technical argument to obtain the estimate

$$\|\underset{\sim}{L}\chi\| \le \rho\lambda^{-1} M_2 \|\chi\|_\lambda \ .$$

Consequently, one has the bound

$$\|(\underset{\sim}{I} - \underset{\sim}{H})^{-1} L\|_\lambda \le \|(\underset{\sim}{I} - \underset{\sim}{H})^{-1}\|_\lambda \cdot \|\underset{\sim}{L}\|_\lambda \le \frac{c}{\lambda}\ , \quad c < \infty\ ,$$

from which it is clear that $[\underset{\sim}{I} - (\underset{\sim}{I} - \underset{\sim}{H})^{-1}\underset{\sim}{L}]^{-1}$ exists on A.

To show that a unique solution exists for our Goursat problem we consider the operator,

$$\begin{aligned} T\underset{\sim}{s}(z,\zeta) \equiv f\Big(z,\zeta,&\underset{\sim}{K}_{00}(s,\chi),\underset{\sim}{K}_{10}(s,\chi)\ ,\\ &\underset{\sim}{K}_{01}(s,\chi),\underset{\sim}{K}_{20}(s,\chi),\underset{\sim}{K}_{11}(s,\chi)\ ,\\ &\underset{\sim}{K}_{02}(s,\chi),\underset{\sim}{K}_{21}(s,\chi),\underset{\sim}{K}_{12}(s,\chi)\Big) \end{aligned}$$

where the $\underset{\sim}{K}_{ij}$ are defined above. The functions $s(z,\zeta)$ are elements in the Banach space $A(\Delta_\rho\ , \Delta_\rho^*)$ of bounded holomorphic functions of two complex variables, with the norm

$$\|s\|_\lambda = \sup\{e^{-\lambda(|z|+|\zeta|)}|s(z,\zeta)| : (z,\zeta) \in (\Delta_\rho, \Delta_\rho^*)\}.$$

It can be shown that the mapping $\underset{\sim}{T}$ has a fixed point in $A(\Delta_\rho, \Delta_\rho^*)$ by using some estimates on the operator norms $\|\underset{\sim}{K}_{ij}(s_k, \chi_k)\|_\lambda$ and the fact that F is holomorphic [CG 4].

The approach above also is useful in treating the vector elliptic equation [CG 5]

(9.17)
$$\Delta\underset{\sim}{u} \equiv \frac{\partial^2 \underset{\sim}{u}}{\partial x^2} + \frac{\partial^2 \underset{\sim}{u}}{\partial y^2} = f(x,y,\underset{\sim}{u},\underset{\sim}{u}_x,\underset{\sim}{u}_y) ,$$
$$\underset{\sim}{u} \equiv (u^{(1)},u^{(2)},\dots,u^{(n)}), \quad \underset{\sim}{f} \equiv (f^{(1)},f^{(2)},\dots,f^{(n)}),$$

with non-linear Goursat data,

(9.18)
$$u_x^{(k)} = h_1^{(k)}(x,\underset{\sim}{u},\underset{\sim}{u}_x,\underset{\sim}{u}_y) \quad \text{on} \quad C_1^{(k)}: y = g_1^{(k)}(x)$$
$$u_y^{(k)} = h_2^{(k)}(y,\underset{\sim}{u},\underset{\sim}{u}_x,\underset{\sim}{u}_y) \quad \text{on} \quad C_2^{(k)}: x = g_2^{(k)}(y) ,$$

$k = 1,2,\dots,n.$

We assume the following about the data:

H_1: $\underset{\sim}{f}$ is holomorphic in its $3n+2$ variables

H_2: $h_j^{(k)}$ is holomorphic in its $3n+1$ variables

H_3: $C_j^{(k)}$ have analytic extensions in a bidisk and

$g_j^{(k)}(0) = 0$

$k = 1,\ldots,n\ ,\quad j = 1,2.$

H_4: No two curves $C_j^{(k)}$ are tangent at the origin.

Under these hypotheses it is possible to reduce the problem to the complex, hyperbolic, Goursat problem

$$
\begin{aligned}
&\underset{\sim}{U}_{z\zeta} = \underset{\sim}{F}(z,\zeta,\underset{\sim}{U},\underset{\sim}{U}_z,\underset{\sim}{U}_\zeta)\ , \\
(9.19)\qquad &\underset{\sim}{U}_z^{(k)} = H_1^{(k)}(z,\underset{\sim}{U},\underset{\sim}{U}_z,\underset{\sim}{U}_\zeta)\ ,\quad \text{on}\quad \zeta = G_1^{(k)}(z)\ , \\
&\underset{\sim}{U}_\zeta^{(k)} = H_2^{(k)}(z,\underset{\sim}{U},\underset{\sim}{U}_z,\underset{\sim}{U}_\zeta)\ ,\quad \text{on}\quad z = G_2^{(k)}\zeta)
\end{aligned}
$$

$(k = 1,2,\ldots,n)$. The non-linearity of the data on the 2n curves leads to a number of technical difficulties; however, the general approach to obtaining an iterative scheme is essentially the same as in the previous case. For further details please see the paper by the author with J. Conlan [CG 5], where estimates are given that ensure the iteration converges.

The higher order Cauchy problem with non-linear data may also be treated in this way [CG 5]. As an example the previous method may be employed for

(9.20) $$\Delta^n u = f\left(x,y,u,u_x,u_y,\ldots,\frac{\partial^{\ell+m}\Delta^p u}{\partial x^\ell \partial y^m},\ldots,\frac{\partial\Delta^{n-1}u}{\partial x}\ \frac{\partial\Delta^{n-1}u}{\partial y}\right)$$

$\ell + m + 2p \leq 2n - 1$, with the data

$u(x,0) = \Phi_0(x,u)$,

$\left(\frac{\partial}{\partial y} u\right)(x,0) = \Phi_1(x,u,u_x,u_y)$,

$(\Delta u)(x,0) = \Phi_2(x,u,u_x,u_y,\Delta u)$,

$\left(\frac{\partial}{\partial y} \Delta u\right)(x,0) = \Phi_3(x,u,u_x,u_y,\Delta u,\Delta u_x,\Delta u_y)$,

.
.
.

$(\Delta^k u)(x,0) = \Phi_{2k}\left(x,u,u_x,u_y,\ldots,\frac{\partial^{p+q}\Delta^j u}{\partial x^p \partial y^q},\ldots\right)$ with $p+q \leq 2(k-j)$,

(9.21)

$\left(\frac{\partial}{\partial y} \Delta^k u\right)(x,0) = \Phi_{2k+1}\left(x,u,u_x,u_y,\ldots,\frac{\partial^{p+q}\Delta^j u}{\partial x^p \partial y^q}\right)$ with

$p + q \leq 2(k - j) + 1$, $k \leq 2n - 2$.

However, for the purposes of complete generality we will concentrate our exposition in the remainder of this chapter on the semi-linear elliptic system (or equation when $i = 1$) [CG 6]

(9.22) $$\Delta^{n(i)} u^{(i)} = f^{(i)}\left(x,y,u^{(1)},\ldots,u^{(m)},\ldots,\frac{\partial^{q+p}\Delta^k u^{(j)}}{\partial x^q \partial y^p},\ldots\right)$$

where $1 \le i,j \le m$, $o \le q,p \le n(j)$, $2k+q+p \le 2n(j)-1$.

We seek solutions of (9.22) with Cauchy data given on the analytic arc $\mathfrak{L}$:

$$u^{(i)}(x,y) = \Phi_o^{(i)}(x,y,\underset{\sim}{u}) \ , \ \text{where} \ \ u \equiv (U^{(1)},\ldots,u^{(m)})^T,$$

$$\frac{\partial u^{(i)}(x,y)}{\partial \nu} = \Phi_1^{(i)}\left(x,y,\underset{\sim}{u}, \frac{\partial \underset{\sim}{u}}{\partial \nu}, \frac{\partial \underset{\sim}{u}}{\partial \sigma}\right),$$

$$\Delta u^{(i)}(x,y) = \Phi_2^{(i)}\left(x,y,\underset{\sim}{u}, \frac{\partial \underset{\sim}{u}}{\partial \nu}, \frac{\partial \underset{\sim}{u}}{\partial \sigma}, \Delta \underset{\sim}{u}\right),$$

$$\vdots$$

$$(9.23) \qquad \Delta^{n(i)-1}u^{(i)}(x,y) = \Phi_{2n(i)-2}^{(i)}\left(x,y,\underset{\sim}{u}, \frac{\partial \underset{\sim}{u}}{\partial \nu}, \frac{\partial \underset{\sim}{u}}{\partial \sigma}, \ldots, \Delta^{n(j)-1}u^{(j)}, \ldots\right.$$

As is customary we introduce complex coordinates $z = x + iy$, $\zeta = x - iy$. This changes the form of (9.22) to

$$(9.24) \qquad U^{(i)}_{(n(i),n(i))}(z,\zeta) = F^{(i)}(z,\zeta,U^{(1)},\ldots,U^{(m)},\ldots,U^{(j)}_{(p,q)},\ldots),$$

where $U^{(j)}_{(p,q)}(z,\zeta) \equiv \dfrac{\partial^{p+q}U^{(j)}(z,\zeta)}{\partial z^p \partial \zeta^q}$, $1 \le i,j \le m$, $0 \le p,q \le n(j)$, $p + q \le 2n(j) - 1$. If we now make the conformal (in $\mathfrak{S}$) transformation, $z = \varphi(t)$, (introduced earlier) which takes $\mathfrak{L}$ into a segment I of the real axis in the t-plane; then (9.24) becomes

$$\left(\frac{1}{\varphi'(t)\bar{\varphi}'(\tau)} \frac{\partial^2}{\partial t \partial \tau} \right)^{n(i)} V^{(i)}(t,\tau)$$

$$= F^{(i)}\left(\varphi(t), \bar{\varphi}(\tau), V^{(1)}, V^{(2)}, \ldots, V^{(m)}, \ldots, \left(\frac{1}{\varphi'(t)} \frac{\partial}{\partial t}\right)^p \left(\frac{1}{\bar{\varphi}'(\tau)} \frac{\partial}{\partial \tau}\right)^q V^{(j)}, \ldots\right)$$

.25) $$= \hat{F}^{(i)}(t, \tau, V^{(1)}, \ldots, V^{(m)}, \ldots, V^{(j)}_{(p,q)}, \ldots).$$

Since $z = \varphi(t)$ is conformal in $\mathfrak{S}$ (and $\zeta = \bar{\varphi}(\tau)$ in $\mathfrak{S}$), $\varphi'(t) \neq 0$ (and $\bar{\varphi}'(\tau) = 0$), hence (3.4) may be written as

.26) $$V^{(i)}_{(m(i),n(i))}(t,\tau) = G^{(i)}(t, \tau, V^{(1)}, \ldots, V^{(m)}, \ldots, V^{(j)}_{(p,q)}, \ldots)$$

where the $G^{(i)}$ are holomorphic functions of their arguments. To see how the data transforms we note, as before, that on $\mathfrak{L}$

$$\frac{\partial u^{(j)}}{\partial \nu} = \frac{i}{|\varphi'(r)|} \left(V^{(j)}_{(1,0)}(r,r) - V^{(j)}_{(0,1)}(r,r) \right),$$

$$\frac{\partial u^{(j)}}{\partial \sigma} = \frac{i}{|\varphi'(r)|} \left(V^{(j)}_{(1,0)}(r,r) + V^{(j)}_{(0,1)}(r,r) \right),$$

$$\Delta u^{(j)} = |\varphi'(r)|^{-2} V^{(j)}_{(1,1)}(r,r),$$

$$\frac{\partial}{\partial \nu}\left(\Delta u^{(j)}\right) = \frac{i}{|\varphi'(r)|} \left(\frac{\partial}{\partial t} - \frac{\partial}{\partial \tau}\right) \left(\frac{V_{(1,1)}(t,\tau)}{\varphi'(t)\bar{\varphi}'(\tau)} \right)_{t=\tau=r},$$

etc. for $r \in I$. By replacing each of the terms $|\varphi'(r)|$ by $\sqrt{\varphi'(t)\bar{\varphi}'(\tau)}$ and $V^{(j)}(r,r)$ by $V^{(j)}(t,\tau)$ the right-hand-sides are seen to have holomorphic extensions into $\mathfrak{S} \times \mathfrak{S}^*$. Therefore, we may rewrite the Cauchy data in the form

$$V^{(i)}(r,r) = \Psi^{(i,0)}(r,\ldots,V^{(j)},\ldots) ,$$

$$V^{(i)}_{(1,0)}(r,r) - V^{(i)}_{(0,1)}(r,r) = \Psi^{(i,1)}(r,\ldots,V^{(j)},\ldots,V^{(j)}_{(1,0)},V^{(j)}_{(0,1)},\ldots),$$

$$V^{(i)}_{(1,1)}(r,r) = \Psi^{(i,2)}(r,\ldots,V^{(j)},V^{(j)}_{(1,0)},V^{(j)}_{(0,1)},V^{(j)}_{(1,1)},\ldots) ,$$

$$V^{(i)}_{(2,1)}(r,r) - V^{(i)}_{(1,2)}(r,r) = \Psi^{(i,3)}(r,\ldots,V^{(j)},V^{(j)}_{(1,0)},V^{(j)}_{(0,1)},V^{(j)}_{(1,1)},V^{(j)}_{(2,1)},V^{(j)}_{(1,2)},\ldots$$

$$\vdots$$

$$V^{(i)}_{(n(i),n(i)-1)}(r,r) - V^{(i)}_{(n(i)-1,n(i))}(r,r) = \Psi^{(i,2n(i)-1)}(r,\ldots,V^{(j)},\ldots,$$

$$V^{(j)}_{(n(j),n(j)-1)},V^{(j)}_{(n(j)-1,n(j))},\ldots) , \tag{9.27}$$

$i = 1,2,\ldots,m$.

Let us define

$$S^{(i)}_{(0,0)}(t,\tau) \equiv \left(\int_0^t dt\right)^{n(i)} \left(\int_0^\tau d\tau\right)^{n(i)} s^{(i)}(t,\tau) + \sum_{j=0}^{n(i)-1} c_j^{(i)} (t\tau)^j , \tag{9.28}$$

where the $c_j^{(i)}$ are constants. If a solution $V^{(i)}(t,\tau)$ exists for (9.26 - 9.27) let us define

(9.29) $$s^{(i)}(t,\tau) \equiv \frac{\partial^{2n(i)} V^{(i)}(t,\tau)}{\partial t^{n(i)} \partial \tau^{n(i)}} .$$

Then

$$V^{(i)}_{(t,\tau)} = S^{(i)}_{(0,0)} + \sum_{j=0}^{n(i)-1} \frac{\tau^j}{j!} \left(\int_0^t dt\right)^{n(i)} \varphi^{(i)}_{n(i),j}(t)$$

(9.30) $$+ \sum_{j=0}^{n(i)-1} \frac{t^j}{j!} \left(\int_0^\tau d\tau\right)^{n(i)} \varphi^{(i)}_{j,n(i)}(\tau) \equiv \underset{\sim}{K}^{(i)}_{(0,0)}(\chi),$$

where

(9.31) $$\chi(t,\tau) \equiv (,,,.\varphi^{(i)}_{n(i),j}(t),\ldots,\ldots,\varphi^{(i)}_{j,n(i)}(\tau),\ldots) ,$$

$$0 \le i \le m , \quad 0 \le j \le n(i) - 1 .$$

We also introduce the notation

$$V^{(i)}_{(p,q)}(t,\tau) = S^{(i)}_{(p,q)}(t,\tau) + \sum_{j=q}^{n(i)-1}{}^{*} \frac{\tau^{j-q}}{(j-q)!} \left(\int_0^t dt\right)^{n(i)-p} \varphi^{(i)}_{n(i),j}(t)$$

(9.32) $$+ \sum_{j=p}^{n(i)-1}{}^{*} \frac{t^{j-p}}{(j-p)!} \left(\int_0^\tau d\tau\right)^{n(i)-q} \varphi_{j,n(i)}(\tau) \equiv \underset{\sim}{K}^{(i)}_{(p,q)}(\chi)$$

where $\sum_{j=k}^{m}{}^{*} = \sum_{j=k}^{m}$ if $k \le m$, and 0 if $k > m$. This allows us to put the Cauchy data in the form

$$V^{(i)}_{(k,k)}(r,r) = S^{(i)}_{(k,k)}(r,r) + \sum_{j=k}^{n(i)-1} \frac{r^{j-k}}{(j-k)!}\left(\int_o^r dt\right)^{n(i)-k} \varphi^{(i)}_{n(i),j}(t)$$

$$\sum_{j=k}^{n(i)-1} \frac{r^{j-k}}{(j-k)!}\left(\int_o^r dt\right)^{n(i)-k} \varphi^{(i)}_{j,n(i)}(t)$$

$$= \hat{\Psi}_{2k}(r, \underset{\sim}{K}^{(j)}_{(0,0)}\chi, \ldots, \underset{\sim}{K}^{(j)}_{(p,q)}\chi, \ldots) , \tag{9.33}$$

$$p + q \leq 2k ,$$

and

$$V^{(i)}_{(k,k-1)}(r,r) - V^{(i)}_{(k-1,k)}(r,r) = S^{(i)}_{(k,k-1)}(r,r) - S^{(i)}_{(k-1,k)}(r,r)$$

$$= \sum_{j=k-1}^{n(i)-1} \frac{r^{j-k+1}}{(j-k+1)!}\left(\int_o^r dt\right)^{n(i)-k} \left[\varphi^{(i)}_{n(i),j}(t) - \varphi^{(i)}_{j,n(i)}(t)\right]$$

$$+ \sum_{j=k}^{n(i)-1} \frac{r^{j-k}}{(j-k)!}\left(\int_o^r dt\right)^{n(i)-k+1} \left[\varphi^{(i)}_{j,n(i)}(t) - \varphi^{(i)}_{n(i),j}(t)\right] . \tag{9.34}$$

$$= \hat{\Psi}_{2k-1}(t, \underset{\sim}{K}^{(j)}_{(0,0)}\chi, \ldots, \underset{\sim}{K}^{(j)}_{(p,q)}\chi, \ldots), p+q \leq 2k-1 .$$

Since

$$\frac{d^{n(i)-1}}{dr^{n(i)-k}} S^{(i)}_{(k,k)}(r,r) = \sum_{p+q \leq n(i)-k} \binom{n(i)-k}{p} S^{(i)}_{(k+p,k+q)}(r,r) ,$$

by differentiating (3.12) $n(i)-k$ times one obtains

$$\sum_{j=k}^{n(i)-1}{}^{*} \varphi^{(i)}_{n(i),j}(r) \frac{r^{j-k}}{(j-k)!} + \sum_{j=k}^{n(i)-1}{}^{*} \varphi^{(i)}_{j,n(i)}(r) \frac{r^{j-k}}{(j-k)!}$$

(9.35) $$=\Omega_{2k}\Big(r, \varphi^{(\ell)}_{n(\ell),j}(r), \ldots \text{ and integrals of the same and } S^{(\ell)}(r,r)\Big),$$

$1 \le \ell \le m$, and $1 \le j \le n(i)-1$.

Similarly, differentiating (9.33) $n(i)-k+1$ times one obtains

$$\sum_{j=k}^{n(i)-1}{}^{*} \varphi^{(i)}_{j,n(i)}(r) \frac{r^{j-k}}{(j-k)!} - \sum_{j=k}^{n(i)-1}{}^{*} \varphi^{(i)}_{n(i),j}(r) \frac{r^{j-k}}{(j-k)!}$$

(9.36) $$= \Omega_{2k+1}\Big(r, \varphi^{(\ell)}_{n(\ell),j}(r), \ldots \text{ and integrals of the same and } S^{(\ell)}(r,r)\Big),$$

$1 \le \ell \le m$, and $1 \le j \le n(i) - 1$.

Consequently, one may solve (9.35 - 9.36) for the

$\varphi^{(i)}_{j,n(i)}(r)$ and $\varphi^{(i)}_{n(i),j}(r)$ to obtain expressions of the form

(9.37) $$\varphi^{(i)}_{j,n(i)}(r) = \Lambda^{(i)}_{j,n(i)}\left(r, \varphi^{(\ell)}_{n(\ell),k}(r), \ldots \text{ and integrals of the same and } S^{(\ell)}(r,r)\right),$$

and

$$\varphi^{(i)}_{n(i),j}(r) = \Lambda^{(i)}_{n(i),j}\left(r, \varphi^{(\ell)}_{n(\ell),k}(r), \ldots \text{and integrals of the same and } S^{(\ell)}(r,r)\right) \tag{9.38}$$

$1 \le \ell \le m$, $1 \le k \le n(i)-1$.

The author with Conlan [CG 6] next introduced the Banach space $\mathfrak{a}(\mathfrak{S})$ of $n \times 2n$ holomorphic matrices $\varphi \equiv \left(\varphi^{(i)}_{n(i),j}; \varphi^{(i)}_{j,n(i)}\right)$ defined over $\mathfrak{S} \subset \mathbb{C}$. The norm is defined as

$$\|\varphi\| = \max\left[\max_{i,j}\{\|\varphi^{(i)}_{n(i),j}\|_\lambda\}, \max_{i,j}\{\|\varphi^{(i)}_{j,n(i)}\|_\lambda\}\right],$$

$$\text{where } \|\varphi^{(i)}_{n(i),j}\|_\lambda = \sup_{t \in \mathfrak{S}}\left\{e^{-\lambda|t|}|\varphi^{(i)}_{n(i),j}(t)|\right\}. \tag{9.39}$$

<u>Problem</u>: Show that

$$\left|\int_0^t \varphi^{(k)}_{n(\ell),k}(t_1)dt_1\right| \le \frac{1}{\lambda} e^{\lambda|t|} \|\varphi^{(k)}_{n(\ell),j}(t)\|_\lambda$$

$$\le \frac{1}{\lambda} e^{\lambda|t|}\|\varphi\|,$$

and more generally that

$$\left|\left(\int_0^t dt\right)^{n(i)-k} \varphi^{(i)}_{n(i),j}(t)\right| \le \frac{\text{const}}{\lambda} e^{\lambda|t|}\|\varphi\|.$$

In order to show that (9.37 - 9.38) has a unique solution in $\mathfrak{C}(\mathfrak{S})$ we utilize the fact that the $\Lambda^{(i)}_{n(i),j}$, and the $\Lambda^{(i)}_{j,n(i)}$ all satisfy a Lipschitz condition in their arguments (independent of r). Since some of these arguments involve integrals of the $\varphi^{(\ell)}_{n(\ell),k}$, $\varphi^{(\ell)}_{k,n(\ell)}$ and the $S^{(\ell)}$, one may obtain analogously as in the above exercise, an estimate of the form

$$(9.40) \qquad \|\varphi - \hat{\varphi}\| \leq \left(L_o + \frac{c}{\lambda}\sum_{\ell=1}^{N} L_\ell\right)\|\varphi - \hat{\varphi}\| ,$$

here φ and $\hat{\varphi}$ are two matrices satisfying (9.37 - 9.38) for a given $S = (S^{(1)},\ldots,S^{(m)})$, and N is the number of terms in (9.37 - 9.38) involving integrals of $\varphi^{\ell}_{n(\ell),k}$ and $\varphi^{\ell}_{k,n(\ell)}$. If $L_o < 1$ then one can easily see that the system (9.37 - 9.38) has a fixed point in $\mathfrak{C}(\mathfrak{S})$. Conlan and Gilbert [CG 6] were then able to establish

<u>Theorem (9.1)</u> <u>If the semi-linear elliptic system</u> (9.22) <u>with nonlinear Cauchy data</u> (9.23) <u>has the following properties</u>:
(1) <u>The functions</u> $\Psi^{(i,K)}$ (K=1,...,n(i)-1), <u>are holomorphic in</u>

$\times C^{\alpha}$, where $\alpha = \frac{1}{4}(K-1)(K+3)$ for K odd and $\alpha = \frac{K^2}{4}$ for K even. Here $\mathfrak{S}$ is a simply-connected domain in C^1. (2) The functions $F^{(i)}$ are holomorphic in $\mathfrak{S} \times \mathfrak{S}^* \times C^{\beta}$, $\beta = [n(i)+1]^2$. (3) The coefficient L_o of (9.39) is less than 1. Then the Cauchy problem (9.22 - 9.23) has a unique solution in $\mathfrak{S}$ if diam($\mathfrak{S}$) is sufficiently small. Furthermore, a constructive procedure may be devised where starting with the initial $s_0^{(i)}(t,\tau)$, $(i=1,\ldots,m)$ the vector $\chi_0(t,\tau)$ is computed as a fixed point for (9.37 - 9.38) and these are used to compute

$$s_1^{(i)}(t,\tau) = \hat{F}^{(i)}(t,\tau,\ldots,\underset{\sim}{K}^{(j)}_{(0,0)},\ldots,\underset{\sim}{K}^{(j)}_{(p,q)},\ldots), 0 \le p,q \le n(i) .$$

The cycle is then repeated ad infinitum.

We consider next the system of semi-linear elliptic equations

$$\Delta^{n(i)} u^{(i)} = f^{(i)}(x,y,\ldots,\frac{\partial^{q+p}}{\partial x^q \partial y^p} \Delta^k u^{(j)},\ldots), \tag{9.41}$$

$$1 \le i, j \le m , \quad 0 \le k \le n(j)-1 , \quad q+p+2k \le 2n(j)-1.$$

(In what follows we will write n instead of $n(i)$ whenever it will cause no confusion.) In this section, however, we will investigate solutions of (9.41) satisfying generalized,

non-linear Goursat data of the form,

$$\left(G_{n,n-1}^{(1)} \frac{\partial}{\partial x} + G_{n,n-1}^{(2)} \frac{\partial}{\partial y} \right) \Delta^{n-1} u^{(i)}$$

$$= h_{n,n-1}^{(i)} \left(x, \ldots, \frac{\partial^{p+q}}{\partial x^q \partial y^p} \Delta^k u^{(j)}, \ldots \right)$$

$$\text{on} \quad C_{n,n-1}^{(i)} : y = g_{n,n-1}^{(i)}(x),$$

$$\left(G_{n,n-2}^{(1)} \frac{\partial^2}{\partial x^2} + G_{n,n-2}^{(2)} \frac{\partial^2}{\partial x \partial y} + G_{n,n-2}^{(3)} \frac{\partial^2}{\partial y^2} \right) \Delta^{n-2} u^{(i)}$$

$$= h_{n,n-2}^{(i)} \left(x, \ldots, \frac{\partial^{q+p}}{\partial x^q \partial y^p} \Delta^k u^{(j)}, \ldots \right)$$

$$\text{on} \quad C_{n,n-2}^{(i)} : y = g_{n,n-2}^{(i)}(x),$$

$$\left(G_{n,0}^{(1)} \frac{\partial^n}{\partial x^n} + G_{n,0}^{(2)} \frac{\partial^n}{\partial x^{n-1} \partial y} + G_{n,0}^{(n+1)} \frac{\partial^n}{\partial y^n} \right) u^{(i)}$$

$$= h_{n,0}^{(i)} \left(x, \ldots, \frac{\partial^{q+p}}{\partial x^q \partial y^p} \Delta^k u^{(j)}, \ldots \right)$$

$$\text{on} \quad C_{n,0}^{(i)} : y = g_{n,0}^{(i)}(x) ,$$

(9.42) where $q + p + 2k \le 2n(j) - 1$,

and

$$\left(a_{n-1,n}^{(1)} \frac{\partial}{\partial x} + a_{n-1,n}^{(2)} \frac{\partial}{\partial y}\right)\Delta^{n-1} u^{(i)}$$

$$= h_{n-1,n}^{(i)}\left(y, \dots, \frac{\partial^{q+p}}{\partial x^q \partial y^p} \Delta^k u^{(j)}, \dots\right)$$

$$\text{on} \quad C_{n-1,n}^{(i)} : x = g_{n-1,n}^{(i)}(y) ,$$

. . .

$$\left(a_{0,n}^{(1)} \frac{\partial^n}{\partial x^n} + a_{0,n}^{(2)} \frac{\partial^n}{\partial x^{n-1} \partial y} + \dots + a_{0,n}^{(n+1)} \frac{\partial^n}{\partial y^n}\right) u^{(i)}$$

$$= h_{0,n}^{(i)}\left(y, \dots, \frac{\partial^{q+p}}{\partial x^q \partial y^p} \Delta^k u^{(j)}, \dots\right)$$

(9.43) on $C_{0,n}^{(i)} : x = g_{0,n}^{(i)}(y)$.

As usual, setting $z = y + iy$, $\zeta = x - iy$, and writing

$\frac{\partial}{\partial z} = D_1$, $\frac{\partial}{\partial \zeta} = D_2$, one obtains

$$\frac{\partial}{\partial x} = D_1 + D_2 \ , \ \frac{\partial}{\partial y} = i(D_1 - D_2) \ , \ \Delta = 4D_1D_2 \ ,$$

$$\frac{\partial^{q+p}}{\partial x^q \partial y^p} \Delta^k = 4^k (i)^p (D_1D_2)^k (D_1 + D_2)^q (D_1 - D_2)^p \ .$$

Hence, transforming to the z, ζ coordinates, equation (9.41) becomes

$$.44) \qquad \frac{\partial^{2n}}{\partial z^n \partial \zeta^n} U^{(i)} = F^{(i)}\left(z, \zeta, \ldots, \frac{\partial^{r+t}}{\partial z^r \partial \zeta^t} U^{(j)}, \ldots\right)$$

where $0 \le r \le k + p + q \le n(j)$, $0 \le t \le k + p + q \le n(j)$, $0 \le r + t \le 2k + p + q \le 2n(j)-1$, where the indices p,q,k refer to the generic term in the right-hand side of (9.41).

The Goursat data (9.42) becomes

$$\left[G^{(1)}_{n,n-1}(D_1 + D_2) + G^{(2)}_{n,n-1} i(D_1 - D_2)\right](D_1D_2)^{n-1} U^{(i)}$$

$$= H^{(i)}_{n,n-1}\left(z, \ldots, D_1^r \, D_2^t \, U^{(j)}, \ldots\right)$$

on $\zeta = G^{(i)}_{n,n-1}(z)$,

$$\Big[\mathcal{G}^{(1)}_{n,n-2}(D_1+D_2)^2 + i\,\mathcal{G}^{(2)}_{n,n-2}(D_1+D_2)(D_1-D_2)$$

$$-\,\mathcal{G}^{(3)}_{n,n-2}(D_1-D_2)^2\Big](D_1D_2)^{n-2}U^{(i)}$$

$$= H^{(i)}_{n,n-2}\Big(z,\ldots,D_1{}^rD_2{}^tU^{(j)},\ldots\Big)$$

on $\zeta = G_{n,n-2}(z)$,

. . .

$$\Big[\mathcal{G}^{(1)}_{n,0}(D_1+D_2)^n + i\,\mathcal{G}^{(2)}_{n,0}(D_1+D_2)^{n-1}(D_1-D_2) + \ldots$$

$$+\,\mathcal{G}^{(n+1)}_{n,0}(i)^n(D_1-D_2)^n\Big]U^{(i)}$$

$$= H^{(i)}_{n,0}(z,\ldots,D_1{}^rD_2{}^tU^{(j)},\ldots)$$

(9.45) on $\zeta = G^{(i)}_{n,0}(z)$, where in the above the indices are bounded by $1 \le r,t \le n(j)$, $r+s \le 2n(j)-1$. One has a similar system derived from (9.42).

For purposes of simplicity of presentation the author with Conlan [CG 6] imposed the following normalization on the coefficients $\mathcal{G}^{(k)}_{r,t}$

$$\mathfrak{a}^{(1)}_{n,n-1} - i\,\mathfrak{a}^{(2)}_{n,n-1} = 0 \;,$$

$$\mathfrak{a}^{(1)}_{n,n-1} + i\,\mathfrak{a}^{(2)}_{n,n-1} = 1 \;,$$

$$\mathfrak{a}^{(1)}_{n,n-2} + i\,\mathfrak{a}^{(2)}_{n,n-2} - \mathfrak{a}^{(3)}_{n,n-2} = 1 \;,$$

$$\mathfrak{a}^{(1)}_{n,n-2} + \mathfrak{a}^{(3)}_{n,n-2} = 0 \;,$$

$$\mathfrak{a}^{(1)}_{n,n-2} - i\,\mathfrak{a}^{(2)}_{n,n-2} - \mathfrak{a}^{(3)}_{n,n-2} = 0 \;,$$

etc.; then (9.45) becomes, using our earlier notation,

$$U^{(i)}_{(n(i),j)} = H^{(i)}_{n(i),j}\left(z,\ldots,U^{(k)}_{(p,q)},\ldots\right) \quad \text{on} \quad \zeta = G^{(i)}_{n(i),j}(z) \;,$$

$$\text{9.47)}\quad U^{(i)}_{(j,n(i))} = H^{(i)}_{j,n(i)}\left(\zeta,\ldots,U^{(k)}_{(p,q)},\ldots\right) \quad \text{on} \quad z = G^{(i)}_{j,n(i)}(\zeta) \;,$$

$0 \le j \le n(i) - 1$, and (k,p,q) runs through the set

$$\mathbf{P} \equiv \{(k,p,q) : 1 \le k \le m,\; 0 \le p,q \le n(k),\; p+q \le 2n(k) - 1\} \;.$$

Our Goursat problem has now been reduced formally to the system (9.44), (9.47). We recall the notation for

$S^{(i)}_{(p,q)}(z,\zeta)$, and $\underset{\sim}{K}^{(i)}_{(p,q)}(\chi,s)$, with

$$\chi(z,\zeta) \equiv (\ldots,\varphi^{(i)}_{n(i),j}(z),\ldots;\ldots,\varphi^{(i)}_{j,n(i)}(\zeta),\ldots),$$

$$(1 \le i \le m,\ 0 \le j \le n(i)-1),$$

as used in (9.34 - 9.37), and let us seek a solution in the form

$$U^{(i)}(z,\zeta) \equiv \underset{\sim}{K}^{(i)}_{(0,0)}(\chi,s)$$

$$\equiv S^{(i)}_{(0,0)} + \sum_{i=0}^{n(i)-1} \frac{\zeta^j}{j!}\left(\int_0^z dz\right)^{n(i)} \varphi^{(i)}_{n(i),j}(z)$$

$$+ \sum_{j=0}^{n(i)-1} \frac{z^j}{j!}\left(\int_o^\zeta d\zeta\right)^{n(i)} \varphi^{(i)}_{j,n(i)}(\zeta) . \tag{9.48}$$

The Goursat data (9.47) can then be written in the form

$$S^{(i)}_{(n(i),k)} + \sum_{j=k}^{n(i)-1} \frac{\zeta^{j-k}}{(j-k)!} \varphi^{(i)}_{n(i),j}(z) = H^{(i)}_{n(i),k}(z,\ldots,\underset{\sim}{K}^{(t)}_{(p,q)},\ldots),$$

on $\zeta = G^{(i)}_{n(i),k}(z)$,

$$S^{(i)}_{(\ell,n(i))} + \sum_{j=\ell}^{n(i)-1}{}^{*} \frac{z^{j-\ell}}{(j-\ell)!} \varphi^{(i)}_{\ell,n(i)}(\zeta) = H^{(i)}_{\ell,n(i)}(\zeta,\ldots,\underset{\sim}{K}^{(t)}_{(p,q)},\ldots),$$

on $z = G^{(i)}_{\ell,n(i)}(\zeta)$,

where (t,p,q) runs through $\mathbb{P}$. We may rewrite (9.49) in the form

$$\varphi^{(i)}_{n(i),k}(z) = H^{(i)}_{n(i),k}(z,\ldots,\underset{\sim}{K}^{(t)}_{(p,q)},\ldots) - S^{(i)}_{(n(i),k)}$$

$$- \sum_{j=k+1}^{n(i)-1}{}^{*} \frac{\zeta^{j-k}}{(j-k)!} \varphi^{(i)}_{n(i),j}(z) \equiv T^{(i)}_{n(i),k}(\chi)(z) ,$$

on $\zeta^{(i)}_{n(i),k}(z)$,

(9.50) and

$$\varphi^{(i)}_{\ell,n(i)}(\zeta) = H^{(i)}_{\ell,n(i)}(\zeta,\ldots,\underset{\sim}{K}^{(t)}_{(p,q)},\ldots) - S^{(i)}_{(\ell,n(i))}$$

$$- \sum_{j=\ell+1}^{n(i)-1}{}^{*} \frac{z^{j-\ell}}{(j-\ell)!} \varphi^{(i)}_{\ell,n(i)}(\zeta) \equiv T^{(i)}_{\ell,n(i)}(\chi)(\zeta)$$

on $z = G^{(i)}_{\ell,n(i)}(\zeta)$.

To treat (4.50) we assume, as we have done earlier, that the data is holomorphic, and hence must satisfy a local Lipschitz condition, i.e. there exist constants $L^{ijk}_{rst} \geq 0$, such that

$$(9.51) \qquad |H^{(i)}_{j,k}(w,\ldots,p^{(r)}_{s,t},\ldots) - H^{(i)}_{j,k}(w,\ldots,\tilde{p}^{(r)}_{s,t},\ldots)|$$

$$\leq \sum_{(r,s,t)\in \mathbf{P}} L^{ijk}_{r\ st}\ \Delta\, p^{(r)}_{s,t} ,$$

where $\Delta p^{(r)}_{s,t} = |p^{(r)}_{s,t} - \tilde{p}^{(r)}_{s,t}|$.

We may rewrite (9.50) in the more compact form

$$(9.52) \qquad \chi(z,\zeta) = (\ldots,T^{(i)}_{n(i),k}(\chi)(z),\ldots;\ldots,T^{(i)}_{k,n(i)}(\chi)(\zeta),\ldots)$$

$$\equiv \mathfrak{J}(\chi)(z,\zeta).$$

To show that (9.52) has a unique solution, consider the mapping

$$(9.53) \qquad \chi^{(r+1)} = \mathfrak{J}(\chi^{(r)})\ ;\ r=0,1,2,\ldots ,$$

and where $\chi^{(0)} = \chi$. We want to show that the mapping $\chi^{(0)} \to \chi^{(2)}$ is a contraction. This will imply that (9.52) has a unique solution. Let

$$(9.54) \qquad \|\Delta\chi\|_\lambda = \sum_{(i,k)\in\mathbb{Q}} \left\{ \|\Delta\varphi^{(i)}_{n(i),k}\|_\lambda + \|\Delta\varphi^{(i)}_{k,n(i)}\|_\lambda \right\},$$

where $\mathbb{Q} = \{(i,k): 1 \le i \le m,\ 0 \le k \le n(i)-1\}$, and $\|\cdot\|_\lambda$ is the λ-norm introduced in (9.39).

We next assume that $|g^{i}_{j,k}(w)| = |w| \le \rho$.

Then by (9.50), (9.51) we have

$$(9.55) \qquad \|\Delta\varphi^{(i)(a+1)}_{n(i),k}\|_\lambda \le \sum_{(n,s,t)\in\mathbb{P}} L^{i,n(i),k}_{r,s,t} \|\Delta K^{(r)(a)}_{s,t}\|_\lambda + \sum_{j=k+1}^{n(i)-1} \frac{\rho^{j-k}}{(j-k)!} \|\Delta\varphi^{(i)(a)}_{n(i),j}\|_\lambda ,$$

It can be shown that this is the same as assuming there exists a conformal mapping taking the curves $C^{(i)}_{n(i),j}$ and $C^{(i)}_{j,n(i)}$, $(0 \le j \le n(i)-1)$ into ray segments initiating at the origin, such that none of the segments coincide where $K^{(r)(a)}_{s\ t} = K^{(r)}_{s,t}(\chi^{(a)})$, $\Delta K = |K(\chi) - K(\tilde{\chi}))|$.

Using our definition for $K^{(r)}_{s,t}(\chi)$ one has

$$(9.56) \qquad \|\Delta K^{(r)(a)}_{s,t}\|_\lambda \le \sum_{j=t}^{n(r)-1}{}^{*} \frac{\rho^{j-t}}{(j-t)!} \left\| \left(\int_o^{|z|} d\xi \right)^{n(r)-s} \Delta\varphi^{(r)(a)}_{n(r),j}(\xi) \right\|_\lambda$$

$$\sum_{j=s}^{n(r)-1}{}^{*} \frac{\rho^{j-s}}{(j-s)!} \left\| \left(\int_0^{|z^*|} d\xi^* \right)^{n(r)-t} \Delta\varphi^{(r)(a)}_{j,n(r)}(\xi^*) \right\|_\lambda .$$

Combining (9.55) and (9.56) we get

$$\|\Delta\varphi^{(i)(a+1)}_{n(i),k}\|_\lambda \le \sum_{(r,s,t)\in \mathbb{P}} L^{i,n(i),k}_{rs\ t} \Big\{ \sum_{j=k}^{n(r)-1}{}^{*} \frac{\rho^{j-t}}{(j-t)!} \Big\| \Big(\int_0^{|z|} d\xi \Big)^{n(r)-s} \Delta\varphi^{(r)(a)}_{n(r),j} \Big\|_\lambda$$

$$+ \sum_{j=s}^{n(r)-1}{}^{*} \frac{\rho^{j-s}}{(j-s)!} \Big\| \Big(\int_0^{|z^*|} d\xi^* \Big)^{n(r)-t} \Delta\varphi^{(r)(a)}_{j,n(r)} \Big\|_\lambda \Big\} + \sum_{j=k+1}^{n(i)-1}{}^{*} \frac{\rho^{j-k}}{(j-k)!} \|\Delta\varphi^{(i)(a)}_{n(i),j}\|_\lambda$$

$$\le \sum_{(r,n(r),t)\in \mathbb{P}} L^{i,n(i),k}_{r,n(r),t} \sum_{j=t}^{n(r)-1}{}^{*} \frac{\rho^{j-t}}{(j-t)!} \|\Delta\varphi^{(r)(a)}_{n(r),j}\|_\lambda$$

$$+ \sum_{(r,s,n(r)\in \mathbb{P}} L^{i,n(i),k}_{r,s,n(r)} \sum_{j=s}^{n(r)-1} \frac{\rho^{j-s}}{(j-s)!} \|\Delta\varphi^{(r)(a)}_{j,n(r)}\|_\lambda$$

$$+ \sum_{j=k+1}^{n(i)-1} \frac{\rho^{j-k}}{(j-k)!} \|\Delta\varphi^{(i)(a)}_{n(i),j}\|_\lambda + \|H^{(i)(a)}_{n(i),k}\|_\lambda \,, \tag{9.57}$$

where $H^{(i)(a)}_{n(i),k}$ stands for all terms involving integrals of the various $\Delta\varphi$'s .

From the definitions of the sets $\mathbb{P}$ and $\mathbb{Q}$ it is clear that

$$\sum_{(r,n(r),t)\in \mathbb{P}} = \sum_{(r,t,n(r)\in \mathbb{P}} = \sum_{(r,t)\in \mathbb{Q}} .$$

Moreover, $\|H^{(i)(a)}_{n(i),k}\|_\lambda \le \frac{1}{\lambda} C\|\Delta\chi^{(a)}\|_\lambda$. Hence putting

$$(9.58)\qquad A^{ik}_{rs} = L^{i,n(i),k}_{r,n(r),s}\ ,\ \tilde{A}^{rk}_{rs} = L^{i,n(i),k}_{r,s,n(r)}\ ,\ A(j-s) = \frac{\rho^{j-s}}{(j-s)!}\ ,$$

we can write (9.57) as

$$(9.59)\qquad \|\Delta\varphi^{(i)(a+1)}_{n(i),k}\|_\lambda \le \sum_{(r,s)\in\mathbb{Q}} \sum_{j=s}^{n(r)-1} \Big\{A^{ik}_{rs}A(j-s)\|\Delta\varphi^{(r)(a)}_{n(r),j}\|_\lambda$$

$$+ \tilde{A}^{ik}_{rs}A(j-s)\|\Delta\varphi^{(r)(a)}_{j,n(r)}\|_\lambda\Big\} + {\sum_{j=k+1}^{n(i)-1}}^{*} A(j-k)\|\Delta\varphi^{(i)(a)}_{n(i),j}\|_\lambda + \frac{C}{\lambda}\|\Delta\chi^{(a)}\|_\lambda\ .$$

If we let

$$\delta^i_j = \begin{cases} 1 & \text{if } i = j \\ 0 & \text{otherwise} \end{cases}\ ,\quad \delta(s,j) = \begin{cases} 1 & \text{if } s \le j \\ 0 & \text{otherwise} \end{cases}$$

then we can treat the last sum in (9.59) as follows:

$$(9.60)\qquad {\sum_{j=k+1}^{n(i)-1}}^{*} A(j-k)\|\Delta\varphi^{(i)(a)}_{n(i),j}\|_\lambda = \sum_{j=0}^{n(i)-1} \delta(k+1,j)A(j-k)\|\Delta\varphi^{(i)(a)}_{n(i),j}\|_\lambda\ ;$$

hence, for any B^i_j , since $i \le m$,

$$\text{(9.61)}\quad \sum_{j=0}^{n(i)-1} B_j^i = \sum_{j=0}^{n(i)-1} B_j^i + \sum_{r\neq i}\sum_{j=0} \delta_r^i B_j^r = \sum_{r=1}^{m}\sum_{j=0}^{n(r)-1} \delta_r^i B_j^r = \sum_{r=1}^{m}\sum_{j=0}^{m(r)-1} \delta_r^i B_j^r$$

Using (9.61) and (9.61) we have

$$\text{(9.62)}\quad \sum_{j=k+1}^{n(i)-1}{}^{*} A(j-k)\|\Delta\varphi_{n(i),j}^{(i)(a)}\|_\lambda = \sum_{(r,j)\in\mathbb{Q}} \delta_r^i\delta(k+1,j)A(j-k)\|\Delta\varphi_{n(r),j}^{(r)(a)}\| .$$

Also, since

$$\text{(9.63)}\quad \sum_{(r,s)\in\mathbb{Q}}\sum_{j=s}^{n(r)-1}{}^{*} A_{rs}^{ik}A(j-s)B_j^r = \sum_{r=1}^{m}\sum_{s=0}^{n(r)-1}\sum_{j=0}^{n(r)-1} \delta(s,j)A_{rs}^{ik}A(j-s)B_j^r$$

$$= \sum_{r=1}^{m}\sum_{j=0}^{n(r)-1}\left[\sum_{r=0}^{n(r)-1} \delta(s,j)A_{rs}^{ik}A(j-s)\right]B_j^r ,$$

we get

$$\text{(9.64)}\quad \sum_{(r,s)\in Q}\sum_{j=s}^{n(r)-1}{}^{*} A_{rs}^{ik}A(j-s)B_j^r = \sum_{r=1}^{m}\sum_{s=0}^{n(r)-1}\sum_{j=0}^{n(r)-1} \delta(s,j)A_{rs}^{ik}A(j-s)B_j^r$$

$$\sum_{(r,j)\in Q}\sum_{s=0}^{n(r)-1} \delta(s,j)A_{rs}^{ik}A(j-s)\|\Delta\varphi_{n(r),j}^{(r)(a)}\|_\lambda ,$$

and similarly for the summation of the $\|\Delta\varphi_{j,n(r)}^{(r)(a)}\|_\lambda$ terms in (9.59). Hence (9.59) can be rewritten in the equivalent form

(9.65) $$\|\Delta\varphi_{n(i),k}^{(i)(a+1)}\|_\lambda \le \sum_{(r,j)\in Q} \left\{B_{rj}^{ik}\|\Delta\varphi_{n(r),j}^{(r)(a)}\|_\lambda + \tilde{B}_{rj}^{ik}\|\Delta\varphi_{j,n(r)}^{(r)(a)}\|_\lambda\right\}$$

$$+ \frac{1}{\lambda} C\|\Delta\chi^{(a)}\|_\lambda ,$$

where

$$B_{rj}^{ik} = \sum_{s=0}^{n(r)-1} \delta(s,j)A_{rs}^{ik}A(j-s) + \delta_r^i\ \delta(k+i,j)A(j-k) ,$$

(9.66)

$$\tilde{B}_{rj}^{\ k} = \sum_{s=0}^{n(r)-1} \delta(s,j)\tilde{A}_{rs}^{\ k}A(j-s).$$

Note that (9.58) and (9.66) completely determine the B_{rj}^{ik} and $\tilde{B}_{rj}^{ik}$ in terms of the Lipschitz constants L_{rst}^{ijk} . In exact analogy to the preceeding, we find

$$\|\Delta\varphi_{k,n(i)}^{(i)(a+1)}\|_\lambda \le \sum_{(n,j)\in Q} \left\{B*_{rj}^{ik}\|\Delta\varphi_{n(r),j}^{(r)(a)}\|_\lambda + \tilde{B}*_{rj}^{ik}\|\Delta\varphi_{j,n(r)}^{(r)(a)}\|_\lambda\right\}$$

(9.67) $$+ \frac{1}{\lambda} C \| \Delta\chi^{(a)} \|_{\lambda} ,$$

where the $B^{*}{}^{ik}_{rj}$ and $\tilde{B}^{*}{}^{ik}_{rj}$ are completely determined in terms of the L^{ijk}_{rst} .

If we set

$$D^{ik}_{rj} = \max \left\{ \left(B^{ik}_{rj} + B^{*}{}^{ik}_{rj} \right) , \left(\tilde{B}^{ik}_{rj} + \tilde{B}^{*}{}^{ik}_{rj} \right) \right\} ,$$

then

(9.68) $$\| \Delta\varphi^{(i)(a+1)}_{n(i),k} \|_{\lambda} + \| \Delta\varphi^{(i)(a+1)}_{k,n(i)} \|_{\lambda} \leq$$

$$\sum_{(n,j) \in Q} B^{ik}_{rj} \left\{ \| \Delta\varphi^{(r)(a)}_{n(r),j} \|_{\lambda} + \| \Delta\varphi^{(r)(a)}_{j,n(r)} \|_{\lambda} \right\} + \frac{2C}{\lambda} \| \Delta\chi^{(a)} \|_{\lambda} .$$

From (9.68) is is clear that

$$\| \Delta\chi^{(a+1)} \|_{\lambda} = \sum_{(i,k) \in Q} \left\{ \| \Delta\varphi^{(i)(a+1)}_{n(i),k} \|_{\lambda} + \| \Delta\varphi^{(i)(a+1)}_{k,n(i)} \|_{\lambda} \right\}$$

$$\leq \left\{ C_1 + \frac{1}{\lambda} C_2 \right\} \| \Delta\chi^{(a)} \|_{\lambda} \leq C_3 \| \Delta\chi^{(a)} \|_{\lambda} ,$$

(9.69) for $C_1 = \max_{(r,j)\in Q}\left\{\sum_{(i,k)\in Q} D^{ik}_{rj}\right\}$

and for some constants C_2, C_3.

Applying (9.68) twice, and using (9.69), we get

$$\|\Delta\varphi^{(i)(2)}_{n(i),k}\|_\lambda + \|\Delta\varphi^{(i)(2)}_{k,n(i)}\|_\lambda \le \sum_{(r,j)\in \mathbb{Q}} D^{ik}_{rj}\left\{\|\Delta\varphi^{(r)(1)}_{n(r),j}\|_\lambda + \|\Delta\varphi^{(r)(1)}_{j,n(r)}\|_\lambda\right\} + \frac{2C}{\lambda}\|\Delta\chi^{(1)}\|_\lambda$$

(9.70)

$$\le \sum_{(r,j)\in\mathbb{Q}} D^{ik}_{rj} \sum_{(r,t)\in\mathbb{Q}} D^{rj}_{st}\left\{\|\Delta\varphi^{(s)}_{n(s),t}\|_\lambda + \|\Delta\varphi^{(s)}_{t,n(s)}\|_\lambda\right\} + \frac{Cu}{\lambda}\|\Delta\chi\|_\lambda ,$$

for some constant C_4. Hence

$$\|\Delta\chi^{(2)}\|_\lambda \le \sum_{(i,k)\in\mathbb{Q}} \sum_{(r,j)\in\mathbb{Q}} \sum_{(s,t)\in\mathbb{Q}} D^{ik}_{rj} D^{rj}_{st}\left\{\|\Delta\varphi^{(r)}_{n(s),t}\|_\lambda + \|\Delta\varphi^{(s)}_{t,n(s)}\|_\lambda\right\} + \frac{C_5}{\lambda}\|\Delta\chi\|_\lambda$$

(9.71)

$$\le \sum_{(s,t)\in\mathbb{Q}} \left\{\left[\sum_{(i,k)\in\mathbb{Q}} D^{ik}_{rj} D^{rj}_{st}\right]\left[\|\Delta\varphi^{(s)}_{n(s)t}\|_\lambda + \|\Delta\varphi^{(s)}_{t,n(s)}\|_\lambda\right]\right\}$$

$$+ \frac{1}{\lambda} C_5 \|\Delta\|_\lambda .$$

Then if $\max_{(s,t)\in\mathbb{Q}}\left[\sum_{(i,k)\in\mathbb{Q}} \sum_{(r,j)\in\mathbb{Q}} D^{ik}_{rj} D^{rj}_{st}\right] \equiv k < 1$, we have

$$(9.72)\qquad \|\Delta\chi^{(2)}\|_\lambda \le \left(k+\frac{C_5}{\lambda}\right)\|\Delta\chi\|_\lambda \le k_1\|\Delta\chi\|_\lambda \,,$$

where $k_1 < 1$ for λ sufficiently large, and so the mapping $\chi \to \chi^{(2)}$ is a contraction.

We summarize the above discussion as

<u>Lemma (9.2):</u> <u>The system of functional equations</u> (9.52) <u>have a solution in the Banach space</u> $\mathfrak{A}^\lambda(\mathfrak{S},\mathfrak{S}^*)$, <u>of elements</u> χ <u>with norm</u> $\|\cdot\|_\lambda$, <u>providing</u>

$$(9.73)\qquad k \equiv \max_{(s,t)\in\mathbb{Q}} \left[\sum_{(i,k)\in\mathbb{Q}} \;\sum_{(r,j)\in\mathbb{Q}} D^{ik}_{rj}\, D^{rj}_{st}\right] < 1 \,.$$

We next turn to showing that the Goursat problem (9.44),(9.52) has a unique solution; we proceed as follows. Let the operator $\underset{\sim}{Q}$ be defined as

$$\underset{\sim}{Q}\, s^{(i)} = F^{(i)}(z,\zeta,\ldots,\underset{\sim}{K}{}^{(j)}_{(r,t)}(\chi,s),\ldots),$$

where χ is understood to be the solution of (9.52) which corresponds to s. If $\hat{\chi}$ is the solution of (9.52) correspondin to a $\hat{s}$, then since the $F^{(i)}$ are holomorphic functions of their arguments (in a certain polycylinder), there exist constants L_{krt} such that

$$|F(z,\zeta,\ldots,\underset{\sim}{K}^{(k)}_{(r,t)}(\chi,s),\ldots) - F(z,\zeta,\ldots,\underset{\sim}{K}^{(k)}_{(r,t)}(\hat{\chi},\hat{s}),\ldots)|$$

$$\leq \sum_{(k,r,t)\in \mathbb{P}} L_{krt}\Delta\underset{\sim}{K}^{(k)}_{(r,t)},$$

with $\Delta\underset{\sim}{K}^{(k)}_{(r,t)} = |\underset{\sim}{K}^{(k)}_{(r,t)}(\chi,x) - \underset{\sim}{K}^{(k)}_{(r,t)}(\hat{\chi},\hat{s})|$;

hence,

$$\|\underset{\sim}{Q}\,s - \underset{\sim}{Q}\,\hat{s}\|_\lambda \leq \sum_{(k,r,t)\in \mathbb{P}} L_{krt}\|\Delta K^{(k)}_{(r,t)}\|_\lambda$$

$$\leq \sum_{(k,n(k),t)\in \mathbb{P}} L_{krt}\sum_{j=t}^{n(k)-1}\frac{\rho^{j-t}}{(j-t)!}\|\Delta\varphi^{(k)}_{n(k),j}\|_\lambda$$

$$+ \sum_{(k,r,n(k))\in \mathbb{P}} L_{krt}\sum_{j=r}^{n(k)-1}\frac{\rho^{j-r}}{(j-r)!}\|\Delta\varphi^{(k)}_{j,n(k)}\|_\lambda$$

$$+ \sum_{(k,r,t)\in \mathbb{P}} L_{krt}\|\Delta S^{(k)}_{(r,t)}\|_\lambda + \|H\|_\lambda \tag{9.74}$$

where H stands for all the terms involving integrals of the various $\Delta\varphi$'s. Using the estimates

$$\|\Delta\varphi^{(i)(a+1)}_{n(i),k}\|_\lambda \le \sum_{(r,s,t)\in \mathbb{P}} L^{i,n(i),k}_{r,s,t}\|\Delta K^{(r)(a)}_{(s,t)}\|_\lambda + \sum_{j=k+1}^{n(i)-1}{}^{*} \frac{\rho^{j-k}}{(j-k)!}\|\Delta\varphi^{(i)(a)}_{n(i),j}\|_\lambda$$

$$\|\Delta K^{(r)(a)}_{(s,t)}\|_\lambda \le \sum_{j=t}^{n(r)-1}{}^{*}\frac{\rho^{j-t}}{(j-t)!}\|\left(\int_0^{|z|} d\xi\right)^{n(r)-s} \Delta\varphi^{(r)(a)}_{n(r),j}(z)\|_\lambda$$

$$+ \sum_{j=s}^{n(r)-1}{}^{*}\frac{\rho^{j-s}}{(j-s)!}\|\left(\int_0^{|\zeta|} d\xi\right)^{n(r)-t} \Delta\varphi^{(r)(a)}_{j,n(r)}(\zeta)\|_\lambda$$

$$(9.75)\qquad + \|\Delta S^{(r)}_{(s,t)}\|_\lambda ,$$

we obtain

$$\|\Delta\varphi^{(i)(a+1)}_{n(i),k}\|_\lambda \le \sum_{(r,s,t)\in \mathbb{P}} L^{i,n(i),k}_{r,s,t}\|\Delta S^{(r)}_{(s,t)}\|_\lambda + \sum_{j=k+1}^{n(i)-1}{}^{*} \frac{\rho^{j-k}}{(j-k)!}\|\Delta\varphi^{(i)(a)}_{n(i),j}\|_\lambda$$

$$+ \sum_{(r,n(r),t)\in \mathbb{P}} L^{i,n(i),k}_{r,n(r),t}\sum_{j=t}^{n(r)-1}\frac{\rho^{j-t}}{(j-t)!}\|\Delta\varphi^{(r)(a)}_{n(r),j}\|_\lambda$$

$$+\sum_{(r,s,n(r))\in\mathbb{P}} L^{i,n(i),k}_{r,s,n(r)} \sum_{j=s}^{n(r)-1} \frac{\rho^{j-s}}{(j-s)!}\|\Delta\varphi^{(r),(a)}_{j,n(r)}\|_\lambda + \|H^{(i)(a)}_{n(i)k}\|_\lambda ,$$

where the last term involves all integrals of the various $\Delta\varphi$. Hence,

$$\|\Delta\varphi^{(i)(a+1)}_{n(i),k}\|_\lambda \le \frac{M_1}{\lambda}\|\Delta S\|_\lambda + \frac{M_2}{\lambda}\|\Delta\chi\|_\lambda + \sum_{j=k+1}^{n(i)-1} \frac{\rho^{j-k}}{(j-k)!}\|\Delta\varphi^{(i)(a)}_{n(i),j}\|_\lambda$$

(9.76)

$$+\sum_{(r,n(r),t)\in\mathbb{P}} L^{i,n(i),k}_{r,n(r),t} \sum_{j=t}^{n(i)-1} \frac{\rho^{j-t}}{(j-t)!}\|\Delta\varphi^{(r)(a)}_{n(r),j}\|_\lambda$$

$$+\sum_{(r,s,n(r))\in\mathbb{P}} L^{i,n(i),k}_{r,s,n(r)} \sum_{j=s}^{n(r)-1} \frac{\rho^{j-s}}{(j-s)!}\|\Delta\varphi^{(r)(a)}_{j,n(r)}\|_\lambda$$

or

$$\|\Delta\varphi^{(i)(a+1)}_{n(i),k}\|_\lambda \le \frac{M_1}{\lambda}\|\Delta S\|_\lambda + \frac{M_2}{\lambda}\|\Delta\chi\|_\lambda$$

(9.77)

$$+\sum_{(r,j)\in\mathbb{Q}} \left\{B^{ik}_{rj}\|\Delta\varphi^{(r)(a)}_{n(r),j}\|_\lambda + \tilde{B}^{ik}_{rj}\|\Delta\varphi^{(r)(a)}_{j,n(r)}\|_\lambda\right\} .$$

Likewise, one obtains

$$\|\Delta\varphi^{(i)(a+1)}_{k,n(i)}\|_\lambda \le \frac{M_1^*}{\lambda}\|\Delta S\|_\lambda + \frac{M_2^*}{\lambda}\|\Delta\chi\|_\lambda$$

$$(9.78) \qquad + \sum_{(r,j)\in\mathbb{Q}} \left\{B^{*ik}_{rj}\|\Delta\varphi^{(r)(a)}_{n(r),j}\|_\lambda + \tilde{B}^{*ik}_{rj}\|\Delta\varphi^{(r)(a)}_{j,n(r)}\|_\lambda\right\} .$$

Here the coefficients B^{ik}_{rj}, $\tilde{B}^{ik}_{rj}$, etc. are as defined in (9.46) etc. . Then one has

$$\|\Delta\chi^{(a+1)}\|_\lambda = \sum_{(i,k)\in\mathbb{Q}} \sum_{(r,j)\in\mathbb{Q}} D^{ik}_{rj}\left\{\|\Delta\varphi^{(r)(a)}_{n(r),j}\|_\lambda + \|\Delta\varphi^{(r)(a)}_{j,n(r)}\|_\lambda\right\}$$

$$+ \frac{C_3}{\lambda}\|\Delta S\|_\lambda + \frac{C_2}{\lambda}\|\Delta\chi^{(a)}\|_\lambda$$

with $C_2 = N\max\{M_2, M_2^*\}$, $C_3 = N\max\{M_1, M_1^*\}$,

$$N = \sum_{i=1}^{m} n(i) - m\,, \quad D^{ik}_{rj} = \max\left\{(B^{ik}_{rj} + B^{*ik}_{rj})\,,\ (\tilde{B}^{ik}_{rj} + \tilde{B}^{*ik}_{rj})\right\}\,,$$

from which it follows that

$$(9.79)\qquad \|\Delta\chi^{(a+1)}\|_\lambda \le \left(C_1 + \frac{1}{\lambda} C_2\right)\|\Delta\chi^{(a)}\|_\lambda + \frac{C_3}{\lambda}\|\Delta S\|_\lambda ,$$

where

$$C_1 = \max_{(r,j)\in\mathbb{Q}} \left\{\sum_{(i,k)\in\mathbb{Q}} D_{rj}^{ik}\right\} .$$

We notice that the coefficient C_1 is the same as occurred earlier in expression (9.69).

If $C_1 < 1$ then for sufficiently large λ, $C_1 + \frac{1}{\lambda} C_2 < 1$ and $\|\Delta\chi^{(a+1)}\|_\lambda \le \frac{C_4}{\lambda}\|\Delta S\|_\lambda$.

Problem: Conclude that

$$(9.80)\qquad \|\underset{\sim}{Q}s - \underset{\sim}{Q}\hat{s}\|_\lambda \le \frac{C_5}{\lambda}\|\Delta S\|_\lambda ,$$

using the previous estimate on $\|\Delta\chi^{(a+1)}\|_\lambda$, and (9.74); consequently, it follows that $\underset{\sim}{Q}$ is a contraction mapping.

We summarize the above discussion as

THEOREM (9.3): We assume the following hypotheses:

H_1 : The functions $f^{(i)}(x,y,\ldots,\frac{\partial^{p+q}\Delta^k u^{(j)}}{\partial x^q \partial y^p},\ldots)$

$1 \le i,j \le m,\ 0 \le k \le n(j)-1,\ q+p+2k \le 2n(j)$

are holomorphic functions of their arguments in some polycylinder.

H_2 : The Goursat data $h^{(i)}_{n,j}(x,\ldots,\frac{\partial^{q+p}\Delta^k u^{(j)}}{\partial x^q \partial y^p},\ldots)$,

$h^{(i)}_{j,n}(x,\ldots,\frac{\partial^{q+p}\Delta^k u^{(j)}}{\partial x^q \partial y^p},\ldots)$, $1 \le i,j \le m,\ q+p+2k \le 2n(j)-1$,

are holomorphic in their respective arguments in some polycylinder.

H_3 : The restriction of the curves $C^{(i)}_{n,j}$, $C^{(i)}_{j,n}$, $1 \le i,j \le n$ to $\mathfrak{S}$, a simply-connected domain containing the origin, are analytic and have an extension to the bidisk $\{|x| \le R\} \times \{|y| \le R\}$ as analytic surfaces. Furthermore, the curves all intersect at the origin.

H_4 : There exists a conformal mapping of $\mathfrak{S}$ into the w-plane which takes each of the analytic curves

$C_{n,j}^{(i)} \cap \mathfrak{S}$ *and* $C_{j,n}^{(i)} \cap \mathfrak{S}$ *into a ray segment initiating at the origin, and such that none of the segments coincide except at the origin.*

Then the generalized Goursat problem has a unique solution in $\mathfrak{S}$ *if the constants* C_1 *of* (9.69) *and* k *of* (9.73) *are less than* 1. *An iterative scheme for solving the problem is suggested by the above fixed point construction.*

X. Elliptic Systems of First Order Differential Equations: General Theory

L. Bers [Be 1] and I. Vekua [Ve 2] have considered the canonical, elliptic systems

$$u_x - v_y + au + bv = f, \quad u_y + v_x + cu + dv = g, \tag{10.1}$$

which generalize the Cauchy-Riemann equations $u_x - v_y = 0$, $u_y + v_x = 0$. We note that a second order differential equation of the form

$$\Delta w + p(x,y) w_x + q(x,y) w_y = 0, \quad \Delta \equiv \frac{\partial^2}{\partial x^2} + \frac{\partial^2}{\partial y^2}, \tag{10.2}$$

can be put into the form of (10.1) by setting $w_x = u$, $w_y = v$. The equation (10.2) then becomes

$$u_x - v_y = 0, \quad u_x + v_y + pu + qv = 0. \tag{10.3}$$

The system (10.1) can be put in a convenient form by introducing the complex quantity

$$U(z) = u(x,y) + i\, v(x,y). \tag{10.4}$$

The system (10.1) may then be written as

$$\frac{\partial U}{\partial \bar{z}} + AU + B\bar{U} = F, \tag{10.5}$$

where $A \equiv \frac{1}{4}(a+d+i[c-b])$, $B \equiv \frac{1}{4}(a-d+i[c+b])$,

and $F \equiv \frac{1}{2}(f+ig)$. (10.6)

Classical solutions of (10.5) exist for only a relatively small class of solutions. As an example, we exhibit a differential equation with continuous coefficients about z=0, but having non continuous first partial derivatives there. The differential equation

$$w_{\bar{z}} = \frac{e^{2i\varphi}}{\ln \frac{1}{r}}, \quad (z = re^{i\varphi}), \tag{10.7}$$

has as its general solution

$$w(z) = -2\, z \ln \ln \frac{1}{r} + \Phi(z), \tag{10.8}$$

where $\Phi(z)$ is an arbitrary analytic function. Moreover,

$$w_z = \tfrac{1}{2}\left(w_x - w_y\right) = -2 \ln \ln \frac{1}{r} + \frac{1}{\ln \frac{1}{r}} + \Phi'(z) \tag{10.9}$$

is not continuous at $z = 0$. It has been shown [Ve 2], however, that the solutions of (10.5) are classical whenever the coefficients of (10.5) are Hőlder continuous.

In this chapter we will investigate distributional solutions of (10.5). We refer to functions of this type as being generalized analytic (GA) functions. We consider both the cases of bounded and unbounded domains $\mathfrak{G}$. When $\mathfrak{G}$ is bounded we ask that the coefficients of (10.5) be of class $L^p(\mathfrak{G})$, $p > 2$. However, for the case of unbounded $\mathfrak{G}$ we introduce a new class, $L^{p,\nu}(\mathfrak{G})$, which we wish the coefficients to lie in. The class $L^{p,\nu}(\mathfrak{G})$ is defined to be the set of complex valued functions having the properties

(i) $f(z)$ is defined on the entire plane $\mathbb{C}$,

(ii) $f \in L^p(\mathbb{C}_0)$ where $\mathbb{C}_0$ is the unit disk,

(iii) $f_\nu(z) \in L^p(\mathbb{C}_0)$ where $f_\nu(z) \equiv |z|^{-\nu} f\left(\frac{1}{z}\right)$ for $z \in \mathbb{C}_0$.

The norm of $f \in L^{p,\nu}$ is given as

$$\|f\|_{p,\nu} \equiv |f|_{p,\mathbb{C}_0} + |f_\nu|_{p,\mathbb{C}_0}, \tag{10.10}$$

where $|f|_{p,\mathfrak{G}}$ is the L^p-norm of f in $\mathfrak{G}$.

The spaces $\mathfrak{C}^m(\mathfrak{G})$ and $\mathfrak{C}^m(\bar{\mathfrak{G}})$ consist of those functions whose derivatives up to the mth order are continuous in $\mathfrak{G}$ and the closure of $\mathfrak{G}$, respectively. We write $\mathfrak{C}^o(\mathfrak{G}) = \mathfrak{C}(\mathfrak{G})$, $\mathfrak{C}^o(\bar{\mathfrak{G}}) = \mathfrak{C}(\bar{\mathfrak{G}})$.

We say $w \in \mathfrak{C}^{m,\alpha}(\bar{\mathfrak{G}})$ where $0 < \alpha \leq 1$, if w and its derivatives up to the mth order are Hölder-continuous in $\bar{\mathfrak{G}}$ with exponent α (i.e., there exists a positive constant

M such that for $z_1, z_2 \in \bar{\mathfrak{G}}$, $|w(z_1) - w(z_2)| \le M|z_1 - z_2|^{\alpha}$, and a similar inequality holds for the derivatives of w to order m).

The space $\mathbb{C}^{m,\alpha}(\mathfrak{G})$ is defined to contain those functions which are in $\mathbb{C}^{m,\alpha}(\bar{\mathfrak{G}}_1)$ for every <u>bounded</u> subdomain $\mathfrak{G}_1$ of $\mathfrak{G}$.

We also use the space $\mathbb{B}^m(\mathbb{C})$, which consists of the functions which, along with their derivatives to order m, are <u>continuous</u> and <u>bounded</u> in the whole complex plane $\mathbb{C}$. If the function and its derivatives are also Hőlder-continuous in $\mathbb{C}$ with exponent α, then we say the function is in $\mathbb{B}^{m,\alpha}(\mathbb{C})$. If $\mathfrak{G}$ is unbounded, but not the entire plane, then we extend f to $\mathbb{C}$ by setting it equal to zero in $\mathbb{C} - \mathfrak{G}$.

The classes $D_z(\mathfrak{G})$ and $D_{\bar{z}}(\mathfrak{G})$ will represent the set of functions having distributional derivatives w.r.t. z and $\bar{z}$ respectively. We say that $U(z)$ is a generalized solution of (10.5) in $\mathfrak{G}$, if for each point of $\mathfrak{G}$ $U \in D_{\bar{z}}(\mathfrak{G})$, (with the exception of a discrete set of points), and (10.5) is satisfied almost everywhere. In this case we refer to $U(z)$ as being of class $\mathfrak{C}(A,B,F;\mathfrak{G})$ with respect to (10.5). The discrete set of points mentioned above are called <u>singularities</u> of the solution. If there are no singularities, our generalized solution will be called regular. It has been shown that when $A, B, F \in L^p(\mathfrak{G})$, $p > 2$, then every generalized solution is regular; indeed, $U(z)$ belongs to the Hőlder class $C^{\alpha}(\mathfrak{G})$, $\alpha = \frac{p-2}{p}$.

---

*Definition: Let $f, g \in L_1(\mathfrak{G})$. If f and g satisfy the relation

$$\iint_{\mathfrak{G}} g \frac{\partial \varphi}{\partial \bar{z}} \, dx\, dy + \iint_{\mathfrak{G}} f \varphi \, dx\, dy = 0 \, ,$$

$$\left(\iint_{\mathfrak{G}} g \frac{\partial \varphi}{\partial z} \, dx\, dy + \iint_{\mathfrak{G}} f \varphi \, dx\, dy = 0 \right) ,$$

Where $\varphi \in C_c^1(\mathfrak{G})$ (C^1 with compact support in $\mathfrak{G}$), then f is said to be the generalized or distributional derivative of g with respect to $\bar{z}$ (w.r.t. z).

Definition: A domain $\mathfrak{G}$ is regular if it is bounded and its boundary Γ consists of a finite number of simple closed curves with piecewise continuous tangent.

We next define the operator $\underset{\sim}{T}$ and list several properties of this operator. Proofs of the stated properties may be found in Chapter 1 of [Ve 2]. Let

$$\underset{\sim\mathfrak{G}}{T}f \equiv -\frac{1}{\pi}\iint_{\mathfrak{G}} \frac{f(\zeta)\,d\xi\,d\eta}{\zeta - z}, \quad \zeta = \xi + i\eta . \tag{10.11}$$

(i) <u>For</u> $\mathfrak{G}$ <u>a bounded domain, and</u> f <u>a function with compact support in</u> $\mathfrak{G}$, <u>the nonhomogeneous Cauchy-Riemann equation</u>

$$\frac{\partial w}{\partial \bar{z}} = f$$

<u>has the solution</u>

$$w(z) = \Phi(z) + (\underset{\sim\mathfrak{G}}{T}f)(z).$$

<u>Here</u> $\Phi(z)$ <u>is analytic in</u> $\mathfrak{G}$ (<u>which we denote by</u> $\Phi \in \mathfrak{a}(\mathfrak{G})$.) <u>Furthermore</u>, <u>if</u> $f \in L^1(\bar{\mathfrak{G}})$, (<u>where</u> $\mathfrak{G}$ <u>is bounded</u>), <u>then</u>

$$\iint_{\mathfrak{G}} \underset{\sim}{T} f \frac{\partial \varphi}{\partial \bar{z}}\, dx\, dy + \iint_{\mathfrak{G}} f\,\varphi\, dx\, dy = 0. \tag{10.12}$$

<u>If</u>

$$\underset{\sim}{\bar{T}} f \equiv -\frac{1}{\pi}\iint_{\mathfrak{G}} \frac{f(\zeta)\,d\xi\,d\eta}{\bar{\zeta} - \bar{z}}, \tag{10.13}$$

<u>in a similar manner it may be shown that</u>

$$\iint_{\mathfrak{G}} \underset{\sim}{\bar{T}} f \frac{\partial \varphi}{\partial z}\, dx\, dy + \iint_{\mathfrak{G}} f\,\varphi\, dx\, dy = 0 . \tag{10.14}$$

(ii) <u>If</u> $f = \frac{\partial g}{\partial \bar{z}} \in L^1(\bar{\mathfrak{G}})$, <u>then</u>

$$g(z) = \Phi(z) + (\underset{\sim}{T}_{\mathfrak{G}} f)(z), \tag{10.15}$$

<u>with</u> $\Phi \in \mathcal{C}(\mathfrak{G})$. <u>Conversely, if</u> $\Phi \in \mathcal{C}(\mathfrak{G})$ <u>and</u> $f \in L^1(\bar{\mathfrak{G}})$, <u>then the function</u> (10.15) <u>is in</u> $D_{\bar{z}}(\mathfrak{G})$ <u>and</u> $\frac{\partial g}{\partial \bar{z}} = f$.

(iii) <u>If</u> $f \in L^1_{loc}(\mathfrak{G})$, $\frac{\partial f}{\partial \bar{z}} \in L^p_{loc}(\mathfrak{G})$ <u>for some</u> p, $p > 1$, <u>then</u> $\frac{\partial f}{\partial z}$ <u>exists and is in</u> $L^p_{loc}(\mathfrak{G})$. <u>Then</u> f_x <u>and</u> f_y <u>exist and are in</u> $L^p_{loc}(\mathfrak{G})$, [Ve 2], pg. 72.

(iv) <u>Let</u> f <u>be a complex valued function which is in</u> $\mathcal{B}^{1,\alpha}(\mathbb{C})$ <u>for some</u> α, $0 < \alpha < 1$, <u>and in</u> $L^p(\mathbb{C})$ <u>for some</u> p, $1 \le p < 2$. <u>Then</u> $(\underset{\sim}{T} f) \in \mathcal{B}^{2,\alpha}(\mathbb{C})$ <u>and moreover</u>

$$(\mathcal{D}_x + i\,\mathcal{D}_y)(\underset{\sim}{T} f) = f\,, \text{ [Ve 2], (pp. 56 - 64).}$$

We consider next those solutions of (10.5) that are in $D_{\bar{z}}(\bar{\mathfrak{G}})$, i.e., those $u \in \mathcal{C}(A,B,F;\bar{\mathfrak{G}})$. This permits us to write (10.5) as the integral equation

$$u(z) - \left(\underset{\sim}{P}_{\mathfrak{G}} u\right)(z) = \Phi(z) + \left(\underset{\sim}{T}_{\mathfrak{G}} \mathbf{F}\right)(z), \tag{10.16}$$

where $\Phi \in \mathcal{C}(\mathfrak{G})$ and

$$\left(\underset{\sim}{P}_{\mathfrak{G}} f\right)(z) \equiv -\underset{\sim}{T}_{\mathfrak{G}}(Af + B\bar{f}). \tag{10.17}$$

We have the following result [Ve 2] pg. 139:

Theorem (10.1). Every function of the class $\mathfrak{C}(A,B,F;\bar{\mathfrak{G}})$ satisfies an integral equation of the form (10.16), where Φ may be uniquely determined from the boundary data. Conversely, if $\Phi \in \mathfrak{C}(\mathfrak{G})$ and the integral equation (10.16) is satisfied by a function $u(z) \ni Au + B\bar{u} \in L^1(\bar{\mathfrak{G}})$, then u is a solution of (10.5). This can be seen by applying the operator $\frac{\partial}{\partial z}$ to both sides of (10.16)

Theorem (10.2): If $A,B,F \in L^{p,2}(\mathfrak{G})$, $p > 2$, then a solution $u(z)$ of (10.5) is in the Hőlder class $C^{\alpha}(\mathfrak{G})$, $\alpha = \frac{p-2}{p}$.

To extablish this result we first prove several lemmas.

Lemma (10.1): [Ve 2] pg. 47: If $f \in L^p(\bar{\mathfrak{G}})$, $1 \le p < 2$, where $\mathfrak{G}$ is a bounded domain, then $g = \left(\underset{\sim}{T}_{\mathfrak{G}} f\right) \in L^{\gamma}(\bar{\mathfrak{G}})$, where γ is an arbitrary real number which satisfies $p < \gamma < \frac{2p}{2-p}$. Furthermore, it holds that the following inequality is valid:

$$\left| \underset{\sim}{T}_{\mathfrak{G}} f \right|_{\gamma,\mathfrak{G}} \le M_{p,\gamma}(\bar{\mathfrak{G}}) \, |f|_{p,\mathfrak{G}} \,. \tag{10.18}$$

For $1 \le p < 2$, $2 < \gamma < \frac{2p}{2-p}$, $g \in L^{\gamma}_{2a}$, i.e., we have

$$\left(\int\!\!\int_{C}\left|g(z+\Delta z)-g(z)\right|^{\gamma}dx\,dy\right)^{1/\gamma}\le M'_{p,\gamma}|f|_{p,\mathfrak{G}}|\Delta z|^{2a}, \tag{10.19}$$

with $a=\frac{1}{\gamma}-\frac{2-p}{2p}>0$.

<u>Remark</u>: By the class $L^p_a(\mathfrak{G})$ is meant the class of L^p functions that are of Hőlder-class a in the L^p-norm, i.e.,

$$\left(\int\!\!\int_{\mathfrak{G}}|f(z+\Delta z)-f(z)|^p\,dx\,dy\right)^{1/p}\le B|\Delta z|^a,\ 0<a\le 1.$$

The $L^p_a(\mathfrak{G})$ norm is given by

$$\|f\|_{p,a,\mathfrak{G}}\equiv|f|_{p,\mathfrak{G}}+|||f|||_{p,a,\mathfrak{G}},$$

with $|||f|||_{p,a,\mathfrak{G}}\equiv\sup\dfrac{\left(\int\!\!\int_{\mathfrak{G}}|f(z+\Delta z)-f(z)|^p\,dx\,dy\right)^{1/p}}{|\Delta z|^a}$

<u>Proof</u>: Our proof follows that given by Vekua ([Ve 2] pg. 47), with minor typographical corrections. If $p<\gamma<\frac{2p}{2-p}$, then one has the obvious identity

$$|T_{\mathfrak{G}}f|\le\frac{1}{\pi}\int\!\!\int_{\mathfrak{G}}|f(\zeta)|^{p/\gamma}\cdot|z-\zeta|^{-2/\gamma+a}|f(\zeta)|^{p(\frac{1}{p}-\frac{1}{\gamma})}|\zeta-z|^{-\frac{2}{q}+a}\,d\xi\,d\eta\,,$$

with $q=\frac{p}{p-1}$, and $a=\frac{1}{\gamma}-\frac{1}{p}+\frac{1}{2}>0$ (as given in the hypothesis).

We next apply the Hölder inequality, noting that

$\frac{1}{\gamma} + \frac{\gamma-p}{p\gamma} + \frac{1}{q} = 1$, and obtain

$$|\underset{\sim}{T}_{\mathfrak{G}} f| \le \frac{1}{\pi}\left(\iint_{\mathfrak{G}} |f(\zeta)|^p |\zeta - z|^{-2+\gamma a} \, d\xi \, d\eta\right)^{1/\gamma} \cdot \left(\iint_{\mathfrak{G}} |f|^p \, d\xi \, d\eta\right)^{1/p-1/\gamma} \cdot$$

$$\cdot \left(\iint_{\mathfrak{G}} |\zeta - z|^{-2+aq} \, d\xi \, d\eta\right)^{1/q} .$$

For $a > 0$ we have

$$\sup_{z \in \mathbb{C}} \left(\iint_{\mathfrak{G}} |\zeta - z|^{-2+aq} \, d\xi \, d\eta\right)^{1/q} \equiv M(aq\,;\,\mathfrak{G}) < \infty .$$

Then

$$\iint_{\mathfrak{G}} |\underset{\sim}{T}_{\mathfrak{G}} f|^{\gamma} \, dx \, dy \le \frac{1}{\pi^{\gamma}} M(aq;\mathfrak{G})^{\gamma} \left(\|f\|_{L_p,\bar{\mathfrak{G}}}\right)^{\gamma-p} .$$

$$\cdot \iint_{\mathfrak{G}} \left[|f(\zeta)|^p \iint_{\mathfrak{G}} |\zeta - z|^{-2+\gamma a} \, dx \, dy\right] d\xi \, d\eta$$

$$\le \frac{1}{\pi^{\gamma}} M(aq;\mathfrak{G})^{\gamma} M(a\gamma;\mathfrak{G})^{\gamma} \left(\|f\|_{L_p,\bar{\mathfrak{G}}}\right)^{\gamma} .$$

This establishes (18) for $p < \gamma < \frac{2p}{2-p}$, $(1 \le p < 2)$.

To obtain the inequality (10.19), set $g(z) \equiv (T_{\underset{\sim}{\mathfrak{G}}} f)(z)$, and estimate the integral similar to the way used before

$$|g(z+\Delta z) - g(z)| \le \frac{|\Delta z|}{\pi} \int\!\!\int_{\mathfrak{G}} \frac{|f(\zeta)|\, d\xi\, d\eta}{|\zeta-z|\cdot|\zeta-z-\Delta z|}$$

$$\le \frac{|\Delta z|}{\pi}\left(\int\!\!\int_{\mathfrak{G}} |f(\zeta)|^p \big(|\zeta-z|\cdot|\zeta-z-\Delta z|\big)^{-2+\gamma z} d\xi\, d\eta\right)^{1/\gamma} \cdot$$

$$\cdot \left(|f|_{p,\mathfrak{G}}\right)^{1-p/\gamma} \left(\int\!\!\int_{\mathfrak{G}} (|\zeta-z|\cdot|\zeta-z-\Delta z|)^{-2+qa} d\xi\, d\eta\right)^{1/q}$$

$$\le M_{p,\gamma}(\mathfrak{G})\cdot|\Delta z|^{2/\gamma}\cdot\left(|f|_{p,\mathfrak{G}}\right)^{1-p/\gamma}\cdot\left(\int\!\!\int_{\mathfrak{G}} |f(\zeta)|^p (|\zeta-z|\cdot|\zeta-z-\Delta z|)^{-2+\gamma a} d\xi\, d\eta\right)^{1/}$$

The term $|\Delta z|^{2/\gamma}$ comes about by using a well-known integral inequality due to Hadamard ([He 1] pg. 198), ([Ve 2] pg. 39), namely

$$\int\!\!\int_{\mathfrak{G}} |\zeta-z|^{-\alpha}\cdot|\zeta-z-\Delta z|^{-\beta} d\xi\, d\eta < M\cdot|\Delta z|^{2-\alpha-\beta}, \quad \alpha+\beta > 2 .$$

To apply this inequality we must check a few items. First we want $\alpha+\beta > 2$, i.e., $4 - 2qa > 2$ or what is the same thing $0 < qa < 1$. We already know from the hypothesis that $qa > 0$. Furthermore, from the definition of a, we have $qa = 1 - q\left(\frac{1}{2} - \frac{1}{\gamma}\right) < 1$ for $\gamma > 2$. We next note that $2 - \alpha - \beta = -2 + 2qa$, and $\frac{-2+2qa}{q} = \frac{2}{\gamma} - 1$, which yields the desired exponent.

To obtain our next inequality we compute with $\alpha = \beta = 2 - \gamma a$. First $\alpha + \beta = 4 - 2\gamma a > 2$, since $1 > \gamma a = 1 - \gamma\left(\frac{2-p}{2p}\right) > 0$ for $1 \le p < 2$. Consequently, one obtains

$$\left(\iint_{\mathfrak{G}} |g(z+\Delta z) - g(z)|^{\gamma}\, dx\, dy\right)^{1/\gamma} \le M_{p,\gamma} |\Delta z|^{2/\gamma} \left(|f|_{p,\bar{\mathfrak{G}}}\right)^{1-p/\gamma} \cdot$$

$$\cdot \left(\iint_{\mathfrak{G}} |f|^{p} \iint_{\mathfrak{G}} \left(|\zeta - z| \cdot |\zeta - z - \Delta z|\right)^{-2+\gamma a} dx\, dy\right)^{1/\gamma}$$

$$\le M_{p,\gamma} |\Delta z|^{2\left(\frac{1}{\gamma} - \frac{1}{p} + \frac{1}{2}\right)} |f|_{p,\bar{\mathfrak{G}}} ,$$

since $\frac{2 + 2(-2+\gamma a)}{\gamma} = -\frac{2}{\gamma} + 2a = 1 - \frac{2}{p}$. The generic constant $M_{p,\gamma}$ is independent of $\mathfrak{G}$, so we may replace the integration domain on the left by $\mathbb{C}$.

<u>Remark</u>: From the inequalities (10.18), (10.19) it follows that $\underset{\sim}{T}_{\mathfrak{G}} f$ is compact in the space $L^{p}(\mathfrak{G})$, $(1 \le p < 2)$, and

maps this space onto the space $L^{\gamma}_{2a}(\mathbb{C})$, where $a = \frac{1}{\gamma} - \frac{1}{p} + \frac{1}{2}$, where γ is an arbitrary real number satisfying $p < \gamma < \frac{2p}{2-p}$.

<u>Lemma (10.2)</u>: <u>If</u> $f \in W^{m,p}(\mathfrak{G})$ (i.e., the linear space of weak L^p-derivatives, up to and including order m, <u>in</u> $\mathfrak{G}$), $m \geq 2$, $1 \leq p < 2$, <u>then in</u> $\mathfrak{G}$, $f \in W^{m-1,\gamma}(\mathfrak{G}) \cap C^{m-2,\frac{\gamma-2}{\gamma}}(\mathfrak{G})$, <u>where</u> $2 < \gamma < \frac{2p}{2-p}$.

<u>Proof</u>: This is almost a direct consequence of Lemma 1. See [Ve 2] for further details.

<u>Lemma (10.3)</u>: <u>If</u> $f \in W^{m,1}(\mathfrak{G})$, $m \geq 3$, then

$$f \in W^{m-1,\gamma}(\mathfrak{G}) \cap W^{m-2,\frac{2\gamma}{2-\gamma}}(\mathfrak{G}) \cap C^{m-3,\frac{2(\gamma-1)}{\gamma}}(\mathfrak{G}),$$

<u>for</u> γ <u>an arbitrary number satisfying</u> $1 < \gamma < 2$.

<u>Proof</u>: This result is again an immediate consequence of Lemma (10.1) by replacing γ (where $p < \gamma < \frac{2p}{2-p}$) by $\frac{2\gamma}{2-\gamma}$ in the hypothesis of that lemma. This follows by a well-known inequality due to Sobolev.

<u>Lemma (10.4)</u>: <u>Let</u> $\mathfrak{G}$ <u>be a bounded domain and</u> $A(z) \in L^p(\bar{\mathfrak{G}})$, $p > 2$. <u>Then the operator</u>

$$\underset{\sim}{P}f = \iint_{\mathfrak{G}} \frac{A(\zeta)f(\zeta)d\xi\,d\eta}{\zeta - z} = -\pi\, \underset{\sim}{T}_{\mathfrak{G}}(Af) \qquad (10.20)$$

is compact in $L^q(\bar{\mathfrak{G}})$ *if* $\frac{1}{2} \le \frac{1}{p} + \frac{1}{q} \le 1$. *Moreover, if* n *is an integer satisfying*

$$n > \frac{2p}{p-2}\left(\frac{1}{p} + \frac{1}{q} - \frac{1}{2}\right) \ge n-1, \tag{10.21}$$

then

$$\|P^k f\|_{\gamma_k, \alpha, \mathfrak{C}} \le M_{p,q,\alpha}(\mathfrak{G})\, |A|_{p,\bar{\mathfrak{G}}} |f|_{q,\bar{\mathfrak{G}}} \tag{10.22}$$

$(k=1,\dots,n)$

$$\|P^{n+1} f\|_{\infty, \beta, \mathfrak{G}} \le M_{p,q,\alpha}(\mathfrak{G})\, |A|_{p,\bar{\mathfrak{G}}} |f|_{q,\bar{\mathfrak{G}}}, \tag{10.23}$$

where

$$\gamma_k = \frac{1}{\frac{1}{q} + \frac{k}{p} - \frac{k}{2} + k\alpha} \quad (k=1,\dots,n) \tag{10.24}$$

$$\beta = 1 - 2\left(\frac{1}{q} + \frac{n+1}{p} - \frac{n}{2} + n\alpha\right), \tag{10.25}$$

α *being an arbitrary number satisfying*

$$0 < \alpha < \frac{p-2}{2p} - \frac{1}{n}\left(\frac{1}{p} + \frac{1}{q} - \frac{1}{2}\right). \tag{10.26}$$

Remark: $\|f\|_{\infty,\beta,\mathfrak{G}}$ is the Hőlder-norm

$$\|f\|_{\infty,\beta,\mathfrak{G}} \equiv \sup_{\mathfrak{G}} |f| + \sup_{z_1, z_2 \in \mathfrak{G}} \frac{|f(z_1) - f(z_2)|}{|z_1 - z_2|^\beta} \tag{10.27}$$

<u>Proof</u>: Since $A \in L^p(\bar{\mathfrak{G}})$, $f \in L^q(\bar{\mathfrak{G}})$, $\frac{1}{2} \leq \frac{1}{p} + \frac{1}{q} \leq 1$, then $Af \in L^{r_1}(\mathfrak{G})$, where $r_1 = \frac{pq}{p+q}$ with $1 \leq \gamma_1 \leq 2$. By Lemma (10.1), if $\frac{1}{\gamma_1} = \frac{1}{q} + \frac{1}{p} - \frac{1}{2} + \alpha$, $\alpha > 0$ but otherwise arbitrarily small, $\underset{\sim}{P}f \in L_{2\alpha}^{\gamma_1}(\mathfrak{G})$. Hence, $A\underset{\sim}{P}f \in L^{r_2}(\mathfrak{G})$ with $r_2 = \frac{p\gamma_1}{p+\gamma_1}$ and $\underset{\sim}{P}^2 f = \pi^2 \underset{\sim}{T}(A\underset{\sim}{P}f) \in L_{2\alpha}^{\gamma_2}(\mathfrak{G})$ with $\frac{1}{\gamma_2} = \frac{1}{q} + \frac{2}{p} - 1 + 2\alpha$. Repeating this process we obtain for the kth iteration $\underset{\sim}{P}^k f \in L_{2\alpha}^{\gamma_k}$ $(k=1,\ldots,n)$ where γ_k is given by (10.24). Finally $A\underset{\sim}{P}^n f \in L^{r_{n+1}}(\mathfrak{G})$ where $r_{n+1} = \frac{p\gamma_n}{p+\gamma_n}$. Hence by (10.24) we have

$$\frac{1}{r_{n+1}} = \frac{1}{q} + \frac{n+1}{p} - \frac{n}{2} + n\alpha$$

$$= \frac{1}{q} + \frac{1}{p} + n\left(\frac{1}{p} - \frac{1}{2} + \alpha\right) .$$

When n satisfies (10.21) Equation (10.26) implies

$$0 < n\alpha < n\left(\frac{1}{2} - \frac{1}{p}\right) - \left(\frac{1}{p} + \frac{1}{q} - \frac{1}{2}\right) ,$$

or

$$0 < n\alpha + \frac{1}{p} + \frac{1}{q} < n\left(\frac{1}{2} - \frac{1}{p}\right) + \frac{1}{2} .$$

Combining this with our expression for $\frac{1}{r_{n+1}}$ above yields

$$\frac{1}{r_{n+1}} < n\left(\frac{1}{p} - \frac{1}{2}\right) + n\left(\frac{1}{2} - \frac{1}{p}\right) + \frac{1}{2} = \frac{1}{2} ,$$

or $r_{n+1} > 2$. We may then conclude that

$$\underset{\sim}{P}^{n+1} f = -\pi \underset{\sim}{T}(A\underset{\sim}{P}^n f) \in C^{\beta}(\mathbb{C})$$

where β is given by (10.25). The remainder of Lemma (10.4) follows from Lemma (10.1).

We now return to the proof of Theorem (10.2).

<u>Proof of Theorem (10.2)</u>: By definition $u \in D_{\bar{z}}(\mathfrak{G})$; hence, it follows from Lemma (10.1) that $u \in L^{\gamma}(\mathfrak{G})$ where $1 < \gamma < 2$. Consequently, we may choose $\gamma \geq \frac{p}{p-1}$. Let $\mathfrak{G}_1$, $\mathfrak{G}_2$ be two subdomains of $\mathfrak{G}$ such that $\mathfrak{G}_1 \subset \bar{\mathfrak{G}}_1 \subset \mathfrak{G}_2 \subset \bar{\mathfrak{G}}_2 \subset \mathfrak{G}$. If we define $\underset{\sim}{P}_2 \equiv \underset{\sim}{P}_{\mathfrak{G}_2}$, $\underset{\sim}{P}_1 \equiv \underset{\sim}{P}_{\mathfrak{G}_1}$, etc. then for $z \in \mathfrak{G}_2$ we have

$$w(z) - \underset{\sim}{P}_2 w = h(z), \quad h(z) = \Phi_0(z) + \underset{\sim}{T}_2 F, \tag{10.28}$$

where Φ_0 is holomorphic in $\mathfrak{G}_2$ and $\underset{\sim}{T}_2 F \in C^{\frac{p-2}{2}}(\mathbb{C})$. It follows from (10.28) by iteration that

$$w = \underset{\sim}{P}_2^n w + h + \underset{\sim}{P}_2 h + \ldots + \underset{\sim}{P}_2^{n-1} h. \tag{10.29}$$

However, by Lemma (10.4) there exists an n such that $\underset{\sim}{P}_2^n w \in C^{\alpha}(\mathbb{C})$, $\alpha = \frac{p-2}{p}$. Consequently, for some n the right-hand side of (10.29) is in $C^{\alpha}(\bar{\mathfrak{G}}_1)$, which proves our theorem.

We next include a discussion of what Vekua refers to as his BASIC LEMMA, [Ve 2], pp. 144 - 145.

Theorem (10.3): If $w(z)$ is a generalized analytic function of class $\mathfrak{C}_{p,2}(A,B;\mathfrak{G})$, $p > 2$ (i.e., the coefficients A,B of the homogeneous equation satisfied by w are of class $L^{p,2}(\bar{\mathfrak{G}})$, $p > 2$). Furthermore, let

$$g(z) = \begin{cases} A(z) + B(z)\dfrac{\overline{w(z)}}{w(z)}, & \text{if } w(z) \neq 0,\ z \in \mathfrak{G} \\ A(z) + B(z), & \text{if } w(z) = 0,\ z \in \mathfrak{G}. \end{cases} \tag{10.21}$$

Then the function

$$\Phi(z) = w(z)\,e^{-\omega(z)}, \tag{10.22}$$

where

$$\omega(z) \equiv \frac{1}{\pi} \iint_{\mathfrak{G}} \frac{g(\zeta)\,d\xi\,d\eta}{\zeta - z} \equiv -\left(T_{\mathfrak{G}}\,g\right)(z) \tag{10.23}$$

if analytic in $\mathfrak{G}$ except for a discrete set of points.

Proof: By construction $g(z) \in L^{p,2}(\bar{\mathfrak{G}})$, $p > 2$. Then by Theorem (1.23) of Vekua [Ve 2] pp. 43 - 45, $w \in \mathfrak{C}^{\alpha}(\mathbb{C})$, $\alpha = \frac{p-2}{p}$. Furthermore $-g = \frac{\partial w}{\partial \bar{z}} \in L^{p}(\bar{\mathfrak{G}})$, $p > 2$. Using a theorem for differentiating analytic functions of functions of class $W^{1,p}(\mathfrak{G})$ we have $e^{\omega(z)} \in W^{1,p}(\mathfrak{G})$, and that $\frac{\partial e^{-\omega}}{\partial \bar{z}} = g e^{-\omega}$.

Formally, we have that

$$\frac{\partial\Phi}{\partial\bar{z}} = e^{-\omega}\left(\frac{\partial w}{\partial z} + wg\right) = e^{-\omega}(-Aw - B\bar{w} + wg)$$

$$\equiv 0[\text{a.e.}],$$

which implies Φ is holomorphic in $\mathfrak{G}$, with the exception of a finite number of points. For the justification that $w \in W^{1,p}(\mathfrak{G})$ (modulo a discrete set of points) see Vekua [Ve 2], pg. 142.

Instead of continuing to present the work of Vekua on systems of the form (10.7) we shall consider elliptic systems of $2r$ equations with $2r$ unknown functions of the independent variables x and y. Our results will contain Vekua's as a special case. Pertinent to our discussion will be the study of hyperanalytic functions which relate to the solutions of elliptic systems as analytic functions relate to generalized analytic functions.

We next present some important work of Douglis [Do 1] on systems of elliptic equations in two variables. His results deal with $2r$ linear equations in $2r$ unknown functions, and in particular to the principal parts of these equations. The general, linear systems of first order for functions of two variables are of the form

$$u_{i,x} + \sum_{j=1}^{n} \left[b_{ij}(x,y) u_{j,y} + c_{ij}(x,y) u_j \right] = g_i(x,y), \tag{10.24}$$

$$u_{i,x} = \frac{\partial u_i}{\partial x}, \quad i = 1,2,\ldots,n.$$

The system is termed elliptic if the characteristic matrix $(\lambda\ \delta_{ij} + b_{ij})$ is non-singular for all real values of λ. The cannonical subsystems of an elliptic system (10.24) can be decomposed such that the dependent variables that occur in the principal parts of a subsystem are absent from the principal parts of the other subsystems; i.e., one obtains sub-systems of the form

$$\begin{aligned}
&u_{j,x} + au_{j,y} - bv_{j,y} + u_{j-1,y} + \ldots = g_i \\
&v_{j,x} + av_{i,y} + bu_{j,y} + v_{j-1,y} + \ldots = h_j \\
&u_{0,x} + au_{0,y} - bv_{0,y} + \ldots \qquad = g_0 \\
&v_{0,x} + av_{0,y} + bu_{0,y} + \ldots \qquad = h_0
\end{aligned} \tag{10.25}$$

$$(j=1,\ldots,r-1).$$

Here the dots represent terms of zeroth order involving any of the dependent variables of the entire system. The coefficients $a,b,g_0,\dots,h_{r-1}$, are in general different for the different subsystems.

Douglis [Do 1] pg. 261 refers to the cannonical subsystems consisting only of first order terms as generalized Beltrami systems. By introducing a certain commutative, associative algebra the Beltrami system can be written in a simplified form. The algebra is generated over the reals by two elements i.e., subject to the multiplication rules

$$i^2 = -1,\ i\,e = e\,i,\ e^r = 0, \tag{10.26}$$

and generates $2r$ independent elements

$$e^k,\ ie^k,\ (k=0,\dots,r-1),\ \text{where}\ \ e^0 = 1.$$

It is clear that the Beltrami system corresponding to (10.25) above may be written as $Dw = 0$, where

$$D = D_x + (a + i\,b + e)\,D_y\ ,\ D_x = \frac{\partial}{\partial x}\ ,\ D_y = \frac{\partial}{\partial y}\ , \tag{10.27}$$

and for w a hypercomplex variable

$$w = \sum_{k=0}^{r-1} \left(u_k + i\,v_k\right) e^k \equiv \sum_{k=0}^{r-1} w_k(x,y)\,e^k\ .$$

The functions $w_k(x,y)$ are referred to as the <u>component functions</u> of w. If the Beltrami equations

$$\left[D_x + (a+ib)\,D_y\right](\xi + i\eta) = 0$$

are satisfied, then we may write

$$\left[D_x + (a+ib+e)\,D_y\right] u$$

$$\equiv\left\{\left[\left(D_x + (a+ib)\,D_y\right)\xi\right]\left(D_\xi + i\,D_\eta\right) + e\left(\xi_y\,D_\xi + \eta_y\,D_\eta\right)\right\} u.$$

Consequently, we may put $Dw = 0$ into the form

$$\mathfrak{D}u \equiv \left[D_x + i\,D_y + e\left(a(x,y)D_x + b(x,y)D_y\right)\right]\sum_{k=0}^{r-1} e^k\,u_k\,(x,y) = 0\;, \tag{10.28}$$

where $a(x,y)$, $b(x,y)$ and $u_k(x,y)$ are complex valued. In the future we will refer to $\mathfrak{D}u = 0$ as the generalized Beltrami system.

We digress for the moment on a discussion of the hypercomplex numbers

$$c = \sum_{k=0}^{r-1} e^k\,c_k\;,$$

where the c_k are complex valued. If the c_k are all real $(k = 0,\ldots,r-1)$ then c is referred to as a <u>real hypercomplex</u>

<u>number</u>. The conjugate of c, denoted by $\bar{c}$, is given by

$$\bar{c} = \sum_{k=0}^{r-1} e^k c_k .$$

c_0 is called the <u>complex part</u> of c and

$$C \equiv \sum_{k=1}^{r-1} e^k c_k = e\left(\sum_{k=0}^{r-2} e^k c_{k+1}\right)$$

is referred to as the <u>nilpotent part</u> of c, namely because $C^r = 0$.

Douglis [Do 1] defined the (algebraic) norm for the hypercomplex numbers

$$|c| = \sum_{k=0}^{r-1} |c_k| .$$

It follows rather easily that for any hypercomplex numbers c and d that

$$|cd| \le |c| \cdot |d|,$$

and

$$|c+d| \le |c| + |d|.$$

We notice further that if the complex part of $c = c_0 + C$ does not vanish, then one may express the multiplicative inverse of c as

$$c^{-1} \equiv \frac{1}{c} = \frac{1}{c_0}\left[1 - \frac{C}{c_0} + \frac{C^2}{c_0^{\,2}} - \dots + (-1)^{r-1}\frac{C^{r-1}}{c_0^{\,r-1}}\right],$$

which implies the inequality

$$\left|\frac{1}{c}\right| \le \frac{1}{|c_0|}\sum_{k=0}^{r-1}\left(\frac{|C|}{|c_0|}\right)^k.$$

When c^{-1} exists we have

$$1 = |c \cdot \tfrac{1}{c}| \le |c|\,|\tfrac{1}{c}|,$$

and thereby

$$\frac{1}{|c|} \le |\tfrac{1}{c}|.$$

Douglis [Do 1] developed his theory of hyperanalytic functions by investigating the classical solutions of $\mathcal{D}u = 0$. We first reproduce some of his results concerning classical solutions before redeveloping these ideas for the case of the generalized solutions (in the sense of Sobolev). What follows in this latter direction is a report on the recent research of the author with Gerald Hile [Hi 1], [GH 1], [GH 2], [Gi 5]. For convenience in exposition we define, at this point, certain spaces of hypercomplex functions. In general, a function w will be said to lie in a given function space if each of its compent functions is in that space; i.e., we say $w \in L^p(\mathfrak{G})$, where $\mathfrak{G}$ is some domain in the plane, if w if defined in $\mathfrak{G}$ and each w_k is in $L^p(\mathfrak{G})$. This criterion is obviously equivalent to the statement $|w| \in L^p(\mathfrak{G})$. The L^p norm of a hypercomplex function w thus will be defined by

$$\|w\|_{p,\mathfrak{G}} \equiv \left(\iint_{\mathfrak{G}} |w|^p \,dx\,dy\right)^{\frac{1}{p}} \tag{10.29}$$

If w is bounded in $\mathfrak{G}$ we have the norm

$$\|w\|_{\infty,\mathfrak{G}} \equiv \sup_{z \in \mathfrak{G}} |w(z)| \tag{10.30}$$

<u>Remark</u>: The other norms that will be of use to us are defined similary with respect to the corresponding norms for the complex case.

Assuming that $a(x,y)$ and $b(x,y)$ are Hőlder continuous Douglis [Do 1], pg. 266, introduced the concept of a generating solution for the operator $\mathcal{D}$. For our purposes, we assume that a and b are defined in the whole plane $\mathbb{C}$ and make the following definition:

Definition: A hypercomplex function t is a generating solution for $\mathcal{D}$ if

(i) t has the form

$$t(z) = z + \sum_{k=1}^{r-1} t_k(z)\, e^k \tag{10.31}$$

$$= z + T(z) \qquad (T(z) \text{ nilpotent})$$

(ii) $T \in \mathcal{B}^2(\mathbb{C})$,

(iii) $\mathcal{D}t = 0$ in $\mathbb{C}$.

We now proceed to confirm the existence of generating solutions.

Theorem (10.4): If a and b are in $\mathcal{B}^{1,\alpha}(\mathbb{C})$ for some α, $0 < \alpha < 1$, and in $L^p(\mathbb{C})$ for some p, $1 \le p < 2$, then there exists a generating solution for $\mathcal{D}$.

Proof:

Define t by $t_o(z) = z$, and for $k = 1,\ldots,r-1$, following Douglis [Do 1], pg. 267 we define recursively

$$t_k = -I\left[\left(\alpha\ \mathscr{D}_x + b\ \mathscr{D}_y\right) t_{k-1}\right].$$

We assume that the first and second order derivatives of t_{k-1} are in $\mathscr{B}^{\alpha}(\mathbb{C})$ (true of course for $k = 1$). Then the function $(a\ \mathscr{D}_x + b\ \mathscr{D}_y)\ t_{k-1}$ satisfies the conditions of property iv of the $\underset{\sim}{T}$-operator. Thus $t_k \in \mathscr{B}^{2,\alpha}(\mathbb{C})$. Furthermore the equation $\mathscr{D}t = 0$ can be written component-wise as

$$\left(\mathscr{D}_x + i\ \mathscr{D}_y\right) t_o = 0$$

$$\left(\mathscr{D}_x + i\ \mathscr{D}_y\right) t_k = -\left(a\ \mathscr{D}_x + b\ \mathscr{D}_y\right) t_{k-1}\ ,\quad k = 1,\ldots,r-1.$$

Thus t is a generating solution.

Remark: Henceforth we will assume that a and b are in $\mathscr{B}^{1,\alpha}(\mathbb{C})$ for some α, $0 < \alpha < 1$, and in $L^p(\mathbb{C})$ for some p, $1 \le p < 2$, so that the existence of a generating solution is assured.

Following are some inequalities concerning the generating solution which will be used later. We denote generic constants by $M(\cdot)$, where inside the parenthesis are listed whatever entities determine M.

$$|t_x(z)|\ ,\ |t_y(z)| \le M(a,b), \tag{10.32}$$

$$\left|\frac{1}{i+eb}\right| \le M(b)\ . \tag{10.33}$$

From

$$\left|\frac{1}{t(\zeta)-t(z)}\right| \le \frac{1}{|\zeta - z|}\sum_{k=0}^{r-1}\left(\frac{|T(\zeta)-T(z)|}{|\zeta - z|}\right)^k, \tag{10.34}$$

we obtain

$$\left|\frac{1}{t(\zeta)-t(z)}\right| \le \frac{M(a,b)}{|\zeta - z|} \qquad (z \ne \zeta), \tag{10.35}$$

since the derivatives of T are bounded. Combining the above inequalities yields

$$\left|\frac{t_x(z)}{i+e\,b\,(z)}\right| \le M(a,b), \tag{10.36}$$

and

$$\left|\frac{t_x(z)}{i+e\,b\,(z)}\cdot\frac{1}{t(\zeta)-t(z)}\right| \le \frac{M(a,b)}{|\zeta - z|} \quad (z \ne \zeta). \tag{10.37}$$

For functions $u, v \in C^1(\mathfrak{G})$ and $\mathfrak{G}$ a regular domain Douglis [Do 1] established the following theorems.

Theorem (Green's Identity) (10.5): *If* $\mathfrak{G}$ *is a regular domain, and* w *and* v *are hypercomplex functions in* $C^1(\bar{\mathfrak{G}})$, *then*

$$\iint_{\mathfrak{G}} \frac{t_x}{i+eb}\,[w(\mathcal{D}v) + (\mathcal{D}w)\,v]\,dx\,dy = -\int_{\Gamma} w\,v\,dt(z). \tag{10.38}$$

Definition: *Let* v *be a hypercomplex function in a domain* $\mathfrak{G}$. *The integral operator* $\underset{\sim}{J}_{\mathfrak{G}}$ *is defined by*

$$(\underset{\sim}{J}_{\mathfrak{G}} v)(z) \equiv \frac{1}{2\pi i} \iint_{\mathfrak{G}} \frac{t_\xi(\zeta)}{i+eb(\zeta)}\,\frac{v(\zeta)}{t(\zeta)-t(z)}\,d\xi\,d\eta\ . \tag{10.39}$$

When $\mathfrak{G}$ is the whole plane $\mathbb{C}$, we simply write $J_{\mathbb{C}} \equiv J$.

Theorem (10.6) (Cauchy Integral Representation): *Let* $\mathfrak{G}$ *be a regular domain. If* $w \in C^1(\mathfrak{G}) \cap C(\bar{\mathfrak{G}})$, *then for* $z \in \mathfrak{G}$,

$$w(z) = \frac{1}{2\pi i}\int_{\Gamma} \cdot \frac{w(\zeta)}{t(\zeta)-t(z)}\,dt(\zeta) + \underset{\sim}{J}_{\mathfrak{G}}\,(\mathcal{D}w)(z). \tag{10.40}$$

Note that if $\mathcal{D}w = 0$ in $\mathfrak{G}$, (10.40) becomes

$$w(z) = \frac{1}{2\pi i}\int_{\Gamma} \frac{w(\zeta)}{t(\zeta)-t(z)}\,dt(\zeta). \tag{10.41}$$

Remark: It follows that a hyperanalytic (HA) function is at least twice continuously differentiable, since for $z \notin \Gamma$ (10.41) may be differentiated under the integral sign as many times as t is continuously differentiable, i.e., all HA-functions are classical solutions of $\mathcal{D}u = 0$.

Definition: Let f be an analytic function of z, and let E be nilpotent. We define the hypercomplex function

$$f(z+E) \equiv \sum_{k=0}^{r-1} \frac{1}{k!} E^k f^{(k)}(z) \tag{10.42}$$

Theorem (10.7) (Douglis): Let f be analytic in a domain $\mathfrak{G}_*$. Let w, be in $\mathbb{C}^1(\mathfrak{G})$ with the values of w_o contained in $\mathfrak{G}_*$. Then in $\mathfrak{G}$,

$$d\,f(w(z)) = f'(w(z))\,d\,w(z),$$

and in particular for $w(z) \equiv t(z)$,

$$\mathcal{D}(f \circ t)(z) = f'(t(z))\,\mathcal{D}\,t(z) = 0.$$

We note that the composition $f \circ t$ is therefore hyperanalytic. The relationship between analytic functions and hyperanalytic functions is given more completely in the next theorem.

<u>Theorem</u> (10.8)(<u>Douglis</u>): <u>If</u> w <u>is hyperanalytic in</u> $\mathfrak{G}$ <u>and in</u> $C(\bar{\mathfrak{G}})$, <u>where</u> $\mathfrak{G}$ <u>is a regular domain</u>, <u>then</u> w <u>can be represented in</u> $\mathfrak{G}$ <u>by</u>

$$w(z) = \sum_{p=0}^{r-1} \sum_{k=0}^{r-1} \frac{1}{k!} (T(z))^k f_p^{(k)}(z) e^p = \sum_{p=0}^{r-1} f_p(t(z)) e^p \tag{10.43}$$

<u>where each</u> f_p <u>is analytic in</u> $\mathfrak{G}$.

Some consequences of the preceding theorems by Douglis may now be developed.

<u>Theorem</u> (10.9): <u>The zeros of a hyperanalytic function are isolated</u>, <u>unless the function is identically zero</u>.

<u>Proof</u>:

Let w be hyperanalytic in a neighborhood of a point z_o, with $w(z_o) = 0$. Since $t(z) - T(z_o)$ is also a generating solution, the above representation yields, in a neighborhood of z_o,

$$w(z) = \sum_{p=0}^{r-1} \sum_{k=0}^{r-1} \frac{1}{k!} (Tz - Tz_o)^k f_p^{(k)}(z) e^p ,$$

and

$$0 = w(z_o) = \sum_{p=0}^{r-1} f_p(z_o) e^p .$$

If we assume w is not identically zero, then some f_p is not identically zero, and since each f_p is analytic, some f_p is nonzero in a deleted neighborhood of z_o. Let $\tilde{p}$ be the smallest p such that f_p has this property. Then in the above representation for $w(z)$ the coefficient of $e^{\tilde{p}}$ is $f_{\tilde{p}}(z)$, and therefore $w(z) \neq 0$ in some deleted neighborhood of z_o.

<u>Theorem</u> (10.10): <u>A hyperanalytic function which is entire and bounded is a constant</u>.

<u>Proof:</u>

Suppose w is hyperanalytic in the whole plane and is bounded by a constant k. Let z_o be any point in the plane, and let c be a circle of radius R with center z_o. By the Cauchy formula when z is inside c,

$$w(z) = \frac{1}{2\pi i} \int_c \frac{w(\zeta)}{t(\zeta) - t(z)} [t_\xi(\zeta) d\xi + t_\eta(\zeta) d\eta]$$

thus

$$w_x(z) = \frac{1}{2\pi i} \int_c \frac{t_x(z)w(\zeta)}{(t\zeta - tz)^2} [t_\xi(\zeta) d\xi + t_\eta(\zeta) d\eta].$$

Now by our previous inequalities, (10.32), (10.35), we have

$$|w_x(Z_o)| \le \frac{1}{2\pi} \int_0^{2\pi} \frac{M(a,b)}{R^2} K R \, d\theta = M(a,b) \frac{K}{R} \to 0 \quad \text{as} \quad R \to \infty.$$

Similarly, $|w_y(z_o)| \to 0$ as $R \to \infty$.

We now consider the operator $\mathcal{D}$ in a generalized or Sobolev sense. To this end we make the following

Definition: Let $\mathfrak{G}$ be a domain in the plane, and let w and v be hypercomplex functions in $L^1_{loc}(\mathfrak{G})$. Then $v = \mathcal{D}w$ in $\mathfrak{G}$ if for all hypercomplex functions φ in $C^1_c(\mathfrak{G})$

$$\iint_{\mathfrak{G}} \frac{t_x}{i + eb} [w(\mathcal{D}\varphi) + v\,\varphi]\, dx\, dy = 0 . \tag{10.44}$$

Because of linearity, it is clear that it is equivalent to ask that (10.44) hold for all complex functions in $C^1_c(\mathfrak{G})$. We are now able to establish

Theorem (10.11): If w and v are in $L^1_{loc}(\mathfrak{G})$, then $v = \mathcal{D}w$ if and only if for all ψ in $C^1_c(\mathfrak{G})$,

$$\iint_{\mathfrak{G}} [w(\mathcal{D}^*\psi) + v\,\psi]\, dx\, dy = 0 , \tag{10.45}$$

where the differential operator $\mathcal{D}^*$ is given by

$$\mathscr{D}^* \psi = (\mathscr{D}_x + i\,\mathscr{D}_y)\,\psi + e\mathscr{D}_x(a\,\psi) + e\,\mathscr{D}_y(b\,\psi).$$

Proof:

If ψ_1 and ψ_2 are continuously differentiable hypercomplex functions, then

$$\mathscr{D}^*(\psi_1\psi_2) = \psi_1(\mathscr{D}\psi_2) + (\mathscr{D}^*\psi_1)\psi_2$$

From Douglis ([Do 1] p. 270), we have

$$\mathscr{D}^*\left(\frac{t_x}{i + eb}\right) = 0$$

Thus if we let $\psi = \frac{t_x}{i + eb}\,\varphi$, we obtain

$$\mathscr{D}^*\psi = \frac{t_x}{i + eb}\,\mathscr{D}\varphi$$

and therefore (10.44) and (10.45) are equivalent.

Theorem (10.12): The Sobolev derivative $\mathscr{D}$ is unique in $L^1_{loc}(\mathfrak{G})$.

Proof: The proof is direct; see [GH 1], or [Hi 1] pg. 21.

Theorem (10.13): Let $w \in L^1_{loc}(\mathfrak{G})$, $\mathscr{D}w = v$, where $v \in L^p_{loc}(\mathfrak{G})$ for some p, $p > 1$. Then for each k, $w_{k,x}$ and $w_{k,y}$ exist in the Sobolev sense and are in $L^p_{loc}(\mathfrak{G})$. Moreover, the following formulas hold:

$$2 \frac{\partial w_0}{\partial \bar{z}} = v_o$$

$$2 \frac{\partial w_k}{\partial \bar{z}} + a\, w_{k-1,x} + b\, w_{k-1,y} = v_k$$

$k = 1, \ldots, r-1.$

Proof:

Let ψ be complex function in $C_c^1(\mathfrak{G})$. By (10.45),

$$\iint_{\mathfrak{G}} [w(\psi_x + i\,\psi_y + e(a\,\psi)_x + e(b\,\psi)_y) + v\,\psi]\; dx\, dy = 0 .$$

Equating powers of e, we obtain the equations

$$\iint_{\mathfrak{G}} [w_o(\psi_x + i\,\psi_y) + v_o\,\psi]\, dx\, dy = 0$$

$$\iint_{\mathfrak{G}} [w_k(\psi_x + i\,\psi_y) + w_{k-1}((a\,\psi)_x + (b\,\psi)_y) + v_k\,\psi]\, dx\, dy = 0$$

$k = 1, \ldots, r-1.$

The conclusion follows by using the Theorem of Vekua [Ve 1] pg. 72 which we have listed as property (iii) of the operator $\underset{\sim}{T}$, and by integrating by parts [Hi 1] pg. 23, [GH 1].

Corollary to Theorem (10.13): If $\mathfrak{D}w = 0$ in $\mathfrak{G}$, then w is continuously differentiable in $\mathfrak{G}$ (and therefore hyperanalytic).

Proof:

This may be established by a straightforward application of the above arguments; combined with Vekua's Theorem [Ve 2] (pp. 56 - 64), listed as property (iv) of the operator $\underset{\sim}{T}$.

The operator $\underset{\sim}{J}$ is the analogue for $\mathfrak{D}$ of the operator $\underset{\sim}{T}$ for the differential operator $\frac{\partial}{\partial \bar{z}}$; consequently, it is important for our theory to investigate the properties of $\underset{\sim}{J}$. For sufficiently smooth functions w in a domain $\mathfrak{G}$, Douglis [Do 1] proved $\mathfrak{D}(\underset{\sim}{J}_{\mathfrak{G}} w) = w$. We establish now the more general result [Hi 1] pg. 26, [GH 1],

Theorem (10.14): Let $\mathfrak{G}$ be any domain. If $v \in L^1(\mathfrak{G})$, then $\underset{\sim}{J}_{\mathfrak{G}} v \in L^1_{loc}(\mathfrak{G})$, and

$$v = \mathfrak{D}(\underset{\sim}{J}_{\mathfrak{G}} v),$$

Proof:

Let $\mathfrak{G}_o$ be a bounded domain inside $\mathfrak{G}$, and denote the characteristic function on $\mathfrak{G}_o$ by $X_{\mathfrak{G}_o}$. Then one has the bound

$$\iint_{\mathfrak{G}} \left(\iint_{\mathfrak{G}} \left| X_{\mathfrak{G}_o}(z) \frac{t_\xi(\zeta)}{i + e\,b(\zeta)} \frac{v(\zeta)}{t(\zeta) - t(z)} \right| d\xi\, d\eta \right) dx\, dy$$

$$\leq M(a,b) \iint_{\mathfrak{G}} \left(\iint_{\mathfrak{G}_0} \frac{1}{|\zeta - z|} \, dx\, dy \right) |v(\zeta)| \, d\xi\, d\eta$$

$$\leq M(a,b) M(\mathfrak{G}_0) \iint_{\mathfrak{G}} |v(\zeta)| \, d\xi\, d\eta \leq M(a,b,\mathfrak{G}_0) \|v\|_{1,\mathfrak{G}}$$

By Fubini's Theorem; the following integral is finite:

$$\iint_{\mathfrak{G}} \left(\iint_{\mathfrak{G}} X_{\mathfrak{G}_0}(z) \frac{t_\xi(\zeta)}{i + e\,b(\zeta)} \frac{v(\zeta)}{t(\zeta) - t(z)} \, dx\, dy \right) d\xi\, d\eta$$

$$= 2\pi i \iint_{\mathfrak{G}_0} (\underset{\sim}{J}_{\mathfrak{G}} v)(z) \, dx\, dy$$

Therefore $\underset{\sim}{J}_{\mathfrak{G}} v \in L^1_{loc}(\mathfrak{G})$.

Next suppose $\varphi \in \mathbf{C}^1_c(\mathfrak{G})$. Apply (10.40) to φ to obtain

$$\iint_{\mathfrak{G}} \frac{t_x}{i + e\,b} \, v\, \varphi \, dx\, dy$$

$$= \iint_{\mathfrak{G}} \frac{t_x}{i + e\,b} \, v\, \underset{\sim}{J}_{\mathfrak{G}} (\mathcal{D}\varphi) \, dx\, dy$$

$$= \iint_{\mathfrak{G}} \frac{t_x(z)}{i + e\,b(z)} \, v(z) \left(\frac{1}{2\pi i} \iint_{\mathfrak{G}} \frac{t_\xi(\zeta)}{i + e\,b(\zeta)} \frac{(\mathcal{D}\varphi)(\zeta)}{t(\zeta) - t(z)} \, d\xi\, d\eta \right) dx\, dy$$

$$= - \iint_{\mathfrak{G}} \frac{t_\xi(\zeta)}{i + e\,b(\zeta)} (\underset{\sim}{J}_{\mathfrak{G}} v)(\zeta) (\mathcal{D}\varphi)(\zeta) \, d\xi\, d\eta.$$

The interchange of orders of integration is justified by Fubini's Theorem and the bound

$$\iint_{\mathfrak{G}} \left(\iint_{\mathfrak{G}} \left| \frac{t_x(z)}{i+e\,b(z)} v(z) \frac{t_\xi(\zeta)}{i+e\,b(\zeta)} \frac{(\mathfrak{D}\varphi)(\zeta)}{t(\zeta)-t(z)} \right| d\xi\, d\eta \right) dx\, dy$$

$$\leq M(a,b) \iint_{\mathfrak{G}} |v(z)| \left(\iint_{\operatorname{supp} \varphi} \frac{|\mathfrak{D}\varphi(\zeta)|}{|\zeta - z|} d\xi\, d\eta \right) dx\, dy$$

$$\leq M(a,b,\varphi) \|v\|_{1,\mathfrak{G}} .$$

Corollary to Theorem (10.14): If $w \in L^1_{loc}(\mathfrak{G})$, $v \in L^1(\mathfrak{G})$, and $v = \mathfrak{D}w$ in $\mathfrak{G}$, then

$$w = \Phi + \underset{\sim}{J}_{\mathfrak{G}} v$$

where Φ is hyperanalytic in $\mathfrak{G}$.

Conversely, if Φ is hyperanalytic in $\mathfrak{G}$, $v \in L^1(\mathfrak{G})$, then $\Phi + \underset{\sim}{J}_{\mathfrak{G}} v \in L^1_{loc}(\mathfrak{G})$, and

$$\mathfrak{D}(\Phi + \underset{\sim}{J}_{\mathfrak{G}} v) = v$$

Proof: The proof is direct; see [Hi 1] pg. 29, or [GH 1]. We next prove an analogue of Theorem (6.1) of [Ve 2] pg. 38.

Theorem (10.15): Let $\mathfrak{G}$ be a bounded domain. If $v \in L^p(\mathfrak{G})$, $2 < p < \infty$, then the function $w = \underset{\sim}{J}_{\mathfrak{G}} v$ satisfies

(i) $|w(z)| \le M(a,b,p,\mathfrak{G}) \|v\|_{p,\mathfrak{G}}$, $z \in \mathbb{C}$,

(ii) $|w(z_1) - w(z_2)| \le M(a,b,p) \quad \|v\|_{p,\mathfrak{G}} |z_1 - z_2|^{\frac{p-2}{p}}$, $z_1, z_2 \in \mathbb{C}$.

Proof: Using the representation

$$w(z) = \frac{1}{2\pi i} \iint_{\mathfrak{G}} \frac{t_\xi(\zeta)}{i + e\,b(\zeta)} \frac{v(\zeta)}{t(\zeta)-t(z)} d\xi\, d\eta$$

we obtain the bound (where $\frac{1}{p}+\frac{1}{q}= 1$)

$$|w(z)| \le M(a,b) \iint_{\mathfrak{G}} \frac{|v(\zeta)|}{|\zeta - z|} d\xi\, d\eta$$

$$\le M(a,b) \|v\|_{p,\mathfrak{G}} \Big(\iint_{\mathfrak{G}} |\zeta - z|^{-q} d\xi\, d\eta \Big)^{1/q}$$

$$\le M(a,b,p,\mathfrak{G}) \|v\|_{p,\mathfrak{G}} .$$

To show (ii) we use

$$w(z_1) - w(z_2) = \frac{t(z_1) - t(z_2)}{2\pi i} \iint_{\mathfrak{G}} \frac{t_\xi(\zeta)}{i + e\,b(\zeta)} \frac{v(\zeta)}{(t\zeta - tz_1)(t\zeta - tz_2)} d\xi\, d\eta .$$

Thus

$$|w(z_1) - w(z_2)| \le M(a,b) |z_1 - z_2| \iint_{\mathfrak{G}} \frac{|v(\zeta)|}{|\zeta - z_1||\zeta - z_2|} d\xi\, d\eta$$

$$\le M(a,b) |z_1 - z_2| \|v\|_{p,\mathfrak{G}} \Big(\iint_{\mathfrak{G}} |\zeta - z_1|^{-q} |\zeta - z_2|^{-q} d\xi\, d\eta \Big)^{1/q} .$$

However from the Hadamard inequality used in Lemma (10.1), one has for $1 < q < 2$, that

$$\iint_{\mathfrak{G}} |\zeta - z_1|^{-q} |\zeta - z_2|^{-q} \, d\xi \, d\eta \le M(p) \, |z_1 - z_2|^{2-2q} .$$

Since $1 + \frac{2-2q}{q} = \frac{p-2}{p}$, (ii) is proved.

Recalling the definition of the space $L_{p,\nu}(\mathfrak{G})$

Theorem (10.16): Let $v \in L^{p,2}(\mathbb{C})$, $2 < p < \infty$. Then the function $w = \underset{\sim}{J}_{\mathbb{C}} v \equiv \underset{\sim}{J}_v$ satisfies,

(i) $|w(z)| \le M(a,b,p) \|v\|_{p,2}$, $z \in \mathbb{C}$

(ii) $|w(z_1) - w(z_2)| \le M(a,b,p) \|v\|_{p,2} |z_1 - z_2|^{\frac{p-2}{p}}$, $z_1, z_2 \in \mathbb{C}$

(iii) For any $R > 1$, a constant $M(a,b,p,R)$ exists such that for $|z| \ge R$,

$$|w(z)| \le M(a,b,p,R) \|v\|_{p,2} |z|^{\frac{2-p}{p}}$$

(iv) $\mathfrak{D}w = \mathfrak{D}(\underset{\sim}{J}v) = v$ in $\mathbb{C}$.

Proof:

We can write $w(z) = \tilde{w}(z) + \hat{w}(z)$, where

$$\tilde{w}(z) = \frac{1}{2\pi i} \iint_{\mathbb{C}_0} \frac{t_\xi(\zeta)}{i + e\,b(\zeta)} \, \frac{v(\zeta)}{t(\zeta) - t(z)} \, d\xi \, d\eta ,$$

$$\hat{w}(z) = \frac{1}{2\pi i} \iint_{C_0} \frac{t_{\xi}(1/\zeta)}{i + e\,b(1/\zeta)} \frac{v(1/\zeta)}{t(1/\zeta) - t(z)} \frac{1}{|\zeta|^4}\, d\xi\, d\eta .$$

By Theorem (10.15), for $z \in C$,

$$|\hat{w}(z)| \le M(a,b,p) \|v\|_{p,C_0}$$

Also,

$$|\hat{w}(z)| \le M(a,b) \iint_{C_0} \frac{|v(1/\zeta)|}{|1/\zeta - z|\,|\zeta|^4}\, d\xi\, d\eta .$$

Thus, if $\frac{1}{p} + \frac{1}{q} = 1$,

$$|\hat{w}(z)| \le M(a,b) \|v^{(2)}\|_{p,C_o} (I_1(z))^{\frac{1}{q}} , \tag{10.45}$$

where

$$I_1(z) = \iint_{C_0} |\zeta|^{-q} |1 - \zeta z|^{-q} d\xi\, d\eta .$$

If $|z| \ge \frac{1}{2}$, then

$$I_1(z) = |z|^{-q} \iint_{C_0} |\zeta|^{-q} |\zeta - \frac{1}{z}|^{-q}\, d\xi\, d\eta .$$

By a known result for an integral of this type ([Ve 2] p. 39) we obtain

$$I_1(z) \le |z|^{-q} M(P) \left|\frac{1}{z}\right|^{2-2q}$$

$$I_1(z) \le M(P)\left(\frac{1}{|z|}\right)^{2-q}$$

$$\le M(P)\ 2^{2-q} \tag{10.46}$$

$$= M(P).$$

If $|z| \le \frac{1}{2}$, then $|1 - z\zeta| \ge \frac{1}{2}$, and

$$I_1(z) \le 2^q \iint_{C_0} |\zeta|^{-q}\, d\xi\, d\eta$$

$$= M(P).$$

Thus

$$|\hat{w}(z)| \le M(a,b,P)\, \|v^{(2)}\|_{p,C_0},$$

and (i) is proved.

For (ii), apply Theorem (10.15) to obtain, for $z_1, z_2 \in C$,

$$|\hat{w}(z_1) - \hat{w}(z_2)| \le M(a,b,P)\, \|v\|_{p,C_0} |z_1 - z_2|^{\frac{P-2}{P}}.$$

Also,

$$\hat{w}(z_1) - \hat{w}(z_2)$$
$$= \frac{t(z_1) - t(z_2)}{2\pi i} \iint_{C_0} \frac{t_\xi(1/\zeta)}{i + e\,b(1/\zeta)}\, \frac{v(1/\zeta)}{|\zeta|^4}\, \frac{1}{t(1/\zeta) - t(z_1)}\, \frac{1}{t(1/\zeta) - t(z_2)}\, d\xi\, d\eta$$

and

$$|\hat{w}(z_1) - \hat{w}(z_2)| \leq M(a,b)|z_1 - z_2| \iint_{C_0} \frac{|v(1/\zeta)|}{|\zeta|^4 |1/\zeta - z_1||1/\zeta - z_2|} d\xi\, d\eta$$

$$\leq M(a,b)|z_1 - z_2| I_2(z_1, z_2),$$

where

$$I_2(z_1, z_2) = \iint_{C_0} \frac{|v(1/\zeta)|}{|\zeta|^2} \frac{1}{|1-\zeta z_1||1-\zeta z_2|} d\xi\, d\eta .$$

But Vekua has shown ([Ve 2], pp. 44 - 45) that I_2 satisfies

$$I_2(z_1, z_2) \leq M(P)\|v^{(2)}\|_{p,C_0} |z_1 - z_2|^{\frac{p-2}{p}} .$$

Thus (ii) is proved.

To prove (iii), we note that for $|z| > 1$,

$$|\hat{w}(z)| \leq M(a,b) \iint_{C_0} \frac{|v(\zeta)|}{|\zeta - z|} d\xi\, d\eta$$

$$\leq M(a,b) \frac{1}{|z| - 1} \iint_{C_0} |v(\zeta)|\, d\xi\, d\eta$$

$$\leq M(a,b,P)\|v\|_{p,C_0} \frac{1}{|z| - 1} .$$

Furthermore, with use of (10.45) and (10.46) we obtain, for $|z| > 1$,

$$|\hat{w}(z)| \leq M(a,b,P) \|v^{(2)}\|_{p,C_o} |z|^{\frac{2-P}{P}} .$$

Hence for $|z| > 1$,

$$|w(z)| \leq M(a,b,P) \|v\|_{p,2} \left[|z|^{\frac{2-P}{P}} + \frac{1}{|z| - 1} \right].$$

But if $R > 1$, $|z| \geq R$, then

$$\frac{1}{|z| - 1} \leq K(R) \; |z|^{\frac{2-P}{P}} ,$$

where $K(R) = \dfrac{R^{1 - \frac{2}{P}}}{R - 1}$. Thus (iii) follows.

The proof of (iv) is the same as that of the last part of Theorem (10.14).

<u>Corollary to Theorem</u> (10.16): <u>If</u> $w \in L^1_{loc}(C)$, $v \in L^{p,2}(C)$, $2 < p < \infty$, <u>and</u> $v = \mathcal{D}w$ in C , <u>then</u>

$$w = \Phi + Jv$$

<u>where</u> Φ <u>is hyperanalytic in</u> C.

Having obtained certain compactness properties concerning the operator $\underset{\sim}{J}$ we shall now, following the approach of Vekua [Ve 2], investigate properties of solutions of the "triangular" system,

$$\mathcal{D}w + \sum_{k=0}^{r-1} e^k \sum_{\ell=0}^{k} \left(A_{k\ell} w_\ell + B_{k\ell} \bar{w}_\ell \right) = 0 \tag{10.47}$$

where w is a hypercomplex function, and the $A_{k\ell}$ and $B_{k\ell}$ are complex valued.

Remark: If we make the identification

$$A_{k\ell} = \tfrac{1}{2}\left(p_{k\ell} + s_{k\ell}\right) + \frac{i}{2}\left(r_{k\ell} - q_{k\ell}\right)$$

$$B_{k\ell} = \tfrac{1}{2}\left(p_{k\ell} - s_{k\ell}\right) + \frac{i}{2}\left(r_{k\ell} + q_{k\ell}\right) \tag{10.48}$$

then we may rewrite Equation (10.) as the system

$$u_{o,x} - v_{o,y} + p_{o,o} u_o + q_{o,o} v_o = 0$$

$$u_{o,y} + v_{o,x} + r_{o,o} u_o + s_{o,o} v_o = 0$$

$$u_{k,x} - v_{k,y} + a\, u_{k-1,x} + b\, u_{k-1,y} + \sum_{\ell=o}^{k} \left(p_{k\ell} u_\ell + q_{k\ell} v_\ell\right) = 0$$

$$u_{k,y} + v_{k,x} + a\, v_{k-1,x} + b\, v_{k-1,y} + \sum_{\ell=0}^{k} \left(r_{k\ell} u_\ell + s_{k\ell} v_\ell\right) = 0$$

$k = 1,\ldots,r-1.$

A characteristic of this system is that the zero order terms in the kth equation involve only the unknown functions $u_o, \ldots, u_k, v_o, \ldots, v_k$. This property will be helpful to us later.

Remark: The solutions of the "complete" system of equations

$$\mathcal{D}u + \sum_{k=0}^{r-1} e^k \sum_{\ell=0}^{r-1} \left(A_{k\ell} w_\ell + B_{k\ell} \bar{w}_\ell \right) = 0 ,$$

we refer to as generalized hyperanalytic (GHA) functions.

We list several results which are necessary to our investigation. See [Hi 1] Chapter V or [GH 1] for further details.

Theorem (10.17): Let $w, v \in L^1_{loc}(\mathfrak{G})$, with $\mathcal{D}w, \mathcal{D}v \in L^p(\mathfrak{G})$, $2 < p < \infty$. Then in $\mathfrak{G}$,

$$\mathcal{D}(w\,v) = (\mathcal{D}w)\,v + w(\mathcal{D}v) .$$

Theorem (10.18): Let $w \in L_{loc}(\mathfrak{G})$, $\mathcal{D}w \in L^p(\mathfrak{G})$, $2 < p < \infty$, and

$$w(z) = w_0(z) + N(z)$$

where w_0 is complex and N nilpotent. Let the values of w_0 lie inside a bounded domain $\mathfrak{G}_*$, and let f be a complex function which is analytic in some domain containing $\bar{\mathfrak{G}}_*$. Then in $\mathfrak{G}$,

$$\mathcal{D} f(w(z)) = f'(w(z))(\mathcal{D}w)(z).$$

We are now in a position to prove an analogue of Vekua's "exponential-bound" theorem, a consequence of Theorem (10.3) for solutions of (10.47).

<u>Theorem</u> (10.19): <u>Let</u> w <u>be continuous and bounded and satisfy</u> (10.47) <u>in the whole plane</u>, <u>where each</u> $A_{k\ell}$, $B_{k\ell} \in L^{p,2}(\mathbb{C})$, $2 < p < \infty$. <u>Then</u> w <u>has the form</u>

$$w(z) = C \exp \omega(z) \tag{10.50}$$

<u>where</u> C <u>is a hypercomplex constant</u>, <u>and</u> ω <u>is a hypercomplex function in</u> $\mathcal{B}^{\alpha}(\mathbb{C})$, $\alpha = \frac{p-2}{p}$. <u>Moreover</u>, <u>near infinity</u>,

$$\omega(z) = 0(|z|^{-\alpha}).$$

<u>Proof</u>: If w is identically zero, we can set $C = 0$, $\omega = 0$. Otherwise, let p be the smallest value of k such that w_k is not identically zero. Since w is continuous and bounded, (10.47) shows that $\mathcal{D}w \in L^{p,2}(\mathbb{C}) \subset L^{p}(\mathbb{C})$. We can use Theorem (10.13) and set the coefficient of e^p in (10.47) equal to zero to obtain

$$\frac{\partial w_p}{\partial x} + i\,\frac{\partial w_p}{\partial y} + A_{p,p}w_p + B_{p,p}\bar{w}_p = 0$$

By the preceding theorem, w_p is bounded away from zero, and therefore the function

$$\left(\sum_{k=p}^{r-1} w_k e^{k-p}\right)^{-1}$$

is bounded and continuous. Define v to be the function

$$v \equiv \left(\sum_{k=p}^{r-1} w_k e^{k-p}\right)^{-1} \sum_{k=p}^{r-1} e^{k-p} \sum_{\ell=0}^{k} \left(A_{k\ell} w_\ell + B_{k\ell} \bar{w}_\ell\right) .$$

Then $v \in L^{p,2}(\mathbb{C})$, and

$$wv = e^p \left(\sum_{k=p}^{r-1} w_k e^{k-p}\right) \cdot v$$

$$= \sum_{k=p}^{r-1} e^k \sum_{\ell=0}^{k} \left(A_{k\ell} w_\ell + B_{k\ell} \bar{w}_\ell\right) = -\mathcal{D}w$$

Now define

$$\omega \equiv -\underset{\sim}{J}v$$

$$\Phi \equiv w \exp(-\omega) = w \exp(\underset{\sim}{J}v)$$

By Theorems (10.17) and (10.18),

$$\mathcal{D}\Phi = \exp(\underset{\sim}{J}v)\,[wv + \mathcal{D}w] = 0 ,$$

thus Φ is a HA function. The stated properties of ω are a result of Theorem (10.16).

Finally, since ω is bounded, $\exp(-\omega)$ is also bounded. By Theorem (10.10) Φ is constant.

Corollary to Theorem (10.19): *If w satisfies the conditions of Theorem (10.19), and $w(z_0) = 0$ for some z_0 in $\mathbb{C}$, then w is identically zero.*

Proof: Since $\exp(z+E)$ is never nilpotent for any hypercomplex number $z+E$, $w(z_0) = 0$ implies $c = 0$ in the representation (10.50).

We now introduce the operator $\underset{\sim}{Q}$, given in terms of the operator $\underset{\sim}{J}$, and the coefficients of the differential equation.

Definition: *The operator $\underset{\sim}{Q}$ is given by*

$$\underset{\sim}{Q}w \equiv \underset{\sim}{J}\left[\sum_{k=0}^{r-1} e^k \sum_{\ell=0}^{k} \left(A_{k\ell} w_\ell + B_{k\ell} \bar{w}_\ell\right)\right]$$

Theorem (10.20): *If each $A_{k\ell}$, $B_{k\ell}$ is in $L^{p,2}(\mathbb{C})$, $2 < p < \infty$, then $\underset{\sim}{Q}$ is compact in the space $\mathcal{B}(\mathbb{C})$ and maps this space into $\mathcal{B}^\alpha(\mathbb{C})$, $\alpha = \frac{p-2}{p}$. Moreover, near infinity,*

$$|\underset{\sim}{Q} w(z)| = O(|z|^{-\alpha}).$$

Proof: If $w \in \mathcal{B}(\mathbb{C})$, then the function

$$v = \sum_{k=0}^{r-1} e^k \sum_{\ell=0}^{k} \left(A_{k\ell} w_\ell + B_{k\ell} \bar{w}_\ell\right)$$

is in $L^{p,2}(\mathbb{C})$. By Theorem (10.16), $\underset{\sim}{Q}w = \underset{\sim}{J}v$ is in $\mathcal{B}^\alpha(\mathbb{C})$. Furthermore, properties (i), (ii), and (iii) of Theorem (10.16) yield

(i) $|\underset{\sim}{Q}w(z)| \le M(a,b,p,A_{k\ell}, B_{k\ell}) \|w\|_{\infty,\mathbb{C}}$,

(ii) $|\underset{\sim}{Q}w(z_1) - \underset{\sim}{Q}w(z_2)| \le M(a,b,p,A_{k\ell}, B_{k\ell}) \|w\|_{\infty,\mathbb{C}} |z_1 - z_2|^\alpha$,

(iii) For $|z| \ge 2$,

$|w(z)| \le M(a,b,p,A_{k\ell}, B_{k\ell}) \|w\|_{\infty,\mathbb{C}} |z|^{-\alpha}$.

(The dependence of M on the $A_{k\ell}$ and $B_{k\ell}$ arises of course from bounds on the $L^{p,2}(\mathbb{C})$ norms of these functions.) Thus a family in $\mathcal{B}(\mathbb{C})$ which is uniformly bounded is mapped onto a family in $\mathcal{B}^\alpha(\mathbb{C})$ which is uniformly bounded and equicontinuous, and uniformly $O(|z|^{-\alpha})$ at infinity. Hence by the Arzela-Ascoli Theorem, $\underset{\sim}{Q}$ is compact in the space $\mathcal{B}(\mathbb{C})$.

Theorem (10.21): *If each* $A_{k\ell}$, $B_{k\ell}$ *is in* $L^{p,2}(\mathbb{C})$, $2 < p < \infty$, *and* v *is in* $\mathcal{B}(\mathbb{C})$, *then the equation*

$$w + \underset{\sim}{Q}w = v$$

has a unique solution in $\mathcal{B}(\mathbb{C})$.

Proof: Since $\underset{\sim}{Q}$ is compact in $\mathcal{B}(\mathbb{C})$, it is sufficient to show the homogeneous equation

$$w + \underset{\sim}{Q}w = 0$$

has only the zero solution. But if $w = -\underset{\sim}{Q}w$, then w is in $\mathcal{B}(\mathbb{C})$ and vanishes at infinity. Furthermore, differentation shows that w satisfies (10.47). Hence w has the representation (10.50), and since w vanishes at infinity, $c = 0$ and $w = 0$.

We can now give an analogue to the "generating pairs" developed by Lipman Bers [Be 3].

Suppose w is continuous and bounded in $\mathbb{C}$ and satisfies (10.47), with $A_{k\ell}$, $B_{k\ell} \in L^{p,2}(\mathbb{C})$, $2 < p < \infty$. By the Corollary to Theorem (10.16),

$$w + \underset{\sim}{Q}w = \Phi$$

Where Φ is hyperanalytic in $\mathbb{C}$. But w, $\underset{\sim}{Q}w \in \mathcal{B}(\mathbb{C})$, and thus Φ is a constant. Therefore

$$(10.51) \qquad w + \underset{\sim}{Q}w = c = c_1 + i\, c_2$$

where c_1 and c_2 are real hypercomplex numbers (i.e., each coefficient of e^k is real). Let F be the unique solution in $\mathcal{B}(\mathbb{C})$ of

$$F + \underset{\sim}{Q}F = 1$$

and let G be the unique solution in $\mathcal{B}(\mathbb{C})$ of

$$G + QG = i$$

Then w, the unique solution in $\mathcal{B}(\mathbb{C})$ to (10.47), is clearly of the form

$$w = c_1 F + c_2 G.$$

Therefore any bounded solution of (10.47) for the same functions $A_{k\ell}$, $B_{k\ell}$ is of this form. The solutions F and G are called the generating pair associated with the coefficients $A_{k\ell}$, $B_{k\ell}$ and the equation (10.47).

We can extend somewhat the definition of generating pairs by following at this point the approach initiated by Bers [Be 3].

Definition: The hypercomplex valued functions F and G are a generating pair for $\mathfrak{G}_o$ if the following hold:

(i) $I_m\{F_o(z)G_o(z)\} > 0$, for $z \in \mathfrak{G}_o$,

(ii) $F_z, F_{\bar{z}}, G_z, G_{\bar{z}}$ exist and are in $C^{\alpha}(\mathfrak{G}_o)$.

Remark: We assume in what follows that $\mathfrak{G}_o$, $\mathfrak{G}$, etc. are regular domains.

Remark: If $w(x)$ is a hypercomplex valued function defined in $\mathfrak{G}$, then for any $z_o \in \mathfrak{G}$ there exist unique real hypercomplex constants λ and $\mu \ni$

$$W(z_o) = \lambda F(z_o) + \mu G(z_o) .$$

Definition: Whenever the following limit exists, we shall refer to it as a "Bers derivative,"

$$\dot{w}(z_o) \equiv \lim_{z \to z_o} \frac{w(z) - \lambda f(z) - \mu G(z)}{z - z_o} .$$

$\dot{w}(z_o)$ is also referred to as an (F,G) - derivative, and is designated by $\dot{w}(z_o) = \frac{d_{(F,G)} w}{dz}$. We list the following lemmas without proof since they may be verified using similar arguments as in the complex valued case studied by Bers [Be 3].

Lemma (10.22): Let $w(z)$ be defined in $\mathfrak{G}$, and for fixed $z_o \in \mathfrak{G}$, set

$$W(z) \equiv w(z) - \lambda F(z) - \mu G(z). \tag{10.52}$$

Then the existence of $\dot{w}(z_o)$ implies the existence of $W_z(z_o)$, $W_{\bar{z}}(z_o)$, and also that $W_{\bar{z}}(z_o) = 0$, with $W_z(z_o) = \dot{w}(z_o)$. Furthermore, if W_z and $W_{\bar{z}}$ exist and are continuous in a neighborhood of z_o, with $W_{\bar{z}}(z_o) = 0$, then $\dot{w}(z_o)$ exists and $\dot{w}(z_o) = W_z(z_o)$.

Definition: For the generating pair $\{F,G\}$ we define the characteristic coefficients

$$a = a_{(F,G)} \equiv -\frac{\bar{F}G_{\bar{z}} - F_{\bar{z}}\bar{G}}{F\bar{G} - \bar{F}G},$$

$$b = b_{(F,G)} = \frac{FG_{\bar{z}} - F_{\bar{z}}G}{F\bar{G} - \bar{F}G},$$

$$A = A_{(F,G)} = -\frac{\bar{F}G_z - F_z\bar{G}}{F\bar{G} - \bar{F}G},$$

$$B = B_{(F,G)} = \frac{FG_z - F_zG}{F\bar{G} - \bar{F}G}.$$

Remark: The hypothesis (i) ensures that the hypercomplex quantity $(F\bar{G}-\bar{F}G)$ has nonvanishing complex part, and hence a multiplicative inverse exists.

Lemma (10.23): Let w be defined in $\mathfrak{G}$. If for fixed $z_o \in \mathfrak{G}$, $\dot{w}(z_o)$ exists, then $w_z(z_o)$, $w_{\bar{z}}(z_o)$ exist and satisfy

(10.53) $$w_{\bar{z}}(z_o) = a(z_o)w(z_o) + b(z_o)\overline{w(z_o)},$$

(10.54) $$\dot{w}(z_o) = w_z(z_o) - A(z_o)w(z_o) - B(z_o)\overline{w(z_o)}.$$

Furthermore, if w_z, $w_{\bar{z}}$ exist, are continuous in a neighborhood of z_o, and (10.53) holds, then $\dot{w}(z_o)$ exists and (10.54) holds. Following Bers [Be 3] we say that w is (F,G) - pseudo hyperanalytic of the first kind, (PHA-1) in $\mathfrak{G}$ if $\dot{w}$ exists everywhere in $\mathfrak{G}$. It can be shown that F and G are PHA-1.

Lemma (10.24): If w_z, $w_{\bar{z}}$ exist and are continuous in a regular domain $\mathfrak{G}$, and if

(10.55) $$w_{\bar{z}} = aw + b\bar{w},$$

then w is (F,G) - PHA-1 in $\mathfrak{G}$ and satisfies $\dot{w}=w_z - AW - B\bar{W}$. Furthermore, if w is (F,G) - PHA-1 in $\mathfrak{G}$, then w_z, $w_{\bar{z}}$ exist and are in $C^{\alpha}(\mathfrak{G})$. Moreover, in $\mathfrak{G}$

$$w_{\bar{z}} = aw + b\bar{w} \ , \quad \dot{w} = w_z - Aw - B\bar{w} \ .$$

<u>Proof</u>: The only part we prove is the conclusion that

w_z , $w_{\bar{z}} \in C^{\alpha}(\mathfrak{G})$. If $(\underset{\sim}{J}_{\mathfrak{G}}u)(z) \equiv -\dfrac{1}{\pi} \displaystyle\int_{\mathfrak{G}} \dfrac{u(\zeta)}{\zeta - z} \, d\xi \, d\eta$, then by (10.14)'

corollary we can show that each component of $w - \underset{\sim}{J}_{\mathfrak{G}}(aw + b\bar{w})$

is analytic in $\mathfrak{G}$. By Theorem (10.15) $\underset{\sim}{J}_{\mathfrak{G}}(aw + b\bar{w}) \in C^{\alpha}(\mathfrak{G})$.

Hence $w \in C^{\alpha}(\mathfrak{G})$, which implies, by well-known properties

of $\underset{\sim}{J}$, that $\underset{\sim}{J}_{\mathfrak{G}}(aw + b\bar{w}) \in C^{1,\alpha}(\mathfrak{G})$, and therefore $w \in C^{1,\alpha}(\mathfrak{G})$.

<u>Definition</u>: <u>If</u> w <u>is</u> (F,G)-PHA-1 <u>in</u> $\mathfrak{G}$, <u>then the function</u>

(10.56) $\qquad \chi \equiv \varphi + i\psi \equiv {}_{*}w \ ,$

<u>where</u> φ <u>and</u> ψ <u>are the unique, real hypercomplex functions</u>
<u>satisfying</u>

(10.57) $\qquad\qquad w = \varphi F + \psi G \ ,$

<u>is said to be</u> (F,G) - <u>pseudohyperanalytic of the second kind</u>
(PHA-2) <u>in</u> $\mathfrak{G}$.

<u>Remark</u>: The functions φ and ψ may be represented by

(10.58) $\qquad \varphi = \dfrac{w\bar{G} - \bar{w}G}{F\bar{G} - \bar{F}G} \ , \qquad \psi = \dfrac{\bar{w}F - w\bar{F}}{F\bar{G} - \bar{F}G} \ .$

The correspondence between w and χ is one-to-one, and is given by

$$\chi = {}_{*}w \equiv \frac{w\bar{G} - \bar{w}G}{F\bar{G} - \bar{F}G} + i\,\frac{\bar{w}F - w\bar{F}}{F\bar{G} - \bar{F}G} \tag{10.59}$$

$$w = {}^{*}\chi \equiv \varphi F + \psi G\ ,$$

with φ and ψ defined by (10.56).

<u>Lemma</u> (10.25): <u>If</u> $w \in B(\mathbb{C})$ <u>and satisfies in</u> $\mathbb{C}$ <u>the equation</u> (10.47) <u>where the coefficients are in</u> $L^{p,2}(\mathbb{C})$, <u>then if</u> $w_o(z)$ <u>vanishes at some</u> z_o, $w_o(z) \equiv 0$.

<u>Proof</u>: From (10.50) we have that $w(z)$ has the form

$$\begin{aligned} w(z) &= c \exp [\omega(z)] \\ &= (c_o + \text{nilpotent})(\exp \omega_o(z) + \text{nilpotent}) \\ &= c_o \exp \omega_o(z) + \text{nilpotent} \\ &= w_o(z) + \text{nilpotent}, \end{aligned}$$

where $w(z) \in \mathbb{B}^{\alpha}(\mathbb{C})$. However, if $w(z_o) = 0$ then one must have $c_o = 0$.

<u>Theorem</u> (10.26): <u>The generating pair</u> $\{F,G\}$ <u>defined in</u> [3] <u>satisfies</u> (i).

If $w \in \beta(C)$ and satisfies (10.47), then $w = cF + dG$ where c and d are real hypercomplex constants. Now if for some $z \in C$ it holds that

$$F_o(z)\overline{G_0(z)} - \overline{F_o(z)}G_o(z) = 0$$

then there exist real constants c_o, d_o, not both zero, such that

$$c_oF_o(z) + d_oG_o(z) = 0.$$

Which means, by the previous lemma, that the complex part of a $\beta(C)$ solution of (10.47) vanishes identically, i.e.

$$c_oF_o(z) + d_oG_o(z) \equiv 0$$

This contradicts the known properties of F and G at ∞, namely $F(\infty) = F_o(\infty) = 1$, $G(\infty) = G_o(\infty) = i$.

<u>Definition</u>: <u>Two generating pairs</u> $\{\tilde{F},\tilde{G}\}$ <u>and</u> $\{F,G\}$ <u>are said to be equivalent if there are real hypercomplex constants</u> a_{ij} $\ni$

(10.60)
$$\tilde{F} = a_{11}F + a_{12}G$$
$$G = a_{21}F + a_{22}G.$$

Remark: The conditions $I_m(\overline{F_o G_o}) > 0$, $I_m(\overline{\tilde{F}_o \tilde{G}_o}) > 0$ imply that the complex part of $(a_{11}a_{22} - a_{21}a_{12}) > 0$ and hence $(a_{11}a_{22} - a_{21}a_{12})^{-1}$ exists.

Theorem (10.27): (i) If two generating pairs defined in the same domain are equivalent, they have the same characteristic coefficients. (ii) If two generating pairs defined in the whole plane and also bounded have the same characteristic coefficients, they are equivalent. (iii) If $\{F,G\}$ and $\{\tilde{F},\tilde{G}\}$ are equivalent, then every $(\tilde{F},\tilde{G})$ - PHA-1 is (F,G)-PHA-1, and

$$\frac{d_{(F,G)}w}{dz} = \frac{d_{(\tilde{F},\tilde{G})}w}{dz}.$$

Proof: Statements (i), and (iii) follow immediately. For the proof of (ii) assume $\{F,G\}$ and $\{\tilde{F},\tilde{G}\}$ are generating pairs for the whole plane, are bounded, and have the same characteristic coefficients. It is possible to find real hypercomplex constants $a_{ij} \ni$ the functions

$$f(z) \equiv \tilde{F}(z) - a_{11}F(z) - a_{12}G(z)$$

$$g(z) = \tilde{G}(z) - a_{21}F(z) - a_{22}G(z)$$

vanish at some point z_o. Moreover,

$$f_{\bar{z}} = af + b\bar{f} \ , \ g_{\bar{z}} = ag + b\bar{g} \ .$$

From Theorem (10.19) it follows since f and $g \in \beta(\mathbb{C})$ that $f \equiv g \equiv 0$.

<u>Remark</u>: If $\{F,G\}$ and $\{\tilde{F},\tilde{G}\}$ are equivalent generating pairs, with

$$\tilde{F} = a_{11}F + a_{12}G, \ \tilde{G} = a_{21}F + A_{22}G,$$

and if

$$w = \varphi F + \psi G = \tilde{\varphi}\tilde{F} + \tilde{\psi}\tilde{G} \ ,$$

then

$$\varphi = a_{11}\tilde{\varphi} + a_{21}\tilde{\psi} \ ,$$

$$\psi = a_{12}\tilde{\varphi} + a_{22}\tilde{\psi} \ .$$

<u>Definition</u>: <u>If</u> $\{F,G\}$ <u>is a generating pair</u>, <u>its adjoint generating pair is given by</u> $\{F,G\}^* = \{F^*,G^*\}$, <u>with</u>

(10.67) $$F^* = \frac{-2\bar{F}}{F\bar{G} - \bar{F}G} \ , \ G^* = \frac{2\bar{G}}{F\bar{G} - \bar{F}G} \ .$$

<u>Remark:</u> This definition makes sense for the hypercomplex case since i, 338,can be verified for the pair $\{F^*,G^*\}$, if it holds for $\{F,G\}$.

Theorem (10.28): If $\{F,G\}$ is a generating pair then

(i) $\{F,G\}^{**} = \{F,G\}$.

(ii) If $\{F^*,G^*\} = \{F,G\}^*$, then

$$a_{(F^*,G^*)} = -a_{(F,G)}, \quad A_{(F^*,G^*)} = -A_{(F,G)}$$

$$b_{(F^*,G^*)} = \overline{-B_{(F,G)}}, \quad B_{(F^*,G^*)} = \overline{-b_{(F,G)}}$$

Proof: The proof is immediate by direct substitution and some minor manipulations.

We may also introduce the concept of successor or predecessor as in the Bers theory. We say that $\{F',G'\}$ is a successor of $\{F,G\}$ and $\{F,G\}$ is a predecessor of $\{F',G'\}$ if

$$a_{(F',G')} = a_{(F,G)},$$
$$b_{(F',G')} = -B_{(F,G)}. \tag{10.62}$$

Remark: If $\{F',G'\}$ is a successor of $\{F,G\}$ then $\{F,G\}^*$ is a successor of $\{F',G'\}^*$.

Proof: This follows immediately from the definitions.

Theorem (10.29): Let $\{F,G\}$ be a generating pair for $\mathfrak{G}$, which is in $C^{1,\alpha}(\mathfrak{G})$. Then $\{F,G\}$ possess both a successor and a predecessor which are also in $C^{1,\alpha}(\mathfrak{G})$.

Proof: Extend F and G to $\mathbb{C}$ by defining them to be zero in $\mathbb{C} - \mathfrak{G}$. From the hypothesis (ii), and (10.62) above, the characteristic coefficients of the successor and predecessor are in $C^{\alpha}(\mathfrak{G}) \cap L^{p,2}(\mathbb{C})$, $p > 2$. Using the results of Theorem (10.27) we conclude the existence of a generating pair in $\mathbb{C}$ which corresponds to the preceeding and succeeding characteristic coefficients. The restriction of these generating pairs to $\mathfrak{G}$ satisfies the hypotheses (i), (ii).

Following Bers, we will call a sequence of generating pairs $\{F^{(\nu)}, G^{(\nu)}\}$ $(\nu = 0, \pm 1, \pm 2, \ldots)$ a generating sequence if $\{F^{(\nu+1)}, G^{(\nu+1)}\}$ is a successor of $\{F^{(\nu)}, G^{(\nu)}\}$. If $\{F^{(o)}, G^{(o)}\} = \{F,G\}$, we will say $\{F,G\}$ is embedded in $\{F^{(\nu)}, G^{(\nu)}\}$, $(\nu = 0, \pm 1, \pm 2, \ldots)$.

Finally it can be shown that the following holds

Theorem (10.30): If $\{F,G\}$ is a generating pair, defined in a bounded regular domain, then $\{F,G\}$ can be embedded in a generating sequence.

XI. Elliptic Systems of First Order Differential Equations : Further Results and Constructive Methods

In this chapter we continue our investigations of solutions of elliptic, first order systems,

$$(11.1) \qquad \mathcal{D}u + \sum_{\ell=0}^{n-1} e^{\ell} \sum_{k=0}^{n-1} (A_{\ell k}u_k + B_{\ell k}\overline{u}_k) = 0,$$

with

$$\mathcal{D}u = \frac{\partial u}{\partial x} + i \frac{\partial u}{\partial y} + e(a(x,y) \frac{\partial u}{\partial x} + b(x,y) \frac{\partial u}{\partial y}) ,$$

where the complex coefficients $A_{\ell k}$, $B_{\ell k} \in L^p(\mathfrak{G})$, $p > 2$, or if $\mathfrak{G}$ is unbounded $\in L^{p,2}(\mathfrak{G})$, $p > 2$. The distributional solutions of (11.1) turn out when $a,b \in C^{1,\alpha}(\mathfrak{G})$. We refer to these functions as <u>generalized</u> <u>hyperanalytic</u> (GHA) functions. This name has been suggested by the work of Douglis [Do 1] concerning strong solutions of $\mathcal{D}u = 0$, which he refers to as <u>hyperanalytic</u> (HA) functions, and the work of Vekua [Ve 2] concerning the distributional solutions of (10.5).

In the last chapter we investigated certain special cases of (11.1) such as $\mathcal{D}u = 0$, and the triangular system

$$(11.2) \qquad \mathcal{D}u + \sum_{\ell=0}^{n-1} e^{\ell} \sum_{k=0}^{\ell} (A_{\ell k} u_k + B_{\ell k} \overline{u}_k) = 0 \quad \text{in } L^p(\mathfrak{G}) .$$

We continue our discussions of the research of G. Hile with the author, by considering a restricted class of (11.2), namely the equation

$$(11.3) \qquad \mathcal{D}w + Aw + B\overline{w} = 0 ,$$

where w, A, and B are hypercomplex functions.

In the special instance where $a(x,y) = b(x,y) = 0$,

$\mathfrak{D}w = 2\frac{\partial w}{\partial \bar{z}}$, and (11.3) becomes the differential equation of pseudohyperanalytic function (PHA) theory.

For (11.3) we can obtain integral representations for solutions w in a regular domain $\mathfrak{G}$. The next lemma and theorem establish a Green's identity for solutions of (11.3).

Lemma (11.1) <u>let</u> $\mathfrak{G}$ <u>be</u> <u>a</u> <u>regular</u> <u>domain</u>, <u>with</u> $u \in C(\overline{\mathfrak{G}})$, $\mathfrak{D}u \in L^p(\mathfrak{G})$ <u>where</u> $2 < p < \infty$. <u>Then</u>

$$\int_\Gamma u(z)\, dt(z) = -\iint_{\mathfrak{G}} \frac{t_x(z)}{i+e\, b(z)} (\mathfrak{D}u)(z)\, dx\, dy.$$

<u>Proof</u>

Let $\{\psi_n\}$ be a sequence in $C_c(\mathfrak{G})$ such that $\psi_n \to \mathfrak{D}u$ in $L^p(\mathfrak{G})$. By Theorem 5.3, $\underset{\sim}{J}_{\mathfrak{G}}\, \psi_n \to \underset{\sim}{J}_{\mathfrak{G}}(\mathfrak{D}u)$ in the whole plane C . In Theorem (10.5) we set $w = \underset{\sim}{J}_{\mathfrak{G}}\, \psi_n$, $v = 1$, to obtain

$$\int_\Gamma (J_{\mathfrak{G}}\, \psi_n)\, dt(z) = -\iint_{\mathfrak{G}} \frac{t_x(z)}{i + e\, b(z)} \psi_n(z)\, dx\, dy$$

Letting $n \to \infty$, we have

$$\int_\Gamma J_{\mathfrak{G}}(\mathfrak{D}u)(z)\, dt(z) = -\iint_{\mathfrak{G}} \frac{t_x(z)}{i+e\, b(z)} (\mathfrak{D}u)(z)\, dx\, dy$$

By Corollary (10.) in $\mathfrak{G}$

$$J_{\mathfrak{G}}(\mathfrak{D}u) = u - \Phi$$

where Φ is hyperanalytic in $\mathfrak{G}$. Furthermore, $\underset{\sim}{J}_{\mathfrak{G}}(\mathfrak{D}u)$, $u \in C(\overline{\mathfrak{G}})$, and thus $\Phi \in C(\mathfrak{G})$. We may use Theorem (10.5) with $w = \Phi$, $v = 1$, to conclude

$$\int_\Gamma \Phi(z)\, dt(z) = 0$$

and the theorem is proved.

<u>Definition</u>: <u>For fixed</u> A <u>and</u> B , <u>we define the operator</u>

$$(11.4) \qquad Cw \equiv \mathfrak{D}w + Aw + B\overline{w}$$

and an associated operator

$$(11.5) \qquad \tilde{C}v \equiv \mathfrak{D}v - Av + B^*\overline{v}$$

where B^* is defined by

$$(11.6) \qquad B^* \equiv \frac{i+eb}{t_x} \frac{\overline{t}_x}{\overline{i + eb}} \overline{B}$$

<u>Theorem (11.3)</u>: <u>Let</u> $\mathfrak{G}$ <u>be a regular domain, and</u> $A, B \in L^p(\mathfrak{G})$, with $2 < p < \infty$. <u>If</u> $w, v \in C(\mathfrak{G})$, <u>and satisfy in</u> $\mathfrak{G}$, $Cw = 0$, $\tilde{C}v = 0$, <u>then the integral</u>

$$\int_\Gamma w(z)\, v(z)\, dt(z)$$

<u>is a real hypercomplex number</u>.

<u>Proof</u>

Since $w, v \in C(\overline{\mathfrak{G}})$,

$$\mathfrak{D}w = -Aw - B\overline{w} \in L^p(\mathfrak{G})$$

$$\mathfrak{D}v = Av - B^*\overline{v} \in L^p(\mathfrak{G})$$

(We remark that the quantity $\frac{i + eb}{t_x} \frac{\overline{t}_x}{\overline{i + eb}}$ is bounded in the whole complex plane, as can be seen by (10.37).) By Lemma (11.1) and Theorem (10.17)

$$\int_\Gamma w(z)\, v(z)\, dt(z) = -\iint_{\mathfrak{G}} \frac{t_x}{i + eb}\, \mathfrak{D}(wv)\, dx\, dy$$

$$= -\iint_{\mathfrak{G}} \frac{t_x}{i + eb}\, (w\mathfrak{D}v + v\mathfrak{D}w)\, dx\, dy$$

$$= \iint_{\mathfrak{G}} \left(\frac{t_x}{i + eb}\, B\bar{w}v + \overline{\frac{t_x}{i + eb}}\, \overline{B\bar{w}v}\right) dx\, dy$$

which is a real hypercomplex number.

Following Vekua's techniques, one may now construct some special solutions of (11.3), [Hi 1], [GH 1].

<u>Theorem</u> (11.4): <u>Let</u> A <u>and</u> B <u>be hypercomplex functions in</u> $L^{p,2}(\mathbb{C})$, <u>where</u> $2 < p < \infty$. <u>Then there exist hypercomplex functions of two complex variables</u>, $X^{(1)}(z,\zeta)$ <u>and</u> $X^{(2)}(z,\zeta)$, <u>with the properties</u>

(i) In $\mathbb{C} - \{\zeta\}$, for $j = 1, 2$,

$$D_z X^{(j)}(z,\zeta) + A(z) X^{(j)}(z,\zeta) + B(z) \overline{X^{(j)}(z,\zeta)} = 0 ,$$

(Here $\mathfrak{D}_z$ denotes our usual differential operator $\mathfrak{D}$ where differentiation is with respect to the variable z rather than ζ.)

(ii) $$X^{(1)}(z,\zeta) = \frac{\exp[\omega^{(1)}(z) - \omega^{(1)}(\zeta)]}{2(t(\zeta) - t(z))}$$

$$X^{(2)}(z,\zeta) = \frac{\exp[\omega^{(2)}(z) - \omega^{(2)}(\zeta)]}{2i(t(\zeta) - t(z))}$$

where for $j = 1, 2$, $\omega^{(j)} \in B_\alpha(C)$, $\alpha = \frac{p-2}{p}$, and

$$\omega^{(j)}(z) = O(|z|^{-\alpha}) \text{ as } |z| \to \infty .$$

Proof

Since the proof for $X^{(1)}$ and $X^{(2)}$ are nearly identical, we give the proof only for $X^{(1)}$. We temporarily fix a point ζ in C , and define a function $\hat{B}$ by

$$\hat{B}(z) = B(z) \frac{\overline{t(z) - t(\zeta)}}{t(z) - t(\zeta)} .$$

We have $\hat{B} \in L^{p,2}(C)$, since

$$|\hat{B}(z)| \le |B(z)| \, |t(z) - t(\zeta)| \, \left| \frac{1}{t(z) - t(\zeta)} \right|$$

$$\le M(a,b) \, |B(z)|$$

Now consider the functional equation

$$(11.7) \quad w(z) + (\underset{\sim}{Q}w)(z) - (\underset{\sim}{Q}w)(\zeta) = 1, \; z \in C$$

where $\underset{\sim}{Q}$ is the operator defined by

$$\underset{\sim}{Q}w = \underset{\sim}{J}(Aw + \hat{B}\overline{w})$$

By Theorem (10.20) is compact in the space $\beta(C)$. If we define an operator $\underset{\sim}{P}$ by

$$(\underset{\sim}{P}w)(z) = (\underset{\sim}{Q}w)(z) - (\underset{\sim}{Q}w)(\zeta) , \; z \in C$$

then (11.7) may be written as

$$(11.8) \quad w(z) + (\underset{\sim}{P}w)(z) = 1, \; z \in C .$$

Moreover, since $\underset{\sim}{Q}$ maps a bounded sequence in $\mathcal{B}(\mathbb{C})$ onto a sequence in $\mathcal{B}(\mathbb{C})$ with a convergent subsequence, P has the same property. Thus P is compact in $\mathcal{B}(\mathbb{C})$. Therefore, in order to show (11.7) has a solution in $\mathcal{B}(\mathbb{C})$, it is sufficient to show the homogeneous equation has only the zero solution. Suppose then that $v \in \mathcal{B}(\mathbb{C})$ and satisfies

$$(11.9) \qquad v(z) + (\underset{\sim}{Q}v)(\zeta) = 0 \ , \ z \in \mathbb{C} \ .$$

Differentiating this equation, we obtain

$$\mathcal{D}v + Av + \hat{B}\bar{v} = 0$$

Since $v(\zeta) = 0$, by Corollary (10.19) $v = 0$.
Thus we may let w be the unique solution in $B(\mathbb{C})$ to (11.7) . Differentiating (11.7), we obtain

$$\mathcal{D}w + Aw + \hat{B}\bar{w} = 0$$

According to Theorem (10.19), w has the form

$$w(z) = C \exp \omega(z)$$

where C is a hypercomplex constant, $\omega \in \mathcal{B}^{\alpha}(\mathbb{C})$, $\alpha = \frac{p-2}{p}$, and $\omega(z) = O(|z|^{-\alpha})$ as $|z| \to \infty$. But since $w(\zeta) = 1$, we conclude $C = \exp[-\omega(\zeta)]$, and

$$w(z) = \exp[\omega(z) - \omega(\zeta)] = w(z, \zeta) \ .$$

we now set

$$X^{(1)}(z, \zeta) = \frac{w(z, \zeta)}{2(t(\zeta) - t(z))} = \frac{\exp[\omega(z) - \omega(\zeta)]}{2(t(\zeta) - t(z))}$$

Then for $z \in \mathbb{C} - \{\zeta\}$.

$$\partial_z X^{(1)}(z,\zeta) = \frac{1}{2(t(\zeta) - t(z))} \partial_z w(z,\zeta)$$

$$= \frac{1}{2(t(\zeta) - t(z))} \cdot (-A(z)\, w(z,\zeta) - \hat{B}(z)\, \overline{w(z,\zeta)}$$

$$= -A(z)\, X^{(1)}(z,\zeta) - B(z)\, \overline{X^{(1)}(z,\zeta)}$$

We now define <u>fundamental</u> <u>kernels</u> associated with fixed A, B in $L^{p,2}$ $(\mathbb{C})$ and the equation (11.3) .

<u>Definition</u>: <u>The</u> <u>fundamental</u> <u>kernels</u> $\Omega^{(1)}$ <u>and</u> $\Omega^{(2)}$, <u>associated</u> <u>with</u> A <u>and</u> B <u>in</u> $L^{p,2}$ $(\mathbb{C})$, <u>are</u>

(11.10) $\quad \Omega^{(1)}(z,\zeta) \equiv X^{(1)}(z,\zeta) + i\, X^{(2)}(z,\zeta)$

(11.11) $\quad \Omega^{(2)}(z,\zeta) \equiv X^{(1)}(z,\zeta) - i\, X^{(2)}(z,\zeta)$

where $X^{(1)}$ and $X^{(2)}$ are the functions described in Theorem (11.4), [Hi 1], [GH 1] .

<u>Theorem</u> (11.6). <u>The</u> <u>fundamental</u> <u>kernels</u> $\Omega^{(1)}$ <u>and</u> $\Omega^{(2)}$ satisfy,

(i) <u>For</u> <u>each</u> ζ in $\mathbb{C}$, in $\mathbb{C} - \{\zeta\}$

(11.12) $\quad \partial_z \Omega^{(1)}(z,\zeta) + A(z)\, \Omega^{(1)}(z,\zeta) + B(z)\, \overline{\Omega^{(2)}(z,\zeta)} = 0$

(11.13) $\quad \partial_z \Omega^{(2)}(z,\zeta) + A(z)\, \Omega^{(2)}(z,\zeta) + B(z)\, \overline{\Omega^{(1)}(z,\zeta)} = 0$

(ii) For fixed ζ, and $j = 1, 2$,

$$|\Omega^{(j)}(z,\zeta)| = O(|z|^{-1}) \text{ as } |z| \to \infty$$

(iii) As $|z - \zeta| \to 0$.

(11.14) $$\left|\Omega^{(1)}(z,\zeta) - \frac{1}{t(\zeta) - t(z)}\right| = o\left(|z - \zeta|^{\frac{-2}{p}}\right)$$

(11.15) $$|\Omega^{(2)}(z,\zeta)| = O\left(|z - \zeta|^{\frac{-2}{p}}\right)$$

Proof

Property (i) is readily verified from (1) of Theorem (11.4). Property (ii) follows from the relations

$$\Omega^{(1)}(z,\zeta) = \frac{\exp[\omega^{(1)}(z) - \omega^{(1)}(\zeta)] + \exp[\omega^{(2)}(z) - \omega^{(2)}(\zeta)]}{2(t(\zeta) - t(z))}$$

$$\Omega^{(2)}(z,\zeta) = \frac{\exp[\omega^{(1)}(z) - \omega^{(1)}(\zeta)] - \exp[\omega^{(2)}(z) - \omega^{(2)}(\zeta)]}{2(t(\zeta) - t(z))}$$

because each $\omega^{(j)}$ is bounded in C, and by (10.35).

To show (iii), first we remark that it is easily seen that the hypercomplex function $\exp(z+E)$ is uniformly Lipschitz continuous wherever $z+E$ remains bounded. Since each $\omega^{(j)}$ is bounded, there is a positive constant K such that

$$\left|\Omega^{(1)}(z,\zeta) - \frac{1}{t(\zeta) - t(z)}\right|$$

$$= \left|\frac{\exp[\omega^{(1)}(z)-\omega^{(1)}(\zeta)]-\exp[0]+\exp[\omega^{(2)}(z)-\omega^{(2)}(\zeta)]-\exp[0]}{2(t(\zeta) - t(z))}\right.$$

$$\leq (K|\omega^{(1)}(z)-\omega^{(1)}(\zeta)| + K|\omega^{(2)}(z)-\omega^{(2)}(\zeta)|)\,\frac{M(a,b)}{2|\zeta-z|} .$$

Because each $\omega^{(j)} \in B^{\alpha}(\mathbb{C})$, $\alpha = \frac{p-2}{p}$, we obtain the desired result for $\Omega^{(1)}$. To obtain the result for $\Omega^{(2)}$; we apply the same analysis to the equation

$$\Omega^{(2)}(z,\zeta)$$

$$= \frac{\exp[\omega^{(1)}(z)-\omega^{(1)}(\zeta)]-\exp[0] + \exp[0]-\exp[\omega^{(2)}(z)-\omega^{(2)}(\zeta)]}{2(t(\zeta) - t(z))}$$

The next three theorems develop an integral formula for solutions of (7.1) , which is useful for constructive methods [Hi 1] [GH 1] .

Theorem (11.7): Let $\mathfrak{G}$ *be a regular domain, and let* A *and* B *be in* $L^{p,2}(\mathbb{C})$ *where* $2 < p < \infty$. *Furthermore, let* w *be in* $C(\overline{\mathfrak{G}})$ *and satisfy in* $\mathfrak{G}$

$$Cw = \mathfrak{D}w + Aw + B\overline{w} = 0$$

If $\tilde{\Omega}^{(1)}$ *and* $\tilde{\Omega}^{(2)}$ *are the fundamental kernels for the* associated equation

$$\tilde{C}v = \mathfrak{D}v - Av + B^{*}\overline{v} = 0 ,$$

then

$$- \frac{1}{2\pi i} \int_{\Gamma} \tilde{\Omega}^{(1)}(\zeta, z)\ w(\zeta)\ dt(\zeta) - \overline{\tilde{\Omega}^{(2)}(\zeta,z)\ w(\zeta)\ dt(\zeta)}$$

$$= \begin{cases} w(z) , & \underline{\text{if}} \quad z \in \mathfrak{G} , \\ 0 , & \underline{\text{if}} \quad z \quad \overline{\mathfrak{G}} . \end{cases}$$

Let $\tilde{X}^{(1)}$ and $\tilde{X}^{(2)}$ be the corresponding solutions of $\tilde{C}v = 0$ as described in Theorem (11.4) . Using Theorem (11.3) , we obtain the formulas, for $j = 1, 2$,

$$\int_\Gamma \widetilde{X^{(j)}}(\zeta,Z)\, w(\zeta)\, dt(\zeta) - \overline{\widetilde{X^{(j)}}(\zeta,z)\, w(\zeta)\, dt(\zeta)}$$

$$= \begin{cases} \displaystyle\int_{|\zeta-z|=\epsilon} \widetilde{X^{(j)}}(\zeta,z)\, w(\zeta)\, dt(\zeta) - \overline{\widetilde{X^{(j)}}(\zeta,z)\, w(\zeta)\, dt(\zeta)} , & \text{if } z \in \mathfrak{G} , \\ 0 , & \text{if } z \notin \overline{\mathfrak{G}} , \end{cases}$$

where ϵ is a sufficiently small positive number. We multiply by i the equation for $j = 2$ and add to the equation for $j = 1$ to obtain

$$\int_\Gamma \widetilde{\Omega^{(1)}}(\zeta,z)\, w(\zeta)\, dt(\zeta) - \overline{\widetilde{\Omega^{(2)}}(\zeta,z)\, w(\zeta)\, dt(\zeta)}$$

$$= \begin{cases} \displaystyle\int_{|\zeta-z|=\epsilon} \widetilde{\Omega^{(1)}}(\zeta,z)\, w(\zeta)\, dt(\zeta) - \overline{\widetilde{\Omega^{(2)}}(\zeta,z)\, w(\zeta)\, dt(\zeta)} , & \text{if } z \in \mathfrak{G} , \\ 0 , & \text{if } z \notin \overline{\mathfrak{G}} . \end{cases}$$

Using (11.12), (11.13) we obtain for z in $\mathfrak{G}$,

$$\lim_{\epsilon\to 0} \int_{|\zeta-z|=\epsilon} \widetilde{\Omega}^{(1)}(\zeta,z)\, w(\zeta)\, dt(\zeta) - \overline{\widetilde{\Omega}^{(2)}(\zeta,z)\, w(\zeta)\, dt(\zeta)}$$

$$= \lim_{\epsilon\to 0} \int_{|\zeta-z|=\epsilon} \frac{w(\zeta)}{t(z)-t(\zeta)}\, dt(\zeta)$$

But Douglis showed [Do 1], (pp. 271-272), using the continuity of w, that the latter limit is $-2\pi i\, w(z)$. Thus the theorem is proved.

Theorem (11.8): Let $A, B \in L^{p,2}(\mathbb{C})$, where $2 < p < \infty$. Let $\Omega^{(1)}, \Omega^{(2)}$ be fundamental kernels for the equation

$$Cw = \partial w + Aw + B\overline{w} = 0$$

and $\widetilde{\Omega}^{(1)}, \widetilde{\Omega}^{(2)}$ fundamental kernels for the equation

$$\widetilde{C}v = \partial v - Av + B^*\overline{v} = 0 .$$

Then, for $z \neq \zeta$.

$$\Omega^{(1)}(z,\zeta) = -\widetilde{\Omega}^{(1)}(\zeta,z)$$

$$\Omega^{(2)}(z,\zeta) = -\overline{\widetilde{\Omega}^{(2)}(\zeta,z)} .$$

Proof

Let z, ζ be fixed, $z \neq \zeta$, and let ϵ be small enough that

$$0 < \epsilon < |z-\zeta| < \frac{1}{\epsilon}$$

Then by the previous theorem, for $j = 1, 2,$

$$X^{(j)}(z,\zeta) = -\frac{1}{2\pi i}\int_{|s-\zeta|=\frac{1}{\epsilon}} \tilde{\Omega}^{(1)}(s,z)\, X^{(j)}(s,\zeta)\, dt(s)$$

$$\overline{-\Omega^{(2)}(s,z)X^{(j)}(s,\zeta)\, dt(s)} + \frac{1}{2\pi i}\int_{|s-\zeta|=\epsilon} \tilde{\Omega}^{(1)}(s,z)X^{(j)}(s,\zeta)dt(s)$$

$$- \overline{\tilde{\Omega}^{(2)}(s,z)X^{(j)}(s,\zeta)dt(s)}$$

Using Theorem (11.6) and the relations (11.8), (11.9), we obtain the estimates

$$|\tilde{\Omega}^{(j)}(s,z)| \;,\; |X^{(j)}(s,\zeta)| = O(|s|^{-1}) \quad \text{as} \quad |s| \to \infty ,$$

$$\left|X^{(1)}(s,\zeta) - \frac{1}{2(t(\zeta) - t(s))}\right| = O(|s - \zeta|^{-2/p}) \quad \text{as} \quad |s - \zeta| \to 0 ,$$

$$\left|X^{(2)}(s,\zeta) - \frac{1}{2i(t(\zeta) - t(s))}\right| = O(|s - \zeta|^{-2/p}) \quad \text{as} \quad |s - \zeta| \to 0 .$$

Letting $\epsilon \to 0$, we therefore obtain

$$X^{(1)}(z,\zeta)=\lim_{\epsilon\to 0} \frac{1}{4\pi i}\Big[\int_{|s-\zeta|=\epsilon} \frac{\tilde{\Omega}^{(1)}(s,z)}{(t(\zeta) - t(s))}\, dt(s) - \overline{\frac{\tilde{\Omega}^{(2)}(s,z)}{(t(\zeta) - t(s))}\, dt(s}$$

$$X^{(2)}(z,\zeta) = \lim_{\epsilon\to 0} \frac{-1}{4\pi}\left[\int_{|s-\zeta|=\epsilon} \frac{\tilde{\Omega}^{(1)}(s,z)}{(t(\zeta)-t(s))}\,dt(s) - \frac{\overline{\tilde{\Omega}^{(2)}(s,z)}}{(t(\zeta)-t(s))}\,dt(s)\right].$$

As in the proof of the previous theorem, we remark that Douglis has shown [Do 1] (pp.271-272) that the above limits are

$$X^{(1)}(z,\zeta) = \frac{1}{4\pi i}\left[-2\pi i\,\tilde{\Omega}^{(1)}(\zeta,z) + 2\pi i\,\overline{\tilde{\Omega}^{(2)}(\zeta,z)}\right]$$

$$= -\tfrac{1}{2}\left[\tilde{\Omega}^{(1)}(\zeta,z) + \overline{\tilde{\Omega}^{(2)}(\zeta,z)}\right]$$

$$X^{(2)}(z,\zeta) = -\frac{1}{4\pi}\left[-2\pi i\,\tilde{\Omega}^{(1)}(\zeta,z) - 2\pi i\,\overline{\tilde{\Omega}^{(2)}(\zeta,z)}\right]$$

$$= -\frac{1}{2i}\left[\tilde{\Omega}^{(1)}(\zeta,z) - \overline{\tilde{\Omega}^{(2)}(\zeta,z)}\right]$$

The relations (11.8), (11.9) complete the proof [Hi 1] [GH 1] .

The next result is merely stated. It is an immediate consequence of the preceding two theorems [Hi 1], [GH 1] .

<u>Theorem (11.9)</u>: <u>Let</u> $\mathfrak{G}$ <u>be a regular domain, and let</u> A <u>and</u> B <u>be in</u> $L^{p,2}(\mathbb{C})$ <u>where</u> $2 < p < \infty$. Furthermore, let w <u>be in</u> $C(\overline{\mathfrak{G}})$ <u>and satisfy in</u> $\mathfrak{G}$

$$Cw = \mathfrak{D}w + Aw + B\overline{w} = 0 .$$

<u>Then</u>

$$\frac{1}{2\pi i}\int_{\Gamma} \Omega^{(1)}(z,\zeta)w(\zeta)dt(\zeta) - \Omega^{(2)}(z,\zeta)\overline{w(\zeta)}\,\overline{dt(\zeta)}$$

$$(11.16) \qquad = \begin{cases} w(z), & \text{if } z \in \mathfrak{G}, \\ 0, & \text{if } z \notin \mathfrak{G}. \end{cases}$$

To develop our next integral formula for solutions of (11.3), we need first two results concerning functions hyperanalytic <u>outside</u> a regular domain.

<u>Theorem (11.10)</u>: <u>Let</u> $\mathfrak{G}$ <u>be a regular domain, and let</u> ψ <u>be a function which is hyperanalytic outside</u> $\overline{\mathfrak{G}}$, <u>continuous up to the boundary</u> Γ, <u>and vanishes at infinity</u>. <u>Then</u>

$$(11.17) \qquad \frac{1}{2\pi i} \int_{\Gamma} \frac{\psi(\zeta)}{t(\zeta) - t(z)}\, dt(\zeta) = \begin{matrix} 0, & \text{if } z \in \mathfrak{G}, \\ -\psi(z), & \text{if } z \notin \overline{\mathfrak{G}} \end{matrix}$$

<u>Proof</u>:

For given z in $\mathbb{C}$, choose R positive such that $R > 2|z|$ and $\mathfrak{G} \subset \{\zeta : |\zeta| \le R\}$. Then applying Theorems (10.5) and (10.6) we have

$$\frac{1}{2\pi i} \int_{|\zeta|=R} \frac{\psi(\zeta)}{t(\zeta) - t(z)}\, dt(\zeta) - \frac{1}{2\pi i} \int_{\Gamma} \frac{\psi(\zeta)}{t(\zeta) - t(z)}\, dt(\zeta)$$

$$= \begin{cases} 0 & \text{if } z \in \mathfrak{G} \\ \psi(z) & \text{if } z \notin \overline{\mathfrak{G}} \end{cases}$$

(We have used an elementary consequence of Theorem (10.5) given in Douglis [Do 1](p. 276), stating that if Φ is hyperanalytic in a regular domain and continuous up to the boundary Γ, then

$$\int_\Gamma \Phi(\zeta)dt(\zeta) = 0.$$

Now, setting $\zeta = Re^{i\theta}$ on $|\zeta| = R$, and using $|\zeta - z| \geq R/2$,

$$\left|\frac{1}{2\pi i}\int_{|\zeta|=R} \frac{\psi(\zeta)}{t(\zeta)-t(z)}\, dt(\zeta)\right| \leq \frac{M(a,b)}{2\pi}\int_0^{2\pi} \frac{|\psi(Re^{i\theta})|}{|\zeta - z|} R\, d\theta$$

$$\leq M(a,b) \sup_{|\zeta|=R} |\psi(\zeta)| \to 0 \text{ as } R \to \infty .$$

Theorem (11.11): Let $\mathfrak{G}$ be a regular domain, and let ψ be a function which is hyperanalytic outside $\overline{\mathfrak{G}}$, continuous up to the boundary Γ, and vanishes at infinity. Let A and B be in $L^{p,2}(\mathbb{C})$ where $2 < p < \infty$, and outside $\overline{\mathfrak{G}}$, $A = B = 0$. Then if $\Omega^{(1)}$ and $\Omega^{(2)}$ are the fundamental kernels for A and B,

$$(11.18) \qquad \frac{1}{2\pi i}\int_\Gamma \Omega^{(1)}(z,\zeta)\, \psi(\zeta)\, dt(\zeta) - \Omega^{(2)}(z,\zeta)\, \overline{\psi(\zeta)\, dt(\zeta)}$$

Proof: For given z in $\mathbb{C}$, choose R as in the proof of the previous theorem. Since $A = B = 0$ outside $\overline{\mathfrak{G}}$. ψ satisfies (11.3) outside $\overline{\mathfrak{G}}$. Thus we can apply Theorem (11.9) to the domain $\{\zeta: |\zeta| \leq R\} - \overline{\mathfrak{G}}$ to obtain

$$\frac{1}{2\pi i}\int_\Gamma \Omega^{(1)}(z,\zeta)\psi(\zeta)dt(\zeta) - \Omega^{(2)}(z,\zeta)\overline{\psi(\zeta)}\;\overline{dt(\zeta)}$$

$$- \frac{1}{2\pi i}\int_\Gamma \Omega^{(1)}(z,\zeta)\psi(\zeta)dt(\zeta) - \Omega^{(2)}(z,\zeta)\overline{\psi(\zeta)dt(\zeta)}$$

$$= \begin{cases} 0, & \text{if } z \in \mathfrak{G}, \\ \psi(z), & \text{if } z \notin \overline{\mathfrak{G}} . \end{cases}$$

Using the same estimates as in the proof of the previous theorem, along with the estimates, for $j = 1,2$,

$$|\Omega^{(j)}(z,\zeta)| = O(|\zeta|^{-1}) \quad \text{as} \quad |\zeta| \to \infty ,$$

we can show that the integral on $|\zeta| = R$ approaches zero as $R \to \infty$.

It is now possible to derive the following integral representation.

Theorem (11.12): *Let* $\mathfrak{G}$ *be a regular domain, and let* A and $B \in L^{p,2}(\mathbb{C})$, where $2 < p < \infty$, and outside $\overline{\mathfrak{G}}$, $A = B = 0$. *Let* $\Omega^{(1)}$ *and* $\Omega^{(2)}$ *be the fundamental kernels for* A *and* B. *If* $w \in C(\overline{G})$ *and satisfies in* $\mathfrak{G}$

$$Cw = \mathcal{D}w + Aw + B\overline{w} = 0 ,$$

then w *has the representation, for* z *in* $\mathfrak{G}$,

$$(11.19) \quad w(z) = \frac{1}{2\pi i} \int_{\Gamma} \Omega^{(1)}(z,\zeta)\Phi(\zeta)dt(\zeta)$$

$$= \mathcal{K}(\Phi,\mathfrak{G}) ,$$

where Φ *is the function, hyperanalytic in* $\mathfrak{G}$ *and continuous in* $\overline{\mathfrak{G}}$, *given in* $\mathfrak{G}$ *by the formula*

$$(11.20) \quad \Phi(z) = \frac{1}{2\pi i} \int_{\Gamma} \frac{w(\zeta)}{t(\zeta) - t(z)} dt(\zeta) .$$

Proof.

Since $\partial w = -Aw - B\overline{w} \in L^p(G)$, we can use Corollary (10.) to conclude that in $\mathfrak{G}$

$$w = \Phi + \underset{\sim}{J}_G(-Aw - B\overline{w})$$

(11.21)

$$= \Phi + \underset{\sim}{J}(-Aw - B\overline{w}) \quad (\text{since } A = B = \text{) outside } \overline{G}) ,$$

where Φ is hyperanalytic in $\mathfrak{G}$. By Theorem (10.16), the function $\underset{\sim}{J}(-Aw - B\overline{w})$ is continuous in C and vanishes at infinity. Moreover, because $A = B = 0$ outside $\overline{\mathfrak{G}}$, $\underset{\sim}{J}(-Aw - B\overline{w})$ is hyperanalytic outside $\overline{\mathfrak{G}}$. Since w is in $C(\overline{\mathfrak{G}})$, Φ must also be in $C(\overline{\mathfrak{G}})$, and thus we can substitute (11.21) for w in formula (1.16) of Theorem (11.11), and use Theorem (11.13) applied to $\underset{\sim}{J}(-Aw - B\overline{w})$, to obtain (11.19). To obtain (11.20), we substitute $\Phi = w + \underset{\sim}{J}(Aw + B\overline{w})$ into the representation (3.4) for Φ , and apply Theorem (11.12) to the function $\underset{\sim}{J}(Aw + B\overline{w})$.

We now apply a Green's theorem to (11.19) to produce still another integral representation.

Theorem (11.13): Under the same hypothesis as in Theorem (11.12), for z in $\mathfrak{G}$ the following representation holds,

$$(11.22) \qquad w(z) = \Phi(z) + \iint_G \Gamma^{(1)}(z,\zeta)\Phi(\zeta)d\xi\, d\eta$$

$$+ \iint_G \Gamma^{(2)}(z,\zeta)\overline{\Phi(\zeta)}d\xi\, d\eta$$

<u>where</u> Φ <u>is as in Theorem</u> (11.12), <u>and</u>

$$\Gamma^{(1)}(z,\zeta) = -\frac{1}{2\pi i}\left[\frac{t_\xi(\zeta)}{i+eb(\zeta)}A(\zeta)\Omega^{(1)}(z,\zeta) - \frac{\overline{t_\xi(\zeta)}}{\overline{i+eb(\zeta)}}B(\zeta)\Omega^{(2)}(z,\zeta)\right.$$

$$\Gamma^{(2)}(z,\zeta) = \frac{1}{2\pi i}\left[\frac{\overline{t_\xi(\zeta)}}{\overline{i+eb(\zeta)}}\overline{A(\zeta)}\Omega^{(2)}(z,\zeta) - \frac{t_\xi(\zeta)}{i+eb(\zeta)}B(\zeta)\Omega^{(1)}(z,\zeta)\right.$$

<u>Proof</u>.

For z in $\mathfrak{G}$ and ϵ a sufficiently small positive number, we can apply Lemma (111) to the domain $\mathfrak{G}_\epsilon = \mathfrak{G} - \{\zeta: |\zeta - z| \le \epsilon\}$, and use the previous theorem to obtain

$$(11.23)\quad w(z) - \frac{1}{2\pi i}\int_{|\zeta-z|=\epsilon} \Omega^{(1)}(z,\zeta)\Phi(\zeta)dt(\zeta) - \Omega^{(2)}(z,\zeta)\overline{\Phi(\zeta)}\,\overline{dt(\zeta)}$$

$$= -\frac{1}{2\pi i}\iint_{G_\epsilon} \frac{t_\xi(\zeta)}{i+eb(\zeta)}\Phi(\zeta)\mathfrak{D}_\zeta\Omega^{(1)}(z,\zeta)d\xi\, d\eta$$

$$+ \frac{1}{2\pi i}\iint_{G_\epsilon} \frac{\overline{t_\xi(\zeta)}}{\overline{i+eb(\zeta)}}\overline{\Phi(\zeta)}\,\overline{\overline{\mathfrak{D}_\zeta\Omega^{(2)}(z,\zeta)}}\,d\xi\, d\eta$$

With the relations (11.14), (11.15) and the analysis used in the proof of Theorem (11.7) we can show that as $\varepsilon \to 0$ the left side of (11.23) becomes $w(z) - \Phi(z)$. Furthermore, from Theorem (7.8) and the relations (11.12), (11.13) we obtain the formulas

$$\mathfrak{D}_\zeta \Omega^{(1)}(z,\zeta) = A(\zeta)\Omega^{(1)}(z,\zeta) - B^*(\zeta)\Omega^{(2)}(z,\zeta)$$

$$\overline{\mathfrak{D}_\zeta \Omega^{(2)}(z,\zeta)} = \overline{A(\zeta)\Omega^{(2)}}(z,\zeta) - \overline{B^*(\zeta)\Omega^{(1)}}(z,\zeta)$$

Substituting these expressions into the right side of (11.23), and using (11.6), we obtain the desired result.

We remark that the representations (11.19) and (11.22) provide a method for obtaining complete families of solutions to (11.3). From Theorem (11.6) it is clear that corresponding to and hyperanalytic function in $\mathfrak{G}$ and continuous in the closure there exists a solution of (11.3) in $\mathfrak{G}$. Consequently by approximating a hyperanalytic function Φ by a sequence of hyperanalytic functions Φ_n, we can generate a sequence of solutions

(11.24) $\quad W_n(z) = \underset{\sim}{\kappa}(\Phi_n, \mathfrak{G}),\ (n=1,2,\ldots),$

which approximate the solution

(11.25) $\quad W(z) = \underset{\sim}{\kappa}(\Phi, \mathfrak{G}).$

Furthermore, using Douglis' result [Do 1] that any hyperanalytic function $\Phi(x)$ may be represented in terms of Γ analytic functions of the generating variable $t(z)$ we have the Theorem (11.14): *Let* $\mathfrak{G}$ *be a regular, simply-connected domain then the set of functions,*

(11.26) $W_{m,p}(z) \equiv \underset{\sim}{\kappa}(t^{mp}c^p;\mathfrak{G})$, $(p=0,1, \ldots,r-1)$, $(m=0,1, \ldots)$

form a complete family of solutions for (11.3) in $\mathfrak{G}$.

Proof:

To each solution $W(z) \in C(\mathfrak{G})$ there exists a unique hyperanalytic function $\phi(z) \in C(\mathfrak{G})$. Since $\phi(z)$ may be represented in $\mathfrak{G}$ as

$$\Phi(z) = \sum_{p=0}^{r-1} f_p(t(z) - T(zo))\, e^p ,$$

where the $f_p(z)$ are analytic, and since the powers of z are complete in the class of analytic functions in $\mathfrak{G}$ the result follows.

We return to the equation (11.2) and investigate it somewhat further by means of the generating pairs approach. We first note that if the functions $A_{k\ell}$, $B_{k\ell}$ in equation (11.2) satisfy further regularity conditions, then so do the generating pairs for that equation.

We list several regularity theorems. See [GH 2] for further details.

Theorem (11.15): *Let* w *be a hypercomplex valued function defined in a domain* $\mathfrak{G}$. *Then*

(i) *if* $\mathcal{D}w$ *is in* $C^{m,\alpha}(\mathfrak{G})$, *where* $m=0$ or 1 , $0 < \alpha < 1$,

then w *is in* $C^{m+1,\alpha}(\mathfrak{G})$

(ii) *if* $\mathfrak{D}w$ *is in* $C(\mathfrak{G})$, *then* w *is in* $C^{\alpha}(\mathfrak{G})$

for any α *in the interval* $0 < \alpha < 1$.

Proof. The proof is obtained using a procedure given previously in these lectures.

Theorem (11.16): *If each* $A_{k\ell}$, $B_{k\ell}$ *is in* $C^{m,\alpha}(\mathfrak{G})$, *where* $m=0$ *or* 1 *and* $0 < \alpha < 1$, *and if* w *is in* $C(\mathfrak{G})$ *and satisfies* (2.1) *in* $\mathfrak{G}$, *then* w *is in* $C^{m+1,\alpha}(\mathfrak{G})$.

Proof: This is a rather direct corrolary of Theorem (11.15).

Theorem (11.17): *If each* $A_{k\ell}$, $B_{k\ell}$ *is in* $C^{m,\alpha}(\mathfrak{G}) \cap L^{p}(\mathfrak{G})$, *where* $m=0$ *or* 1, $0 < \alpha < 1$, $2 < p < \infty$, *and* $\mathfrak{G}$ *is a bounded domain, then there exists a generating pair* (F,G) *in* $\mathfrak{G}$ *for* (11.2) *satisfying* $F,G \in C^{m+1,\alpha}(\mathfrak{G})$.

Proof.

By the previous theorem, *any* generating pair is in $C^{m+1,\alpha}(\mathfrak{G})$. Existence already has been shown in Chapter X.

It now is possible to develop analogues of Bers' results involving (F,G)-integrals. The proofs of the next few theorems are with minor variations the same proofs given by Bers [Be 3].

Definition: *If* (F,G) *is a generating pair in* $\mathfrak{G}$, Γ *a rectifiable curve in* $\mathfrak{G}$ *leading, say, from* z_0 *to* z_1, *and* w *a continuous hypercomplex valued function defined in* $\mathfrak{G}$, *we define*

$$(11.27)\quad {}^*\int_\Gamma w\, d_{(F,G)}z = \mathrm{Re}\int_\Gamma G^*w\, dz + i\,\mathrm{Re}\int_\Gamma F^*w\, dz\ ,$$

$$(11.28)\quad \int_\Gamma w\, d_{(F,G)}z = F(z_1)\mathrm{Re}\int_\Gamma G^*w\, dz + G(z_1)\mathrm{Re}\int_\Gamma F^*w\, dz\ .$$

The function w is (F,G)-integrable in $\mathfrak{G}$ if for every closed curve Γ contained in a simply-connected subdomain of $\mathfrak{G}$,

$$(11.29)\quad \int_\Gamma w\, d_{(F,G)}z = 0\ .$$

Remark: Condition (11.29) may be replaced by

$$(11.30)\quad \mathrm{Re}\int_\Gamma G^*w\, dz = \mathrm{Re}\int_\Gamma F^*w\, dz = 0\ .$$

Theorem (11.18): If w is (F,G)-pseudohyperanalytic of the first kind in $\mathfrak{G}$, then $\dot{w}$ is (F,G)-integrable in $\mathfrak{G}$, and

$$(11.31)\quad {}^*\int_{z_0}^{z} \dot{w}\, d_{(F,G)}z = \chi(z) - \chi(z_0)$$

$$(11.32)\quad \int_{z_0}^{z} \dot{w}\, d_{(F,G)}z = w(z) - \varphi(z_0)F(z) - \psi(z_0)G(z)$$

where $\chi = \varphi + i\psi = {}_*w \pmod{F,G}$.

Proof.

By direct computation we may show [GH 2], that

$$(11.33)\quad {}^*\int_{z_0}^{z} \dot{w}\, d_{(F,G)}z = \int_{z_0}^{z} d\varphi + i\int_{z_0}^{z} d\psi\ .$$

(11.34) $\int_{z_0}^{z} \dot{w}\, d_{(F,G)}z = F(z)\int_{z_0}^{z} d\varphi + G(z)\int_{z_0}^{z} d\psi$.

These relations give (11.31) and (11.32).

Theorem (11.19): If v is continuous and (F,G)-intergrable in a connected, simply-connected domain $\mathfrak{G}$, then there exists an (F,G)-pseudohyperanalytic function of the first kind, w , in $\mathfrak{G}$ such that $v = \frac{d_{(F,G)}w}{dz}$.

Proof. The proof is obtained by using the necessary and sufficient conditions for a function to be (PHA)-2; see [GH 2] for further details.

Theorem (11.20) : Let (F',G') be a successor of (F,G) in $\mathfrak{G}$, and let w be (F',G')-pseudohyperanalytic of the first kind in $\mathfrak{G}$. Then w is (F,G)-integrable in $\mathfrak{G}$.

Proof.

It is sufficient to show that if $\mathfrak{G}_1$ is a domain inside $\mathfrak{G}$, $\bar{\mathfrak{G}}_1 \subset \mathfrak{G}$, bounded by a piecewise smooth Jordan curve Γ , then

(11.35) $\operatorname{Re}\int_{\Gamma} G^* w\, dz = \operatorname{Re}\int F^* w\, dz = 0$.

Using a complex form of Green's formula, and the definitions of successor and predecessor, we have

$$\int_{\Gamma} G^* w \, dz = 2i \iint_{\mathfrak{G}_1} (G^* w)_{\bar{z}} \, dxdy$$

$$= 2i \iint_{\mathfrak{G}_1} \{(-A_{(F,G)} G^* - \overline{D_{(f,G)}} \, \overline{G^*}) w + G^*(A_{(F,G)} w - D_{(F,G)} w - D_{(F,G)} \bar{w}\} dx$$

$$= -4i \iint_{\mathfrak{G}_1} \mathrm{Re}(G^* D_{(F,G)} \bar{w}) \, dxdy$$

which is a purely imaginary hypercomplex number. A similar formula holds with G^* replaced by F^*.

For the remaining results of this chapter we place slightly stronger conditions on our generating pairs (F,G). We assume that F and G are in $C^{2,\alpha}(\mathfrak{G})$, and consequently the coefficients A, B, C, D are in $C^{1,\alpha}(\mathfrak{G})$. Under these conditions we observe that if w is (F,G)-pseudohyperanalytic of the first kind in $\mathfrak{G}$, then according to our regularity Theorems, w is in $C^{2,\alpha}(\mathfrak{G})$. Moreover φ, and ψ are in $C^{2,\alpha}(\mathfrak{G})$ [GH 2]. We use these observations in obtaining the results that follow.

<u>Lemma (11.21)</u>: <u>If</u> (F,G) <u>is</u> <u>a</u> <u>generating</u> <u>pair</u> <u>in</u> $\mathfrak{G}$, with F and G in $C^{2,\alpha}(\mathfrak{G})$, then any successor, (F', G'), or predecessor, (F^{-1}, G^{-1}), also has its members in $C^{2,\alpha}(\mathfrak{G})$.

<u>Proof</u>.

The functions F^1 and G^1 are in $C^{2,\alpha}(\mathfrak{G})$ because of the relation between the characteristic coefficients of successors and predecessors. Similarly F^* and G^* are in $C^{2,\alpha}(\mathfrak{G})$. Since $(F^{-1}, G^{-1})^*$ is a successor of

$(F,G)^*$, $(F^{-1})^*$ and $(G^{-1})^*$ are in $C^{2,\alpha}_{\alpha}$. But $F^{-1} = (F^{-1})^{**}$, $G^{-1} = (G^{-1})^{**}$.

Theorem (11.22): If F and G are in $C^{2,\alpha}(\mathfrak{G})$, w is (F,G)-pseudohyperanalytic of the first kind in $\mathfrak{G}$, and (F^1,G^1) is a successor of (F,G) in $\mathfrak{G}$, then

$\dot{w} = \dfrac{d_{(F,G)}w}{dz}$ is (F^1,G^1)-pseudohyperanalytic of the first kind in $\mathfrak{G}$.

Proof.

Differentiation of $\dot{w} = F\varphi_z + G\psi_z$ gives

$$(\dot{w})_{\bar{z}} = F_{\bar{z}}\varphi_z + G_{\bar{z}}\psi_z + F\varphi_{z\bar{z}} + G\psi_{z\bar{z}} .$$

Next we differentiate $F\varphi_{\bar{z}} + G\psi_{\bar{z}} = 0$ and combine with the above to obtain

$$(\dot{w})_{\bar{z}} = F_{\bar{z}}\varphi_z + G_{\bar{z}}\psi_z - F_z\varphi_{\bar{z}} - G_z\psi_{\bar{z}} .$$

Since F and G satisfy $\dfrac{\partial w}{\partial \bar{z}} = Aw + B\bar{w}$,

$F_z = CF + D\bar{F}$ and $G_z = CG + D\bar{G}$. Where $C = \dfrac{F_z\bar{G} - G_z\bar{F}}{F\bar{G} - \bar{F}G}$,

and $D = \dfrac{-F_zG + G_zF}{F\bar{G} - \bar{F}G}$ we may combine terms to obtain

$$(\dot{w})_{\bar{z}} = A_{(F,G)}\dot{w} - D_{(F,G)}\bar{\dot{w}} .$$

As a consequence of the preceding two results, if F and G are in $C^{2,\alpha}(\mathfrak{G})$ and (F,G) is embedded in a generating sequence in $\mathfrak{G}$, then any (F,G)-pseudoanalytic function of the first kind in $\mathfrak{G}$ has derivatives of all orders, defined by the formulas

$$w^{(o)} = w ,$$

$$w^{(n+1)} = \frac{d_{(F^n,G^n)}w^{(n)}}{dz}$$

Theorem (11.22): *If* (F,G) *is a generating pair in* $\mathfrak{G}$, *with* F *and* G *in* $C^{2,\alpha}(\mathfrak{G})$, *and* (F^{-1},G^{-1}) *is a predecessor of* $(F,G,)$, *then a continuous function* w *is* $(F,G,)$-*pseudohyperanalytic of the first kind in* $\mathfrak{G}$ *if and only if it is* (F^{-1},G^{-1})-*integrable in* $\mathfrak{G}$.

Proof. The proof is obvious.

Finally, we give another characterization of pseudohyperanalytic functions of the second kind. If (F,G) is a generating pair with F,G in $C^{2,\alpha}(\mathfrak{G})$, we set

$$-i\,\frac{G}{F} = \sigma + i\tau$$

$$-i\,\frac{F}{G} = \tilde{\sigma} + i\,\tilde{\tau}$$

where σ, τ, $\tilde{\sigma}$, $\tilde{\tau}$ are real hypercomplex valued. The equation

$$F\varphi_{\bar{z}} + G\psi_{\bar{z}} = 0$$

is equivalent to the following system involving real hypercomplex valued functions,

$$\varphi_x = \tau\psi_x + \sigma\psi_y \tag{11.36}$$

$$\varphi_y = \sigma\psi_x + \tau\psi_y \tag{11.37}$$

We further define

$$\alpha = \frac{\sigma_x + \tau_y}{\sigma}\ , \qquad \beta = \frac{\sigma_y - \tau_x}{\sigma}$$

$$\tilde{\alpha} = \frac{\tilde{\sigma}_x + \tilde{\tau}_y}{\tilde{\sigma}}\ , \qquad \tilde{\beta} = \frac{\tilde{\sigma}_y - \tilde{\tau}_x}{\tilde{\sigma}}$$

One can differentiate (11.36) and (11.37) to obtain

$$\varphi_{xx} + \varphi_{yy} + \tilde{\alpha}\varphi_x + \tilde{\beta}\varphi_y = 0 \tag{11.38}$$

$$\psi_{xx} + \psi_{yy} + \alpha\psi_x + \beta\psi_y = 0 \tag{11.39}$$

These considerations lead to the following theorem:

Theorem (11.23): *If (F,G) is a generating pair in a connected, simply-connected domain $\mathfrak{G}$, and F,G are in $C^{2,\alpha}(\mathfrak{G})$, then a real hypercomplex valued function $\varphi[\psi]$ is the real [imaginary] part of an (F,G)-pseudohyperanalytic function of the second kind in $\mathfrak{G}$ if and only if it is in $C^{2,\alpha}(\mathfrak{G})$ and satisfies (11.38) [(11.39)].*

Proof.

One implication already has been proved. If ψ is in $C^{2,\alpha}(\mathfrak{G})$ and satisfies (11.39), we define

$$\varphi(z) = \int_{z_0}^{z} (\tau\psi_x + \sigma\psi_y)dx + (-\sigma\psi_x + \tau\psi_y)dy$$

(where z_0 is fixed in $\mathfrak{G}$).

This integral is independent of the path, is in $C^{2,\alpha}(\mathfrak{G})$, and we have

$$F\varphi_{\bar{z}} + G\varphi_{\bar{z}} = 0.$$

In a similar manner we prove the other part of this implication.

Added in proof:

Alternate ways of investigating elliptic systems of the first order have been developed using matrix notation instead of the hypercomplex algebra. The reader is directed to the folowing works in this area:

a) Bojarski, B. B.: The theory of generalized analytic vectors, Annales Polonici Mathematici, 17, (1966), pp. 281-320. (In Russian)

b) Pascali, Dan: Vecteurs analytiques généralizés, Revue Romaine de Mathématiques Pures et Appliqueés, 10 (6), (1965) pp. 779-808.

c) __________: Sur la representation de premiere espéce des vecteurs analytiques généralisés, Revue Romaine de Mathématiques Pures et Appliquées, 12 (5), (1967), pp. 685-689.

REFERENCES

[Ag 1] Agmon, S.: Lectures on Elliptic Boundary Value Problems, Van Nostrand, Princeton, 1965.

[An 1] Anselone, P.: Collectively Compact Operator Approximation Theory and Applications to Integral Equations, Prentice-Hall, Englewood Cliffs, 1971.

[At 1] Atkinson, K.: Extensions of the Nyström method for the numerical solution of linear integral equations of the second kind, MRC Technical Summary Report #686, August, 1966.

[At 2] ________ : The numerical solution of Fredholm integral equations of the second kind with singular kernels, Numerische Math. 19, 248 - 259, (1972).

[At 3] ________ : A Survey of Numerical Methods for the Solution of Fredholm Integral Equations of the Second Kind, (book manuscript).

[AG 1] Aziz, A.K. and Gilbert, R.P. : A generalized Goursat problem for elliptic equations, J. reine u. angew. Mathematik, 222, 1 - 13, (1966).

[AGH 1] Aziz, A.K., Gilbert, R.P., and Howard, H.C.: A second order, nonlinear elliptic boundary value problem with generalized Goursat data, Ann. Math. Pura Appl., 72, 325 - 341, (1966).

[Be 1] Bergman, S.: Integral Operators in the Theory of Linear Partial Differential Equations, Ergeb. Math. u. Grenzgeb., 23, Berlin, 1960.

[Be 2] ________ : Solutions of linear partial differential equations of the fourth order, Duke Math. J., 11, 617 - 649, (1944).

[Be 3] Bers, L. : Theory of pseudo-Analytic Functions, Lecture Notes, Institute for Mathematics and Mechanics, New York University, New York, 1953.

[Br 1] Bremermann, H. : Lectures on Several Complex Variables, lecture notes, Univ. of Washington.

[Br 2] Browder, F.E. : Approximations by solutions of partial differential equations, Amer. J. Math., 84, 134-160, (1962).

[BS 1] Bergman, S., and Schiffer, M.: Kernel Functions and Elliptic Differential Equations in Mathematical Physics, Academic Press, New York, 1953.

[Co 1] Colton, D. : Cauchy's problem for almost linear elliptic equations in two independent variables, J. Approx. Theory, 3, 66 - 71, (1970).

[Co 2] ________ : Cauchy's problem for almost linear elliptic equations in two independent variables II, J. Approx. Theory, (to appear).

[Co 3] ________ : Bergman Operators for elliptic equations in three independent variables, Bull. Amer. Math. Soc., 77, 752 - 756, (1971).

[Co 4] ________ : Bergman operators for elliptic equations in four independent variables, SIAM J. Math. Anal., (to appear).

[Co 5] ________ : Integral operators for elliptic equations in three independent variables, I, Applicable Analysis (to appear).

[Co 6] ________ : Integral operators for elliptic equations in three independent variables, II, Applicable Analysis,(to appear).

[Co 7] ________ : Uniqueness theorems for axially symmetric partial differential equations, J. Math. Mech., 18, 921 - 930, (1969).

[Co 8] ________ : Jacobi polynomials of negative index and a nonexistence theorem for the generalized axially symmetric potential equation, SIAM J. Appl. Math., 16, 771 - 776.

[CG 1] Colton, D., and Gilbert, R.P. : Integral operators and complete families of solutions for $\Delta^2 u + A\Delta u + Bu = 0$, Arch. Rat. Mech. and Anal., 43, 62 - 78, (1971).

[CG 2] ______________ : An integral operator approach to Cauchy's problem for $\Delta^2 u + F(x)u = 0$, SIAM J. Math. Anal., 2, 113 - 132, (1971).

[CG 3] ______________ : Function theoretic methods in the theory of boundary value problems for generalized metaharmonic functions, 75, 948 - 952, (1969).

[CG 4] Conlan, J., and Gilbert, R.P. : Initial value problems for fourth order, semi-linear elliptic equations, Applicable Analysis (to appear).

[CG 5] ______________ : Non-linear initial data for, second and higher order, semi-linear elliptic equations, J. reine u. angewandte Math. (to appear).

[CG 6] ______________ : Iterative schemes for almost linear elliptic systems with nonlinear initial data, J. reine u. angewandte Math. (to appear).

[Di 1] Dienes, P. : The Taylor Series, Dover Publications New York, 1957.

[Do 1] Douglis, A.: A function-theoretic approach to elliptic systems of equations in two variables, Comm. Pure Appl. Math., 6, 259 - 289, (1953).

[Do 2] __________ : On uniqueness in Cauchy problems for elliptic systems of equations, Comm. Pure Appl. Math., 13, 593 - 607, (1960).

[Er 1] Erdélyi, A.: Singularities of generalized axially symmetric potentials, Comm. Pure Appl. Math. 9, 403 - 414, (1956).

[Er 2] Erdélyi, A., et al. : Higher Transcendental Functions, McGraw Hill, New York, 1953.

[Ga 1] Garabedian, P.R. : Partial Differential Equations, John Wiley, New York, 1964.

[Gi 0] Gilbert, R.P. : Singularities of Three Dimensional Harmonic Functions, Ph.D. thesis, Carnegie-Mellon University, 1958.

[Gi 1] ________ : Function Theoretic Methods in Partial Differential Equations, Academic Press, New York, 1969.

[Gi 2] ________ : Construction of solutions of boundary value problems, SIAM J. Math. Anal., 1, 96 - 114, (1970).

[Gi 3] ________ : Integral operator methods for approximating solutions of Dirichlet problems, in Iterationsverfahren, Numerische Mathematik, Approximationstheorie, Birkhauser-Verlag, Bassel, 1970.

[Gi 4] ________ : A method of ascent, Bull. Amer. Math. Soc., 75, 1286 - 1289, (1969).

[Gi 5] ________ : Pseudo hyperanalytic function theory, Proceedings of the Conference "Funktionentheoretische Eigenschaften von Lősungen partieller Differentialgleichungen" ed. by St. Ruscheweyh , (to appear).

[GH 1] Gilbert, R.P., and Hile, Gerald: Generalized hyperanalytic function theory, Bull. Amer. Math. Soc., 78, 998 - 1001, (1972).

[GH 2] ________ : Generalized hyperanalytic function theory: II, Trans. Amer. Math. Soc., (to appear).

[GH 3] ________ : Hypercomplex function theory in the sense of L. Bers, (to appear).

[GK 1] Garabedian, P.R., and Korn, D.G. : Numerical design of transonic airfoils, in Numerical Solution of Partial Differential Equations-II, ed. by B. Hubbard, Academic Press, 1970.

[GK 2] Gilbert, R.P., and Kukral, D.: Function theoretic methods for higher order elliptic equations in three independent variables, Bull. Amer. Math. Soc., 79, 96 - 100, (1973).

[GK 3] ______________ : A function theoretic methods for $\Delta_3^2 u + Q(x)u = 0$, to appear).

[GK 4] ______________ : A function Theoretic method for $\Delta_4^2 u + Q(x)u = 0$, Annali di Math. Pura ed Appl., (to appear).

[GK 5] ______________ : Function theoretic methods for elliptic equations with more than four variables, Rend. Sem. Mat. Torino, (to appear).

[GK 6] ______________ : A function theoretic method for $\Delta_3 u + f(x_1, x_2)u = 0$, (to appear).

[GL 1] Garabedian, P.R., and Lieberstein, H.M.: On the numerical calculation of detached bow shock waves in hypersonic flow, J. Aero. Sci., 25, 109 - 118, (1958).

[He 1] Hellwig, G.: Partial Differential Equations, Blaisdell, New York, 1964.

[He 2] Henrici, P.: Zur Funktionentheorie der Wellengleichung, Commentarii Mathematici Helvetici, 27, 235 - 293, (1953).

[He 3] __________ : On the domain of regularity of generalized axially symmetric potentials, Proc. Amer. Math. Soc. 8, 29 - 31, (1957).

[He 4] __________ : Complete systems of solutions for a class of singular elliptic partial differential equations, in Boundary Value Problems in Differential Equations, Univ. of Wis. Press, Madison, 19 - 34, 1960.

[Hi 1] Hile, G.: Hypercomplex Function Theory Applied to Partial Differential Equations, Ph.D. thesis, Indiana University, 1972.

[Ho 1] Hörmander, L.: Linear Partial Differential Operators, Springer Verlag, Berlin, 1964.

[Hu 1] Huber, A.: On uniqueness of generalized axially symmetric potentials, Ann. Math., 60, 351 - 385, (1954).

[Hu 2] ________ : Some results on generalized axially symmetric potentials, Proc. Conf. Part. Diff. Equat., Univ. Maryland, 147-155, (1955).

[Jo 1] John, F.: Plane Waves and Spherical Means Applied to Partial Differential Equations, Wiley, New York, 1955.

[Ke 1] Kellogg, O.D.: Foundations of Potential Theory, Dover, New York, 1953.

[Kr 1] Kreyszig, E.: Kanonische Integral operatoren zur Erzeugung harmonisher Funktionen von vier Veranderlichen, Archive der Mathematik, XIV, 193 - 203, (1963).

[Ku 1] Kukral, D.: Constructive Methods for Determining the Solutions of Higher Order Elliptic Partial Differential Equations, Ph.D. thesis, Indiana University, 1972.

[Ku 2] ________ : On a Bergman-Whittaker type operator in five or more variables, Proc. Amer. Math. Soc., 39, 122 - 124, (1973).

[La 1] Lax, P.: A stability theory of abstract differential equations and its applications to the study of local behaviors of solutions of elliptic equations, Comm. Pure App. Math. 8, 747 - 766, 1956.

[Ma 1] Malgrange, B.: Existence et approximation des solutions des equations aux derivees partielles et des equations de convolutions, Ann. Inst. Fourier, 6, 271 - 355, (1956)

[Mu 1] Muskhelishvili, N.: Singular Integral Equations, Noordhoff, Groningen, 1953.

[MT 1] Marić, V. and Tjong, B.: On an integral operator, Indian, J. Math., 12, 163 - 168, (1970).

[Pa 1] Parter, S.: On the existence and uniqueness of symmetric axially symmetric potentials, Arch. Rat. Mech. Anal. 20, (1965).

[Pa 2] Payne, L.: Some general remarks on improperly posed problems for partial differential equations, in Lecture Notes in Math. #316, Springer-Verlag, New York, 1972.

[Pl 1] du Plessis, N.: Runge's theorem for harmonic functions, J. London Math. Soc., 1, 404 - 408, (1969).

[PW 1] Protter, M., and Weinberger, H.: Maximum Principles in Differential Equations, Prentice Hall, Englewood Cliffs, 1967.

[Ta 1] Taylor, A.: Introduction to Functional Analysis, Wiley, New York, 1958.

[Tj 1] Tjong, B.: Operators Generating Solutions of $\Delta\psi + F(x,y,z)\psi = 0$, Ph.D. thesis, Univ. Kentucky, 1968.

[Ve 1] Vekua, I.: New Methods for Solving Elliptic Equations, Wiley, New York, 1967.

[Ve 2] ________ : Generalized Analytic Functions, Addison Wesley, Reading, 1962.

[We 1] Weinacht, R.: Fundamental solutions for a class of singular equations, Contr. Diff. Equats., 3, 43-55, 1964.

[We 2] ________ : A mean value theorem in generalized axially symmetric potential theory, Acc. Naz. Lincei 38, 610 - 613, (1965).

[We 3] Weinstein, A.: Generalized axially symmetric potential theory, Bull. Amer. Math. Soc., 59, 20-38, (1953).

[We 4] ________ : Singular partial differential equations and their applications, in Fluid Dynamics and Applied Mathematics, Gordon & Breach, New York, 1961.

Vol. 215: P. Antonelli, D. Burghelea and P. J. Kahn, The Concordance-Homotopy Groups of Geometric Automorphism Groups. X, 140 pages. 1971. DM 16,–

Vol. 216: H. Maaß, Siegel's Modular Forms and Dirichlet Series. VII, 328 pages. 1971. DM 20,–

Vol. 217: T. J. Jech, Lectures in Set Theory with Particular Emphasis on the Method of Forcing. V, 137 pages. 1971. DM 16,–

Vol. 218: C. P. Schnorr, Zufälligkeit und Wahrscheinlichkeit. IV, 212 Seiten. 1971. DM 20,–

Vol. 219: N. L. Alling and N. Greenleaf, Foundations of the Theory of Klein Surfaces. IX, 117 pages. 1971. DM 16,–

Vol. 220: W. A. Coppel, Disconjugacy. V, 148 pages. 1971. DM 16,–

Vol. 221: P. Gabriel und F. Ulmer, Lokal präsentierbare Kategorien. V, 200 Seiten. 1971. DM 18,–

Vol. 222: C. Meghea, Compactification des Espaces Harmoniques. III, 108 pages. 1971. DM 16,–

Vol. 223: U. Felgner, Models of ZF-Set Theory. VI, 173 pages. 1971. DM 16,–

Vol. 224: Revètements Etales et Groupe Fondamental. (SGA 1). Dirigé par A. Grothendieck XXII, 447 pages. 1971. DM 30,–

Vol. 225: Théorie des Intersections et Théorème de Riemann-Roch. (SGA 6). Dirigé par P. Berthelot, A. Grothendieck et L. Illusie. XII, 700 pages. 1971. DM 40,–

Vol. 226: Seminar on Potential Theory, II. Edited by H. Bauer. IV, 170 pages. 1971. DM 18,–

Vol. 227: H. L. Montgomery, Topics in Multiplicative Number Theory. IX, 178 pages. 1971. DM 18,–

Vol. 228: Conference on Applications of Numerical Analysis. Edited by J. Ll. Morris. X, 358 pages. 1971. DM 26,–

Vol. 229: J. Väisälä, Lectures on n-Dimensional Quasiconformal Mappings. XIV, 144 pages. 1971. DM 16,–

Vol. 230: L. Waelbroeck, Topological Vector Spaces and Algebras. VII, 158 pages. 1971. DM 16,–

Vol. 231: H. Reiter, L^1-Algebras and Segal Algebras. XI, 113 pages. 1971. DM 16,–

Vol. 232: T. H. Ganelius, Tauberian Remainder Theorems. VI, 75 pages. 1971. DM 16,–

Vol. 233: C. P. Tsokos and W. J. Padgett. Random Integral Equations with Applications to stochastic Systems. VII, 174 pages. 1971. DM 18,–

Vol. 234: A. Andreotti and W. Stoll. Analytic and Algebraic Dependence of Meromorphic Functions. III, 390 pages. 1971. DM 26,–

Vol. 235: Global Differentiable Dynamics. Edited by O. Hájek, A. J. Lohwater, and R. McCann. X, 140 pages. 1971. DM 16,–

Vol. 236: M. Barr, P. A. Grillet, and D. H. van Osdol. Exact Categories and Categories of Sheaves. VII, 239 pages. 1971. DM 20,–

Vol. 237: B. Stenström, Rings and Modules of Quotients. VII, 136 pages. 1971. DM 16,–

Vol. 238: Der kanonische Modul eines Cohen-Macaulay-Rings. Herausgegeben von Jürgen Herzog und Ernst Kunz. VI, 103 Seiten. 1971. DM 16,–

Vol. 239: L. Illusie, Complexe Cotangent et Déformations I. XV, 355 pages. 1971. DM 26,–

Vol. 240: A. Kerber, Representations of Permutation Groups I. VII, 192 pages. 1971. DM 18,–

Vol. 241: S. Kaneyuki, Homogeneous Bounded Domains and Siegel Domains. V, 89 pages. 1971. DM 16,–

Vol. 242: R. R. Coifman et G. Weiss, Analyse Harmonique Non-Commutative sur Certains Espaces. V, 160 pages. 1971. DM 16,–

Vol. 243: Japan-United States Seminar on Ordinary Differential and Functional Equations. Edited by M. Urabe. VIII, 332 pages. 1971. DM 26,–

Vol. 244: Séminaire Bourbaki – vol. 1970/71. Exposés 382–399. IV, 356 pages. 1971. DM 26,–

Vol. 245: D. E. Cohen, Groups of Cohomological Dimension One. V, 99 pages. 1972. DM 16,–

Vol. 246: Lectures on Rings and Modules. Tulane University Ring and Operator Theory Year, 1970–1971. Volume I. X, 661 pages. 1972. DM 40,–

Vol. 247: Lectures on Operator Algebras. Tulane University Ring and Operator Theory Year, 1970–1971. Volume II. XI, 786 pages. 1972. DM 40,–

Vol. 248: Lectures on the Applications of Sheaves to Ring Theory. Tulane University Ring and Operator Theory Year, 1970–1971. Volume III. VIII, 315 pages. 1971. DM 26,–

Vol. 249: Symposium on Algebraic Topology. Edited by P. J. Hilton. VII, 111 pages. 1971. DM 16,–

Vol. 250: B. Jónsson, Topics in Universal Algebra. VI, 220 pages. 1972. DM 20,–

Vol. 251: The Theory of Arithmetic Functions. Edited by A. A. Gioia and D. L. Goldsmith VI, 287 pages. 1972. DM 24,–

Vol. 252: D. A. Stone, Stratified Polyhedra. IX, 193 pages. 1972. DM 18,–

Vol. 253: V. Komkov, Optimal Control Theory for the Damping of Vibrations of Simple Elastic Systems. V, 240 pages. 1972. DM 20,–

Vol. 254: C. U. Jensen, Les Foncteurs Dérivés de $\varprojlim$ et leurs Applications en Théorie des Modules. V, 103 pages. 1972. DM 16,–

Vol. 255: Conference in Mathematical Logic – London '70. Edited by W. Hodges. VIII, 351 pages. 1972. DM 26,–

Vol. 256: C. A. Berenstein and M. A. Dostal, Analytically Uniform Spaces and their Applications to Convolution Equations. VII, 130 pages. 1972. DM 16,–

Vol. 257: R. B. Holmes, A Course on Optimization and Best Approximation. VIII, 233 pages. 1972. DM 20,–

Vol. 258: Séminaire de Probabilités VI. Edited by P. A. Meyer. VI, 253 pages. 1972. DM 22,–

Vol. 259: N. Moulis, Structures de Fredholm sur les Variétés Hilbertiennes. V, 123 pages. 1972. DM 16,–

Vol. 260: R. Godement and H. Jacquet, Zeta Functions of Simple Algebras. IX, 188 pages. 1972. DM 18,–

Vol. 261: A. Guichardet, Symmetric Hilbert Spaces and Related Topics. V, 197 pages. 1972. DM 18,–

Vol. 262: H. G. Zimmer, Computational Problems, Methods, and Results in Algebraic Number Theory. V, 103 pages. 1972. DM 16,–

Vol. 263: T. Parthasarathy, Selection Theorems and their Applications. VII, 101 pages. 1972. DM 16,–

Vol. 264: W. Messing, The Crystals Associated to Barsotti-Tate Groups: With Applications to Abelian Schemes. III, 190 pages. 1972. DM 18,–

Vol. 265: N. Saavedra Rivano, Catégories Tannakiennes. II, 418 pages. 1972. DM 26,–

Vol. 266: Conference on Harmonic Analysis. Edited by D. Gulick and R. L. Lipsman. VI, 323 pages. 1972. DM 24,–

Vol. 267: Numerische Lösung nichtlinearer partieller Differential- und Integro-Differentialgleichungen. Herausgegeben von R. Ansorge und W. Törnig, VI, 339 Seiten. 1972. DM 26,–

Vol. 268: C. G. Simader, On Dirichlet's Boundary Value Problem. IV, 238 pages. 1972. DM 20,–

Vol. 269: Théorie des Topos et Cohomologie Etale des Schémas. (SGA 4). Dirigé par M. Artin, A. Grothendieck et J. L. Verdier. XIX, 525 pages. 1972. DM.50,–

Vol. 270: Théorie des Topos et Cohomologie Etale des Schémas. Tome 2. (SGA 4). Dirigé par M. Artin, A. Grothendieck et J. L. Verdier. V, 418 pages. 1972. DM 50,–

Vol. 271: J. P. May, The Geometry of Iterated Loop Spaces. IX, 175 pages. 1972. DM 18,–

Vol. 272: K. R. Parthasarathy and K. Schmidt, Positive Definite Kernels, Continuous Tensor Products, and Central Limit Theorems of Probability Theory. VI, 107 pages. 1972. DM 16,–

Vol. 273: U. Seip, Kompakt erzeugte Vektorräume und Analysis. IX, 119 Seiten. 1972. DM 16,–

Vol. 274: Toposes, Algebraic Geometry and Logic. Edited by. F. W. Lawvere. VI, 189 pages. 1972. DM 18,–

Vol. 275: Séminaire Pierre Lelong (Analyse) Année 1970–1971. VI, 181 pages. 1972. DM 18,–

Vol. 276: A. Borel, Représentations de Groupes Localement Compacts. V, 98 pages. 1972. DM 16,–

Vol. 277: Séminaire Banach. Edité par C. Houzel. VII, 229 pages. 1972. DM 20,–

Vol. 278: H. Jacquet, Automorphic Forms on GL(2). Part II. XIII, 142 pages. 1972. DM 16,–

Vol. 279: R. Bott, S. Gitler and I. M. James, Lectures on Algebraic and Differential Topology. V, 174 pages. 1972. DM 18,–

Vol. 280: Conference on the Theory of Ordinary and Partial Differential Equations. Edited by W. N. Everitt and B. D. Sleeman. XV, 367 pages. 1972. DM 26,–

Vol. 281: Coherence in Categories. Edited by S. Mac Lane. VII, 235 pages. 1972. DM 20,–

Vol. 282: W. Klingenberg und P. Flaschel, Riemannsche Hilbertmannigfaltigkeiten. Periodische Geodätische. VII, 211 Seiten. 1972. DM 20,–

Vol. 283: L. Illusie, Complexe Cotangent et Déformations II. VII, 304 pages. 1972. DM 24,–

Vol. 284: P. A. Meyer, Martingales and Stochastic Integrals I. VI, 89 pages. 1972. DM 16,–

Vol. 285: P. de la Harpe, Classical Banach-Lie Algebras and Banach-Lie Groups of Operators in Hilbert Space. III, 160 pages. 1972. DM 16,–

Vol. 286: S. Murakami, On Automorphisms of Siegel Domains. V, 95 pages. 1972. DM 16,–

Vol. 287: Hyperfunctions and Pseudo-Differential Equations. Edited by H. Komatsu. VII, 529 pages. 1973. DM 36,–

Vol. 288: Groupes de Monodromie en Géométrie Algébrique. (SGA 7 I). Dirigé par A. Grothendieck. IX, 523 pages. 1972. DM 50,–

Vol. 289: B. Fuglede, Finely Harmonic Functions. III, 188. 1972. DM 18,–

Vol. 290: D. B. Zagier, Equivariant Pontrjagin Classes and Applications to Orbit Spaces. IX, 130 pages. 1972. DM 16,–

Vol. 291: P. Orlik, Seifert Manifolds. VIII, 155 pages. 1972. DM 16,–

Vol. 292: W. D. Wallis, A. P. Street and J. S. Wallis, Combinatorics: Room Squares, Sum-Free Sets, Hadamard Matrices. V, 508 pages. 1972. DM 50,–

Vol. 293: R. A. DeVore, The Approximation of Continuous Functions by Positive Linear Operators. VIII, 289 pages. 1972. DM 24,–

Vol. 294: Stability of Stochastic Dynamical Systems. Edited by R. F. Curtain. IX, 332 pages. 1972. DM 26,–

Vol. 295: C. Dellacherie, Ensembles Analytiques, Capacités, Mesures de Hausdorff. XII, 123 pages. 1972. DM 16,–

Vol. 296: Probability and Information Theory II. Edited by M. Behara, K. Krickeberg and J. Wolfowitz. V, 223 pages. 1973. DM 20,–

Vol. 297: J. Garnett, Analytic Capacity and Measure. IV, 138 pages. 1972. DM 16,–

Vol. 298: Proceedings of the Second Conference on Compact Transformation Groups. Part 1. XIII, 453 pages. 1972. DM 32,–

Vol. 299: Proceedings of the Second Conference on Compact Transformation Groups. Part 2. XIV, 327 pages. 1972. DM 26,–

Vol. 300: P. Eymard, Moyennes Invariantes et Représentations Unitaires. II. 113 pages. 1972. DM 16,–

Vol. 301: F. Pittnauer, Vorlesungen über asymptotische Reihen. VI, 186 Seiten. 1972. DM 18,–

Vol. 302: M. Demazure, Lectures on p-Divisible Groups. V, 98 pages. 1972. DM 16,–

Vol. 303: Graph Theory and Applications. Edited by Y. Alavi, D. R. Lick and A. T. White. IX, 329 pages. 1972. DM 26,–

Vol. 304: A. K. Bousfield and D. M. Kan, Homotopy Limits, Completions and Localizations. V, 348 pages. 1972. DM 26,–

Vol. 305: Théorie des Topos et Cohomologie Etale des Schémas. Tome 3. (SGA 4). Dirigé par M. Artin, A. Grothendieck et J. L. Verdier. VI, 640 pages. 1973. DM 50,–

Vol. 306: H. Luckhardt, Extensional Gödel Functional Interpretation. VI, 161 pages. 1973. DM 18,–

Vol. 307: J. L. Bretagnolle, S. D. Chatterji et P.-A. Meyer, Ecole d'été de Probabilités: Processus Stochastiques. VI, 198 pages. 1973. DM 20,–

Vol. 308: D. Knutson, λ-Rings and the Representation Theory of the Symmetric Group. IV, 203 pages. 1973. DM 20,–

Vol. 309: D. H. Sattinger, Topics in Stability and Bifurcation Theory. VI, 190 pages. 1973. DM 18,–

Vol. 310: B. Iversen, Generic Local Structure of the Morphisms in Commutative Algebra. IV, 108 pages. 1973. DM 16,–

Vol. 311: Conference on Commutative Algebra. Edited by J. W. Brewer and E. A. Rutter. VII, 251 pages. 1973. DM 22,–

Vol. 312: Symposium on Ordinary Differential Equations. Edited by W. A. Harris, Jr. and Y. Sibuya. VIII, 204 pages. 1973. DM 22,–

Vol. 313: K. Jörgens and J. Weidmann, Spectral Properties of Hamiltonian Operators. III, 140 pages. 1973. DM 16,–

Vol. 314: M. Deuring, Lectures on the Theory of Algebraic Functions of One Variable. VI, 151 pages. 1973. DM 16,–

Vol. 315: K. Bichteler, Integration Theory (with Special Attention to Vector Measures). VI, 357 pages. 1973. DM 26,–

Vol. 316: Symposium on Non-Well-Posed Problems and Logarithmic Convexity. Edited by R. J. Knops. V, 176 pages. 1973. DM 18,–

Vol. 317: Séminaire Bourbaki – vol. 1971/72. Exposés 400–417. IV, 361 pages. 1973. DM 26,–

Vol. 318: Recent Advances in Topological Dynamics. Edited by A. Beck, VIII, 285 pages. 1973. DM 24,–

Vol. 319: Conference on Group Theory. Edited by R. W. Gatterdam and K. W. Weston. V, 188 pages. 1973. DM 18,–

Vol. 320: Modular Functions of One Variable I. Edited by W. Kuyk. V, 195 pages. 1973. DM 18,–

Vol. 321: Séminaire de Probabilités VII. Edité par P. A. Meyer. VI, 322 pages. 1973. DM 26,–

Vol. 322: Nonlinear Problems in the Physical Sciences and Biology. Edited by I. Stakgold, D. D. Joseph and D. H. Sattinger. VIII, 357 pages. 1973. DM 26,–

Vol. 323: J. L. Lions, Perturbations Singulières dans les Problèmes aux Limites et en Contrôle Optimal. XII, 645 pages. 1973. DM 42,–

Vol. 324: K. Kreith, Oscillation Theory. VI, 109 pages. 1973. DM 16,–

Vol. 325: Ch.-Ch. Chou, La Transformation de Fourier Complexe et L'Equation de Convolution. IX, 137 pages. 1973. DM 16,–

Vol. 326: A. Robert, Elliptic Curves. VIII, 264 pages. 1973. DM 22,–

Vol. 327: E. Matlis, 1-Dimensional Cohen-Macaulay Rings. XII, 157 pages. 1973. DM 18,–

Vol. 328: J. R. Büchi and D. Siefkes, The Monadic Second Order Theory of All Countable Ordinals. VI, 217 pages. 1973. DM 20,–

Vol. 329: W. Trebels, Multipliers for (C, α)-Bounded Fourier Expansions in Banach Spaces and Approximation Theory. VII, 103 pages. 1973. DM 16,–

Vol. 330: Proceedings of the Second Japan-USSR Symposium on Probability Theory. Edited by G. Maruyama and Yu. V. Prokhorov. VI, 550 pages. 1973. DM 36,–

Vol. 331: Summer School on Topological Vector Spaces. Edited by L. Waelbroeck. VI, 226 pages. 1973. DM 20,–

Vol. 332: Séminaire Pierre Lelong (Analyse) Année 1971-1972. V, 131 pages. 1973. DM 16,–

Vol. 333: Numerische, insbesondere approximationstheoretische Behandlung von Funktionalgleichungen. Herausgegeben von R. Ansorge und W. Törnig. VI, 296 Seiten. 1973. DM 24,–

Vol. 334: F. Schweiger, The Metrical Theory of Jacobi-Perron Algorithm. V, 111 pages. 1973. DM 16,–

Vol. 335: H. Huck, R. Roitzsch, U. Simon, W. Vortisch, R. Walden, B. Wegner und W. Wendland, Beweismethoden der Differentialgeometrie im Großen. IX, 159 Seiten. 1973. DM 18,–

Vol. 336: L'Analyse Harmonique dans le Domaine Complexe. Edité par E. J. Akutowicz. VIII, 169 pages. 1973. DM 18,–

Vol. 337: Cambridge Summer School in Mathematical Logic. Edited by A. R. D. Mathias and H. Rogers. IX, 660 pages. 1973. DM 42,–

Vol: 338: J. Lindenstrauss and L. Tzafriri, Classical Banach Spaces. IX, 243 pages. 1973. DM 22,–

Vol. 339: G. Kempf, F. Knudsen, D. Mumford and B. Saint-Donat, Toroidal Embeddings I. VIII, 209 pages. 1973. DM 20,–

Vol. 340: Groupes de Monodromie en Géométrie Algébrique. (SGA 7 II). Par P. Deligne et N. Katz. X, 438 pages. 1973. DM 40,–

Vol. 341: Algebraic K-Theory I, Higher K-Theories. Edited by H. Bass. XV, 335 pages. 1973. DM 26,–

Vol. 342: Algebraic K-Theory II, "Classical" Algebraic K-Theory, and Connections with Arithmetic. Edited by H. Bass. XV, 527 pages. 1973. DM 36,–